Everything Forever

Everything Forever

Learning to See Timelessness

Gevin Giorbran

Enchanted Puzzle
Publishing

Seattle

Science / Personal Growth / Spirituality

First Edition

Soft Cover
ISBN 10: 0-9791861-0-2
ISBN 13: 978-0-9791861-0-3

Hard Cover
ISBN 10: 0-9791861-1-0
ISBN 13: 978-0-9791861-1-0

Library of Congress Catalog Control Number: 2006910726

Published by Enchanted Puzzle Publishing, Seattle Wa.
http://www.enchantedpuzzle.com

This book contains fractal art from some of the best fractal artists in the world. Sincerest thanks to all the contributing Artists and Photographers, including:

Kerry Mitchell Paul DeCelle Damien Jones	www.fractalus.com
Doug Harrington	www.fractalarts.com
Mike Levin	www.mikelevin.com
Charles Beck	www.enchantedpuzzle.com/beck/
Carol Taylor	www.caroltaylorquilts.com
Ken Libbrecht	www.snowcrystals.com
Michel Meynsbrughen	www.sxu.hu/gallery/clafouti/

Author's Website:

everythingforever.com

General thanks to NASA scientists, photographers, and artists for your work and inspiration.

Dedicated to my Grandparents

Special Thanks to: Mary, Robert, Quentin, Emma, Dennett, John and Rob.

Foreword

If someone picks up this book and reads only this one page I want them to be left knowing what is to come in the future. In 1998 astrophysicists discovered that the expansion of our universe is accelerating. We know we are accelerating away from the dense and hot conditions of the big bang, but what is the universe accelerating towards? Recently physicists are beginning to state openly that time ends in the future with our universe evolving into empty space. Of course empty space is the ultimate zero, the bottom end to all physics. If our universe reaches zero all space will be stretched perfectly flat and no matter will remain. A single unified space will then extend infinitely in all directions. So is this final space the ultimate nothingness? Actually many physicists and mathematicians think of zero as the most ordered state of all possibilities. Zero is balance. Zero is perfect symmetry. But what is this ultimate zero doing in our future? The answer is that zero is timelessness. Absolute zero is the timeless quantum superposition of all the universes that exist. Zero is the great sum of all. An ultimate zero has always existed, and will always exist. Zero is the native state of existence, or what the physicist David Bohm, Einstein's favorite student, called *Implicate order*. It sounds odd at first but we are inside zero.

Today in science the second law of thermodynamics suggests our single universe is becoming increasingly disordered with time. Many scientists claim our universe is winding down and dying of disorder. It is certainly true that entropy, the measure of spent energy, is always increasing. However, half of the second law is wrong. Our universe is not becoming increasingly disordered with time. Quite the contrary, we are headed for zero, and zero is a powerful kind of order.

The timeless zero in our future is the internal complexity of everything and the outer simplicity of nothing at the same time. There cannot be the simplicity of the single whole without all the inner complexity of universes that enfold into and create zero. What zero is not, is nonexistence. As Parmenides said, nonexistence cannot be. There is no state more extreme, either *less than* or *more than* the perfect zero. Zero is the default setting of reality.

The big mystery of "why is there something rather than nothing?" is answered simply by understanding that nothing still exists. All possible moments of time and all possible universes physically exist simultaneously, because all are merely fragments of a physically real zero. In the same way all colors exist in white light, or just as all positive and negative numbers sum to zero, all the moments of time sum up to construct a greater balanced whole we call zero. Zero is like a whole pie that can be sliced up infinitely many ways, but always remains a single whole. It is a difficult mental switch to adjust to, but everything we know is less than zero, not simply more than nothing. And so our beautiful universe is not dying. The very surprising purely scientific truth, as explained in this book, is that our universe is in the process of merging with the timeless sum of all, with the infinite whole, with everything forever.

This separation between past, present and future is only an illusion...

Albert Einstein

~~~

Just as we envision all of space as really being out there, as really existing, we should also envision all of time as really being out there, as really existing too.

Brian Greene

~~~

Nothing ever happened in the past; it happened in the Now.
Nothing will ever happen in the future; it will happen in the Now.

Eckhart Tolle

~~~

Ultimately, all moments are really one, therefore now is an eternity.

David Bohm

~~~

Our actual universe evolves to empty space.

Sean Carroll

~~~

Eternity is a long time, especially towards the end.

Woody Allen
~~~

Introduction

There is a wide assortment of excellent books out there if one is interested in the science of quantum mechanics or superstrings theory. What the shelves are missing are books on the science of timelessness, true even though the three most remarkable physicists of the last century, Albert Einstein, Richard Feynman, and Stephen Hawking, each concluded from their own individual accomplishments in science that the actual Universe exists apart from our sense of time. Each scientist developed their own unique way of understanding timelessness. The renowned Stephen Hawking, who holds Newton's chair at Cambridge has been the most adamant and regularly refers to another mode of time in which the Universe has no beginning or end. None of these scientists have said that time is purely an illusion. It appears more accurate to say that in the same way the permanent pages of a book tell a changing story, the past, present, and future moments of our lives all exist simultaneously in another kind of time. Today, Hawking and others call this other realm *imaginary time*, even though this other form of time should probably be considered more solidly existent and more tangibly real than our own time.

What all scientists agree on is that we have begun to enter the golden age of astronomy and cosmology. Quite suddenly we have reached a period when the most important questions physicists have asked over the past one hundred years are finally being answered by hi-tech probes and the Hubble space telescope, as they extract the needed information from distant galaxies and as they map the echo left over from the big bang. Already this golden age has produced startling revelations about our existence. For example, the Wilkinson Microwave Anisotropy Probe (WMAP) determined with unparalleled precision that the large-scale cosmos is spatially flat. The geometry of the overall cosmos shows no indication of being curved into a figure eight or any kind of closed circular volume that would allow the cosmos to be spatially limited. And so, it appears the stars and the galaxies, the physical cosmos we live in, extends outward in every direction infinitely without end.

Many scientific minded philosophers have in the past imagined the greater Universe might be timeless and infinite, as far back as Parmenides and as recent as Giordano Bruno, and many today in and out of science are convinced that quantum theory indicates an infinity of parallel worlds within the inner space of particles and energy. But this probe has in fact transported us into a very different age of learning, both for science and all of humanity. In scientifically concluding the cosmos is infinite we are no longer discussing various scenarios of how the cosmos might be, we are finally discussing and exploring one scenario of how the cosmos actually is. Consequently we are now being led toward a much deeper and very profound understanding of the cosmic big picture. However, there is one recognizable stumbling block… the second law.

The sixties was distinctly a time when people began to question and challenge established ways of seeing the world. Many recognize that movement was greatly influenced by two scientific theories that had finally gained wide acceptance, the big bang model and Darwin's theory of evolution. Both theories provided insight into how our world changes over time, and both greatly influenced the youth of that period, myself included. However, informative as both theories were, unfortunately there was no lasting change, as there was no pot of gold at the end of the rainbow. The theory of evolution and the big bang both expose details about the past, but neither revealed what the universe is evolving towards. Consequently the knowledge bestowed from these new comprehensions, although wonderfully educational, fell short of exposing any sense of deeper meaning or purpose to the evolution of the cosmos.

The reason both theories failed to provide any type of enlightenment as to what the universe and life are about, can be summed up in five words: *the second law of thermodynamics*. The most psychologically disturbing law found in science is without question the second law which claims that everything in the universe evolves from an ordered state to a more disordered state as time evolves. The second law has been written about extensively, it is one of the two most basic laws of nature, but the underlying conclusion which everyone must draw is always the same. In moving toward disorder the universe is winding down, it is dying. So the grand lesson of science has become that the long-term evolution of the cosmos has no ultimate purpose or goal. Our beautiful universe is dying. This conclusion is forced upon every person who learns the second law, and in fact the second law hangs over science and humanity like a black cloud.

It is easy to imagine how much more interested people would be in science today had we instead discovered that our universe is evolving into something meaningful, and not simply dying of disorder. Imagine instead that scientists had discovered some deep purpose to time. Imagine scientists had found the order of the universe is ever increasing, moving us steadily toward some incomprehensible perfection. We all occasionally stop and contemplate the world, and how we scientifically view the universe's future effects us on many levels. Knowing the universe has a future goal, knowing time has an innate purpose, would at least subtly influence each one of us, and eventually it would undoubtedly change humanity.

Could the second law be wrong? Actually, today the second law stands as one of the most fiercely defended laws in science. It describes the most basic way that the cosmos changes with time. Most believe it will never be overturned. Only there is one thing to consider. Something totally unexpected happened recently in science, and it is something that promises to dramatically change how we view the distant future. In the summer of 1999, NASA officials and a team of scientists in a television broadcast announced one of the most startling discoveries ever made, a discovery comparable even to when Edwin Hubble first discovered the galaxies are expanding away from one another. NASA scientists using the Hubble space telescope had carefully verified the discovery originally made in 1998 and were ready to officially announce the findings. On NASA television a large group of scientists announced, "the expansion of the universe is presently accelerating."

What does this discovery mean? Since the Big Bang was first discovered it was thought that all expansion was slowing down, decelerating ever since time began 13.7 billion years ago. But careful measurements of galactic distances measuring the brightness of a special type of supernova revealed distinctly that expansion is no longer in decline. After slowing for nearly eight billion years, the deceleration of expansion turned to acceleration approximately six billion years ago. Apparently there are two phases to the life of the cosmos, one where expansion slows as time moves away from the point of the big bang, and one where expansion accelerates. What are we accelerating towards? The universe is moving directly toward the opposite extreme from which time began, the state of absolute zero. What is absolute zero? Absolute zero is the timeless whole of all universes.

Science is now in an unprecedented adjustment period. Old questions must be reconsidered, such as, what is the future like? What is the final result of time? What is absolute zero? Could this acceleration change our bleak outlook of the distant future? The discovery that the expansion of the cosmos is now accelerating was not a complete surprise to me personally, as I had written three books between 1994 and 1997, all prior to the '98 discovery, in each book explaining that time is moving toward absolute zero. Although I agreed with the big bang model I departed from the conclusions of mainstream scientists who argued that time will never reach the ground state of zero.

The primordial vacuum of science, the inexplicable emptiness of eastern philosophy, the classic idea of nothingness, creation itself, cannot be found in the direction of the past. The ground state of zero exists in the direction of the future, and very plainly and evidently so, once one considers without assumptions what we know of the universe from basic physics and cosmology. We know in science that the universe is cooling and expanding toward absolute zero, not away from it. Furthermore, there is no evidence of a "creation from nothing" in our past, only increasing density and energy. Most scientists know this to be true, and yet today we continue to project nothing into our past (and reject its obvious presence in the future) based on assumptions that our existence necessarily begins in the past.

In the now famous Big Rip scenario three physicists led by Robert Caldwell mention the possibility that time ends at the "ultimate singularity". More recently the bright and popular physicist Sean Carroll of Caltech in a presentation given to other physicists has stated "our actual universe evolves to empty space" as if this is plainly evident. What is the universe accelerating toward? The simple answer is that, as if shot from Robin Hood, the arrow of time has turned on its rocket boosters and is flying straight at the perfect zero center of the target. We can expect it to become increasingly commonplace for scientists to openly state that time ends at a ground state of absolute zero or empty space. If we take a small step backward and look at the big picture, a final end of time at the ultimate singularity of zero derives from accelerating expansion as equally obvious as the big bang past derives from expansion. And once the goal of our universe is seen and understood, the really big picture finally starts to make sense.

The reason acknowledging the true location of zero is vitally important to science is because the void of empty space in our future isn't really empty. The zero of physics isn't a cancellation of everything. Absolute zero is the sum of everything, the sum of all universes, all possible states, and all life. Zero is Einstein's timelessness. The big bang most certainly happened, and time does begin in a highly ordered state, but the tiny Alpha singularity in our past is merely one of two special types of order in nature. Present in our future there exists another type of order; the true state of highest order, i.e., perfect balance, neutrality, unity, perfect symmetry, the great infinite whole. As impossible as something this profound seems, an ultimate state of oneness really does exist as a physically real stage in the life of our own expanding space-time, directly in the future. Our universe is literally in the process of merging together with all other universes in the greater multiverse, and this book contains the map to prove it.

In Terry Gilliam's movie *Time Bandits*, a small band of God's helpers steal a map of the Universe which allows them to travel through special portals that bridge different periods of time. Seeking gold and jewels, the bandits invade periods of history which in the movie are portrayed as different regions of a larger timeless Universe displayed on the map. Turning that story line into non-fiction, in this book we are going to sneak a peak at God's map. We are going to map the timeless realm of all possibilities (sorry, portals not included). And once we cross into this timeless realm, the panoramic view of the big picture unlocks a real magical chest of gold and jewels, in the form of ultimate knowledge about why the universe is this way.

In science today a completely new way of seeing the universe is emerging. Science tends to study the small, the constructing parts of a system, and so the direction of learning is from the bottom-up. Scientists have managed genuine miracles in discovering the tiny building blocks of the larger world. But rarely do scientists ever attempt to view the greater whole Universe from a top-down perspective. There has been one major exception to this rule in the recent past; the physicist David Bohm.

In his younger years David Bohm was a student and close friend of Albert Einstein. As a physicist Bohm made major contributions to the development of nuclear physics and quantum theory, but in his later years Bohm encountered a book written by Jiddu Krishnamurti, an eastern philosopher, and Bohm was surprised to find there were many ideas about wholeness in this book that related to his own ideas about quantum theory. Bohm later was led to write *Wholeness and the Implicate Order*, a book in which Bohm claims that there are two kinds of order in nature.

Bohm laid a foundation but never realized the full extent of his own claim, but he was certainly correct. Still unbeknownst to the science of today, there are in fact two distinct and separate types of order in nature, rather than simply order and disorder. One order exists in extreme in our past, the other kind of order exists in extreme in our future. And so the universe isn't dying. Rather our universe is evolving away from the powerful influences of one type of order toward a more powerful other type of order in the future.

Having spent a lifetime exploring the idea of eternity and the infinite whole, and having mapped the timeless realm, I discovered profoundly that we exist caught in between two great powers. The emergence of orderliness and life, all the intricate becoming of nature, the systemization we know as the forces of nature, all result from the natural struggle between these two great powers. The theory of two orders is an entirely new science and of course anything new always sounds incredibly complex, but in fact the fifth chapter which explains the two orders is extremely straightforward and simple, it can be explained to a grade school student. The two orders could instead have been discovered in the age of philosophers, by Plato or Aristotle. But somehow it was overlooked, and consequently here we are today having trouble fitting all the pieces of the cosmic puzzle together.

It turns out that there is a very good reason the expansion of the universe is accelerating. All time in every universe moves toward the balance of a universal zero. Although it is a bit startling to clearly recognize that time has both a beginning AND an end, in discovering timelessness we also find that our single cosmos is like a story in a great book that tells an infinite number of stories. All the stories; my story, your story, exists forever. We are led finally to imagine a deeper level of reality, even from a purely scientific perspective, where all life across infinite worlds exists eternally unified within an implicate ground state of zero, forming an omniscience ever present in our own future. A bit too profound I know to be good hard science, but here comes a truly extraordinary way of seeing the Universe.

Contents

Everything Forever

Time is one enormous moment
Where children play
not knowing of a tomorrow
where people walk along an ocean
and gaze in wet air

This sense of separation and loss
is all illusion
though old men tell of the past
as if it is gone somewhere else
to children who listen
as if it used to be

We all walk here in time
not yet knowing
as we ponder the mystery
and animals listen
that all in this same moment
the world begins
and the world ends
while these waves
crash upon the shore
regardless

And now as I touch your hand
time will stand still
and trap something there forever
for us to view from some heaven
as we are forever born
into an endless moment

It is unity that enchants me.

Giordano Bruno

Fractal Art: © Kerry Mitchell

Part One

The Beginning of Timelessness

I dream myself awake. I have come to accept the fact that I will wake up early in the morning full of leftover thoughts after lucidly dreaming of the mysteries remaining in my mind. Awake I think of the patterns which are becoming ever more evident, patterns I now recognize in nature and in us, in everything from art to politics. After about an hour I ease back into quiet sleep. The writer Gerhard Staguhn once wrote, "Whenever man tries to probe into the universe's dimension of time, he will finally be confronted with eternity. Where he tries to understand the dimension of space, he will be finally confronted with infinity." Such exposures have become true for my own journey, and are steadily becoming true for science, but no one, myself included, ever believed the timeless infinite could be so fathomable.

Most people comfortably assume that the past no longer exists and the future only becomes real as time evolves to it. When someone refers to the beginning of time we assume they are referring to the beginning of the existence of the universe. Usually we imagine the whole of physical reality moves along with us through time. Yet that assumption might be like someone reading a book and believing that once a page is turned it no longer exists, or someone believing the pages that haven't been read yet do not exist until one turns the page. It might be the same as believing that nothing existed until the book was suddenly opened to the first page. There is actually no valid reason to assume the past and future do not exist, just because we can't turn the pages forward or backward at will, or we can't yet read all the pages at once, to verify all of them are always there.

Having long ago learned to escape time in various ways, it's easy to forget how convinced most people are that the past and future don't exist. Many stop at the question of whether the past still exists, but are quite convinced there isn't anything out there yet in the future. What is it that makes us believe being here in this moment has any influence on the reality of other times we've known or might experience? Why does the here and now make the past seem as if it doesn't exist any longer? And why does the present make the future seem nonexistent? We know how real this moment is, and we know the other moments we've experienced were just as real. Why can't they all be real at the same time?

That strong sense that we possess of experiencing any given moment as if it is all of reality is only naturally true of all the moments that we experience. It couldn't be any different. Our experience is always of moments which if they were any different then we would just experience that different place. We would be someplace else in time. There is nothing about our existence in this moment that suggests that we can't be in both places, thinking we only exist in each place. What we define as our self can be here and there simultaneously without creating any existential crisis. It is only the definition of each moment that makes each moment seem exclusively real. And if we think about it, that principle alone, the reality of each moment, actually indicates that all the moments are real, far more than it indicates only one moment can be real.

Imagine that you could somehow experience two places simultaneously. You would still sense change and time taking place, just in two places at once. Your existence in time wouldn't be upset, just your sense of place would be unusual. Your sense of position in one single place would feel disturbed. At first you would surely assume both versions of you existed at the same time, as if one was across town from the other. "Why is there two of me in the same time?" you would wonder. But if you looked at a calendar or a clock and realized that one of your experiences was in the future, then your ordinary sense of time would suddenly shatter. Your time advanced future self would sense a 'now' in the past. Your time retarded past self would sense a 'now' in the future. You would become time dizzy at not being able to tell which of the two is the true and actual present? One experience seems to be in the past of the other, while the other experience seems to be in the future of the other. Your sense of a common now for both would conflict with the whole notion of a past and future. Facing such a dilemma would actually leave you with an improved sense of time, because to resolve the paradox you would have to realize that only your experience of one place makes it seem as if other places and times don't exist.

If all the moments we have ever experienced in our life simultaneously exist we wouldn't experience the world any differently. We are limited to experiencing each moment as if it is the only one that is real and furthermore, the only way that we can sense that we exist at all is if we experience a series of individual moments. The dynamic relationships within time make us conscious. So we aren't just experiencing one moment. Our conscious experience overlaps many moments.

Of course most everyone is unfamiliar with these kinds of ideas, and freshly encountering the rationality of such ideas rarely convinces anyone outright that the existence of the universe doesn't evolve. Timelessness is an interesting concept, but what proof is there? Our ordinary notion of time is simple, straightforward, and practical. It is the least of what we know for certain. It isn't as profound as the idea of timelessness. But what is truly real? Do we each exist somewhere else? Are we each being born at this same time? Do the dinosaurs still exist, just somewhere else in another place we can't readily visit? Could we look through a window at the past or the future and see it existing as real as we exist? Could we walk through a door and visit other times? Is there any way to know for certain?

The objective world simply is; it does not happen.

Herman Weyl

~~~

If we accept multiple universes then we no longer need worry about what really happened in the past, because every possible past is equally real. Therefore, to avoid... insanity, we can, with clear consciences, arbitrarily define reality as that branch of the past that agrees with our memories.

Joseph Gerver

~~~

What has been will be again, what has been done will be done again; there is nothing new under the sun. Is there anything of which one can say, "Look! This is something new?" It was here already, long ago; it was here before our time.

King Solomon

~~~

But according to conventional physics, we inhabit a universe where time and space are frozen into a single unchanging space-time. All the events that have happened or will ever happen are marked by points in this "block" of space-time, like bubbles suspended in ice. Past and future have the same footing, and there's no flow.

Stephen Battersby

~~~

I have realized that the past and future are real illusions, that they exist in the present, which is what there is and all there is.

Alan Watts

~~~

Why is [the now] the most precious thing? Firstly, because it is the *only* thing. It's all there is. The eternal present is the space within which your whole life unfolds, the factor that remains constant. Life is now. There was never a time when your life was not now, nor will there ever be. Secondly, the Now is the only point that can take you beyond the limited confines of your mind. It is your only point of access to the timeless and formless realm of Being.

Eckhart Tolle

~~~

"You have the sight now Neo, you are looking at the world without time."

The Oracle in *The Matrix*

~~~

The ultimate stuff of the universe is mind stuff.

Sir Arthur Eddington
~~~

Time has no independent existence apart from the order of events by which we measure it.

Albert Einstein

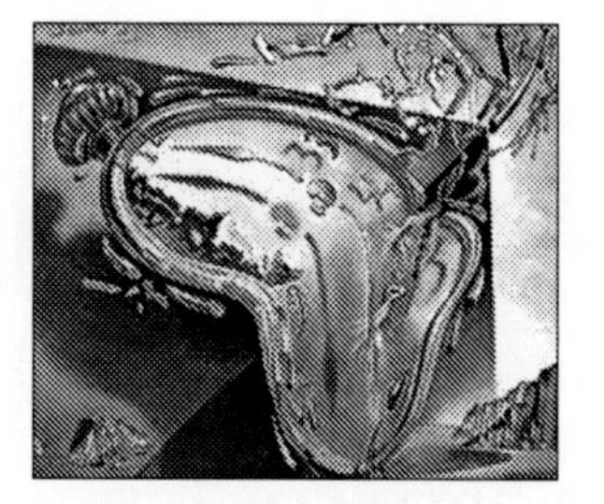

Chapter One

Time is Imaginary

The People of Timelessness

Surprising as it may be to most non-scientists and even to some scientists, Albert Einstein concluded in his later years that the past, present, and future all exist simultaneously. In 1952, in his book *Relativity*, in discussing Minkowski's Space World interpretation of his theory of relativity, Einstein writes:

> Since there exists in this four dimensional structure [space-time] no longer any sections which represent "now" objectively, the concepts of happening and becoming are indeed not completely suspended, but yet complicated. It appears therefore more natural to think of physical reality as a four dimensional existence, instead of, as hitherto, the evolution of a three dimensional existence.

Einstein's belief in an undivided solid reality was clear to him, so much so that he completely rejected the separation we experience as the moment of now. He believed there is no true division between past and future. His most descriptive testimony to this faith came when his lifelong friend Besso died. Einstein wrote a letter to Besso's family, saying that although Besso had preceded him in death it was of no consequence, "...for us physicists believe the separation between past, present, and future is only an illusion, although a convincing one."

Most everyone knows that Einstein proved that time is relative, not absolute as Newton claimed. With the proper technology, such as a very fast spaceship, one person is able to experience several days while another person simultaneously experiences only a few hours or minutes. The same two people can meet up again, one having experienced days or even years while the other has only experienced minutes. The person in the spaceship only needs to travel near to the speed of light. The faster they travel, the slower their time will pass relative to someone planted firmly on the Earth. If they were able to travel at the speed of light, their time would cease completely and they would only exist trapped in timelessness. Einstein could hardly believe there were physicists who didn't believe in timelessness, and yet the wisdom of Einstein's convictions had very little impact on cosmology or science in general. The majority of physicists have been slow to give up the ordinary assumptions we make about time.

The two most highly recognized physicists since Einstein made similar conclusions and even made dramatic advances toward a timeless perspective of the universe, yet they also were unable to change the temporal mentality ingrained in the mainstream of physics and society. Einstein was followed in history by the colorful and brilliant Richard Feynman. Feynman developed the most effective and explanatory interpretation of quantum mechanics that had yet been developed, known today as *Sum over Histories.*

Just as Einstein's own Relativity Theory led Einstein to reject time, Feynman's *Sum over Histories* theory led him to describe time simply as a direction in space. Feynman's theory states that the probability of an event is determined by summing together all the possible histories of that event. For example, for a particle moving from point A to B we imagine the particle traveling every possible path, curved paths, oscillating paths, squiggly paths, even backward in time and forward in time paths. When summed the vast majority of all these directions add up to zero, and all that remains is the comparably few paths that abide by the laws and forces of nature. Sum over histories indicates the direction of our ordinary clock time is simply a path in space which is more probable than the more exotic directions time might have taken otherwise.

Other worlds are just other directions in space, some less probable, some equally as probable as the one direction we experience. Feynman's summing of all possible histories could be described as the first timeless description of a multitude of space-time worlds all existing simultaneously. In a recent paper entitled *Cosmology From the Top Down*, Professor Stephen Hawking of Cambridge writes; "Some people make a great mystery of the multi universe, or the Many-Worlds interpretation of quantum theory, but to me, these are just different expressions of the Feynman path integral."

Hawking, the most popular physicist since Einstein, who has battled against what is known as Lou Gehrig's disease for some thirty years, has expanded upon both Einstein's and Feynman's theories supporting timelessness. Hawking demystified the black hole, and wrote books so enjoyable that he has managed to educate billions of people about modern physics and cosmology. From his wheelchair, presently unable to communicate without his computer, Hawking still actively lectures while he professionally holds Newton's chair as Lucasian professor of mathematics at Cambridge University in England. As if such miracles were commonplace, Hawking has introduced what could be said to be the scientific theory of forever.

Hawking and James Hartle developed the *No Boundary Proposal*, a theory which extends other theories such as *Sum Over Histories.* The no boundary proposal is a model of the early universe during the big bang which includes a second reference of time, called *Imaginary Time* which has no beginning or end. In this mode of time we could in fact reach back and touch the original conditions of the early universe, because they still exist in a common time to all moments. Hawking explains that what we think of as real time has a beginning at the Big Bang, some ten to twenty billion years ago, but in imaginary time the universe simply exists.

People often think from the tag imaginary that this other mode of time isn't real. Quite the contrary, clock time could be said to be imaginary compared to this ultimate mode of time, since in imaginary time our clock time is totally indistinguishable from directions in space. In his most popular book *A Brief History of Time* Hawking writes:

> Quantum theory introduces a new idea, that of imaginary time. Imaginary time may sound like science fiction, and it has been brought into Doctor Who [an English Star Trek]. But never the less, it is a genuine scientific concept. One can picture it in the following way. One can think of ordinary, real, time as a horizontal line. On the left, one has the past, and on the right, the future. But there's another kind of time in the vertical direction. This is called imaginary time, because it is not the kind of time we normally experience. But in a sense, it is just as real, as what we call real time.

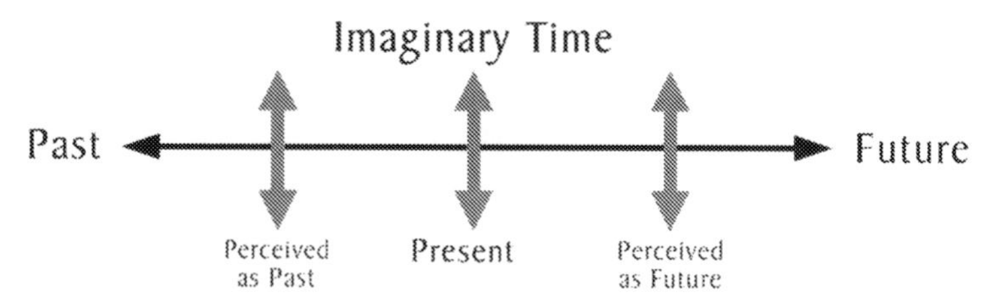

Figure I.I: All moments share an Imaginary Time reference which has no beginning or end.

The simple lines in this image above effectively portray imaginary time occurring at right angles to our ordinary sense of time. Of course since the moments of past, present, and future all exist simultaneously in this other mode of time, the duration of each moment of time would seem to be ceaseless and eternal. The existence of the universe in imaginary time doesn't have a past or a future, instead all times exist in one enormous moment of now. Hawking writes:

> One could say: "The boundary condition of the universe is that it has no boundary." The universe would be completely self-contained and not affected by anything outside itself. It would neither be created nor destroyed. It would just BE.

According to Hawking the universe doesn't have a boundary point where it suddenly begins existing. The first moment isn't any different than the second in respect to existence. Both moments exist forever in imaginary time. It takes very little reasoning to figure out that if the universe exists in an unseen way without beginning or end, at right angles to regular time, then that reference to time is simply more elementary and even more real than ordinary clock time. The term imaginary applies more accurately to our time.

Hawking himself writes:

> This might suggest that the so-called imaginary time is really the real time, and that what we call real time is just a figment of our imaginations. In real time, the universe has a beginning and an end at singularities that form a boundary to space-time and at which the laws of science break down. But in imaginary time, there are no singularities or boundaries. So maybe what we call imaginary time is really more basic, and what we call real is just an idea that we invent to help us describe what we think the universe is like.

Using the no boundary proposal, we can imagine the whole of time by imaginatively placing oneself inside a giant hollow globe. If we look up we see the North Pole from the inside. Within this globe of time, looking up is looking into the past, but not as if it no longer exists, instead one can actually touch the past since it is a place existing permanently. The North Pole, the beginning of time, is just a single position upon the rounded surface like the first page of a book. Looking down one even sees the future. And in this globe, looking down we see the South Pole, the end of what we call time.

If the universe exists in another time reference where conditions are permanent or static, suddenly it doesn't matter that we humans so convincingly observe a beginning to time, since the imaginary time reference applies regardless of our sense of where we are in time. The universe could be said to exist before our clock time began, and after our clock time ends. The past and future exist now. Obviously, imaginary time relates directly to existence. Imaginary time relates to the whole, to all that can be imagined. It also easily relates to numbers and ideas and the concepts we think with, which we already sense exist forever.

The only reason this can be so disorienting at first is because we are splitting time into two separate dimensions. We are splitting in two the more common meaning of the word time. Here one time dimension is related purely to the existence of each moment, so it is the omni-directional time we exist within. The other time dimension, the time we measure with clocks, is here limited to being change which is necessarily a construction of many moments in the first dimension bound together in some way that creates a second time dimension. Each moment is necessarily a time frame, which is a sort of fixed pattern of matter and space. Somehow those frames or spaces are fused together, creating a false sense that existence is changing and transforming, when change is actually observed only by whatever moves from one time frame to another.

Another English theoretical physicist, Julian Barbour, believes that time simply doesn't exist. Barbour, an independent theoretician not affiliated with any University, is never the less highly respected in the upper physics community.

And Barbour has extensively explored the concept of timelessness and the illusion of motion, and is perhaps the first person since Ludwig Boltzmann to set his focus directly on modeling the timeless world of all possible states. Barbour's version of timelessness, Platonia, named in respect of Plato's allegory of the cave which describes a world of illusion, theorizes that the set of all possible nows can be reduced to the patterns created by all the possible triangular positions of only three particles. In an interview with John Brockman, Barbour describes his version of the wedge model and shows his passion for describing timelessness:

> What really intrigues me is that the totality of all possible Nows of any definite kind has a very special structure. You can think of it as a landscape, or country. Each point in the country is a Now. I call it Platonia, because it is timeless and created by perfect mathematical rules.

I so strongly agree with and respect how Barbour has introduced to science the base assumption that what exists in timelessness is describable and it is shaping the world we experience. Barbour is convinced that there is a distinct shape to a timeless realm of all possibilities which is exclusively responsible for guiding the path of time and fashioning the physical universe we experience. Such a view is presently uncommon but it can be appreciated as the only possible explanation. When a respected scientist clearly emphasizes a perspective as Barbour has done it opens the doors for others.

Another popular physicist, the string theorist Brian Greene, author of the book and PBS television series *The Elegant Universe*, has stated the following in his most recent book *The Fabric of the Cosmos*. "Just as we envision all of space as really being out there, as really existing, we should also envision all of time as really being out there, as really existing too." It appears we have finally reached a new era of taking the idea of timelessness seriously. This means however that we have to begin to look at the universe differently. We have to learn to think differently and ask different questions. The most important question is a big one. How can a universe simply exist?

Because imaginary time behaves like another direction in space, histories in imaginary time can be closed surfaces, like the surface of the Earth, with no [existential] beginning or end.

Stephen Hawking

~~~

When I consider the small span of my life absorbed in the eternity of all time, or the small part of space which I can touch or see engulfed by the infinite immensity of spaces that I know not and that know me not, I am frightened and astonished to see myself here instead of there...now instead of then.

Blaise Pascal

~~~

We all operate within a framework of concepts that make sense of the world to us, which we use to formulate our goals, hopes, and dreams, and to seek ways to overcome problems and obstacles as we build our lives. Certainly the universe out there has much to say about all this, but it's hard to figure out what it says when our scientific description exists for us as a remote framework without clearly articulated connections to the concepts which we operate in daily life. So we live in a disconnected state: abstract and evolving knowledge of the grand universe on one hand, and the immediate need for a guide to our individual choices on the other hand. How do we bring these together, so that we can guide our immediate choices from a perspective that is informed by and connected to the big picture?

Todd Duncan

~~~

Time past and time future,
what might have been and what has been,
point to one end, which is always present.

T.S. Eliot

~~~

Part of metaphysics moves, consciously or not, around the question of knowing why anything exists - why matter, or spirit, or God, rather than nothing at all? But the question presupposes that reality fills a void, that underneath Being lies nothingness, that de jure there should be nothing, that we must therefore explain why there is de facto something.

Henry Bergson

Fractal Art: © Kerry Mitchell

Thou canst not recognize not-being (for this is impossible), nor couldst thou speak of it, for thought and being are the same thing.

Parmenides

Chapter Two

Why the Universe Exists Timelessly

A Journey Beyond Nothing

As we move backward through the semi-illusion of time we watch the universe de-evolve, we pass the dinosaurs and the emergence of life on this planet, then view the Earth de-form into clouds of stellar materials produced from exploded stars which themselves de-explode and then de-coalesce toward becoming a dense uniform opaque plasma. As time accelerates backwards space itself collapses inward, as if it is being vacuumed away, moving all the material in the universe ever nearer, with all finally crashing into a beginning point. As the universe crashes inward it seems obvious that we must be closing in on some sort of birth. We seem to be moving backward toward what must inevitably be a distinct creation event, where the somethingness of matter arises from a primordial nothing.

Be this moment an act of omnificent magic, a fortunate accident, or something completely inexplicable, considering the universe is expanding it appears evident that somehow all that we know, has been, and everything that shall follow in the wake of the present, came to be all at once at one moment of time in our past. It seems evident that somehow something impossibly erupted to create a beginning, even if all the laws of nature as they are known today in science forbid such an event. The first law states that energy is neither created nor destroyed. Furthermore, every ounce of logic, be it intuitive or mathematical, demands that something cannot be created out of absolutely nothing. A zillion zeros still add up to zero. And logically, if something comes from nothing, then it wasn't really nothing to begin with then was it. And yet the universe is here, and all is expanding away from one single place and one single time, before which there is no possibility of time as we perceive time.

Every bone in a reasonable person's body screams that this sudden creation event could not have happened by itself. A universe cannot just pop into existence. The existence of a universe and our own existence requires a cause. And so we ask, does this impossibility of 'something coming from nothing' mean that the universe absolutely had to have been created? Did a powerful being of some kind (usually assumed to be named God) create the first moment of our universe? It is almost a relief to consider this possibility in the face of such a paradoxical dilemma, except we actually know that this solution only suspends and relocates the mystery. All the same questions we ask about how the universe came to be, must then be diverted to this being called God. The inference of some seems to be that God is so powerful that God is beyond needing an explanation, yet realistically the same old questions apply. How long has this being existed? How did God begin from nothing? If it has existed forever, then how can it just exist? Why does God exist rather than nothing at all?

We usually know better than to try to explain the existence of the human world as a product of a human act, at least not logically. We don't imagine the Universe created the Universe. We don't even pretend that God created God. By definition the first thing cannot come from itself or anything else. So how then did the very first thing begin if it really didn't exist before it suddenly existed? In truth there isn't a proper answer to these questions. The answer to why we exist isn't answered by explaining the impossible. Rather the great mystery of why we are here is answered by recognizing our own inevitability.

A lot of people who believe in God believe God has existed forever, which leads to the question, could something just exist eternally, without beginning or end? But then if so, if that door is actually open, if it is possible for something complex and powerful like a god to have existed forever, could such reasons for being able to innately exist forever also apply to a seemingly more simple universe? Is it possible that the seed of the big bang existed forever before undergoing the transformation we know as the big bang? Is it possible that everything, even we ourselves, exist forever in each moment apart from our sense of time, making time ultimately an illusion. This would mean that the past, present and even the future, all exist simultaneously.

Presently it doesn't seem possible to us that things might simply exist. Why? Because a universe is complicated, God is complicated, while nothingness in comparison is simple. Nothingness wouldn't need an explanation. Complication requires a reason for being so. In fact there is only one principle idea that holds us back from believing that there are things, or beings, or realms of time and place, which exist forever without cause, without beginning or end. That reason is our expectation that a pure and total nothingness is more primary, more basic, and simpler, than every other possibility. The true root question, the one that applies to everything equally, both God and the universe, is why does anything exist rather than nothing at all? Yet that question assumes that nothing is basic and primary.

And so, if we could somehow make that question go away, if we could realize we are making some type of mistake, and realize that 'nothing' isn't really simpler or more primordial to everything else, then we might actually be able to, in the same realization, understand clearly why something like a God or a cosmos should exist timelessly. Then we would no longer need to battle the paradox of how something came from nothing, because then we would know why there was never an absolute nothingness to begin with.

What Nothing Really is

Why is there this existence we are taking part in instead of nothing at all? So let's focus now on that question. We should not merely ask the question, but study the question. How did something come from nothing? What are we asking with that question? Most of us think we know what *somethingness* is, but what exactly is *nothing*? Let's walk right up to it and find out.

Imagine we are transported all the way back to the beginning of time. We are standing at the very precipice of the birth of the world, the birth of being itself. It would be a bit like kneeling down and crawling out to the tip of a cliff. Out beyond the edge of the cliff there is nothing at all. So now you crawl out, and you put your hand out to the surface of the beginning, to the origin of everything that will ever think or be. Imagine you can touch the very beginning, the originating moment. Now push through it. Reach beyond that outermost edge. Reach into the blankness beyond and touch the original void. Touch the simplicity. Imagine it, imagine the nothingness, the abyss that would have been prior to existence, and try even to understand it. Understand its nature. What words best describe it? Is it frightening or menacing, or is it vibrant with all the potential of being? Is it thick or dark, warm or cold? It surely must at least be simple, as simple as simple can be.

Can you bring words to what you sensed? What words describe the complete blankness? Actually if you are able to imagine something, or describe something, or feel anything, you need to realize that you haven't yet gone far enough beyond the edge of real existing things. Actually if we are able to imagine or describe anything at all, or feel anything, we cannot be all the way beyond the edge of somethingness. Nothing is nothing at all. We must move beyond what ordinary words can describe. So try one more time. Let your mind drift beyond the edge of time, beyond all descriptions, beyond all senses. And yes now we can't see it, there it isn't, just beyond the edge of rational thought itself, hidden there in a blackness darker than black, a quiet beyond silence, a stillness beyond rest. Oh my, there "isn't" the absolute void.

Are you still here! You didn't disappear? And you didn't get sucked in? But did you feel it? Did you at least sense it? "NO!" What do you mean "NO"! We were right there! How could that be? I wonder what went wrong. You must not have a very good imagination! No wait, maybe you do, maybe that is the problem. Maybe your imagination is getting in the way because what we are trying to imagine isn't cold or dark, or a void or an abyss, it isn't quiet or simple, and it's hardly anything to be afraid of, because it doesn't exist. Maybe this nothing is unimaginable because there is nothing to imagine. Indeed if you came up with any sense of what is beyond the cliff, then you sort of missed the point.

The thought exercise above reveals a sort of anomaly in how we see the world, and it reveals something about the world that anyone can appreciate regardless of education or religious beliefs. We cannot actually imagine or describe nothingness, that is, if we are referring to a nothing prior to existence. We can describe the type of nothing that is common in our lives, the nothing that we encounter everyday. There is nothing here or there. There is nothing to talk about. There is nothing in the refrigerator. That type of nothing is something empty, something lacking substance, something uniform or plain or simple. But the other nothing that is prior to existence is a special case in terms of semantics and meaning. By definition, words simply can't describe it, so it is different than everything else that we define with words and everything else imaginable.

A fact about reality we are discovering here is that there are two very different nothings, and presently the two are entangled together when they don't belong together. In other words, there is actually something wrong with the word nothing as we use it today. If we carefully study the definition of the word nothing we can discover two very different definitions of nothing. One definition of nothing is a physically real condition that has no discernable form or substance, such as a white canvas, or a uniform void in empty space. This type of nothing is real and exists, and is actually quite ordinary. An empty refrigerator has nothing in it. A white artist canvas has nothing painted on it. The real nothing is always a place or a space that is uniformly undefined, where there are no distinct things. There is just one thing, like one color, or just space alone, so we call it nothing. But the other definition of 'nothing', the one we were just a moment ago trying to touch and describe is nonexistence, which is a very difficult concept to understand when defined separately from the real nothing, which is the very reason we confuse the two. We confuse the two out of need, because one we can describe, the other we cannot.

When the dictionary defines nothing as 'something that does not exist', it is reasonably obvious that the syntax of the phrase makes no real sense. How can 'nothing' be a *something* which does not exist? In fact simply using any word in an attempt to mean non-existence creates a sort of riddle. How do we make a word refer to something that doesn't exist? What word can represent a form that isn't a form; a thing that isn't a thing? What language can define a concept that has no reality or meaning?

Of course we cannot solve the great old riddle of how something came from nonexistence. It's the ultimate oxymoron, and the ultimate contradiction in terms. We cannot even refer to a state of nonexistence when there is no such state, and no such form, to refer to. Any attempt to describe it isn't describing it. Any word representing it, isn't representing it. Non-existence can only really be defined as something that cannot be defined with a word. It can only refer to something that cannot be referred to. Obviously there is a vexing fundamental problem here. Any attempt to define a nonexistence using any meaningful idea or thought, by using the meaning that otherwise defines all language, that defines our reality, is predestined to fail.

Nonexistence cannot be. It cannot exist. It cannot even be meant. And that predicament, that total paradox, is very different from the real nothing that exists and can be talked about. And the fact that we confuse these two concepts is the very reason we don't yet clearly understand why we exist. We exist because there is no alternative. There never was a non-existence in the past and there never will be a non-existence. Existence is the default setting of reality. Existence belongs here. It has always been.

The Real Nothing

Imagine you are standing in a white world, like the commercials or movies portraying heaven. In this world there is nothing but white everywhere. The oneness of white extends away from you in every direction. You try to look out into the distance, but because there is just the one color you can't tell if the space of this world extends out forever or if its edge remains just out of reach. As you reach out your hand, you realize that your physical body provides the only sense of distance here. Your body is all that exists in a giant field of nothingness. There is no length or width beyond your body. There is no distance to anywhere else, because there isn't anything else to measure a distance to. So if your body happens also to turn white, then suddenly all sense of dimension is erased. The very meaning of place and distance is lost. Soon even the one color of white will disappear from your experience. You will soon become blind to white, because you don't have any other color to judge the meaning of this one color against. Soon, for you, this endless white world becomes nothing at all.

If you were born into this one color dimension you wouldn't ever be able to see it, you would not even know it was right there in front of you, since you would not have any other color or shade of gray to reference it by. Someone who is blind, for example, doesn't see black or darkness, because even if they did temporarily upon initially going blind, the black quickly loses meaning for them because it is just one color, and without differentiation the mind interprets such a world as a perceptual nothing. And in fact the mind is correct, because this is the real nothing that exists in physical reality. The real nothing is just singular form. A real nothing is a singularity, and a singularity is all a real nothing can ever be.

Within a singularity, all distances and locations lose meaning because once there is a perfect unity, a oneness, then every object, every distance, every place, is the same as any other. Singularities are commonplace. Any single color is a singularity. A perfect blue sky is a singularity. The most common everyday example of a singularity is the ordinary empty space we travel through, which is why we typically refer to it as nothing. Never the less, singularities can have content. Most everyone has heard the idea of a polar bear in a snow storm. Singularities can even be full instead of empty. Suppose we take everything from a household refrigerator, put it all in a big stove pot, add some water, and begin stirring. After we cook all this awhile all the distinct parts begin to break down and blend together evenly into a soup. If we keep heating and stirring this stew for five or six hours, or two or three days, eventually all the many ingredients will unify into a single paste-like substance. Many have become one. All the ingredients of the refrigerator are still in there, within the one, they have just transformed into a singularity.

As we shall see, there are extreme cosmological singularities in our distant past and our distant future. Singularities are an interesting novelty of reality because, in the same way all the fruits and vegetables, the condiments, the juices and milk in the refrigerator all vanish in creating the paste, all the physical properties of our universe suddenly vanish into thin air at the stage of becoming a singularity. If we imagine the infinity of all possible universes unified into an ultimate singularity, it would still have no size or properties. In fact, if all possible universes in the entire multiverse of worlds are at some ultimate level unified into a whole, the totality becomes something we perceive as nothing at all. The great unified whole is the white world. It can be imagined the size of a pin head small enough to fit in the palm of your hand, or an endless space stretching out forever. It can be said to exist in any point of space, as well as every place in space, here, there, and everywhere.

Photo: Point Reyes Beach © Mike Levin

The paradox of limits lies in the fact that limits combine two opposite functions: setting apart and joining.

Piet Hut

~~~

In the theory of relativity, the concept of time begins with the Big Bang the same way as parallels of latitude begin at the North Pole. You cannot go further north than the North Pole.

Kari Enqvist

~~~

If your position is everywhere, your momentum is zero.

William Lipscomb

~~~

A region of space might be expanding or contracting. If it is expanding stuff dilutes away until we get empty space. If it's contracting it will ultimately collapse to a black hole. But that black hole will eventually evaporate, leaving empty space.

Sean Carroll

~~~

Consider the most obvious question of all about the initial state of the universe: Why is there an initial state at all?

Lawrence Sklar

~~~

When you have eliminated the impossible, whatever remains, however improbable, must be the truth.

Sir Arthur Conan Doyle
~~~

If we extrapolate this prediction [of contraction] to its extreme, we reach a point when all distances in the universe have shrunk to zero. An initial cosmological singularity therefore forms a past temporal extremity to the universe. We cannot continue physical reasoning, or even the concept of spacetime, through such an extremity. For this reason most cosmologists think of the initial singularity as the beginning of the universe.

Paul Davies
Physicist and Author

Chapter Three

Cosmic Boundaries

The Timeless Extremes that Shape Reality

An idea can stretch the mind's awareness beyond dreams and yet the same idea can limit the imagination of every genius who has ever lived. Why? Because there are distinct boundaries to what is ultimately possible, and those same boundaries work to limit our imaginations. Even living here inside an infinite Universe there are still ideas which we simply cannot think beyond. Such places are found in our very own time and space. In fact ultimate boundaries shape the flow of time and virtually all that we observe. They literally shape reality, and make the universe a sensible place. One of these boundaries is already fully recognized by science. The other as yet is hardly noticed and remains completely unappreciated. We all are at least slightly aware of the first cosmic boundary.

Scientists have long known that the space of our cosmos is stretching outward like the outer surface of a balloon being filled with air. The visible cosmos is expanding as if it is being inflated, and this sends all the distant galaxies whirling away from us. The large-scale bodies of stars known as galaxies are moving away from one another, but they are not moving away from a center. Rather all the space between the galaxies is expanding everywhere in the cosmos equally. Consequently, galaxies twice as far away are speeding away twice as fast.

It is a really simple conclusion scientists are forced to make. If we turn time backward the inevitable result of letting all the air out of the balloon is that all the matter in the universe collapses back into the same space. If time were reversed all the stars and galaxies, rather than expand outward, would collapse inward on themselves. In our past the whole cosmos becomes ever more dense and hot, as every star and galaxy in the heavens is drawn nearer together. What this invariably means is that time only turns back so far. After thirteen point seven billion years of tracing time backwards the collapse is complete, the volume of the cosmos disappears and all material objects are collapsed and condensed into a single solitary place, an extreme called the *Alpha State.*

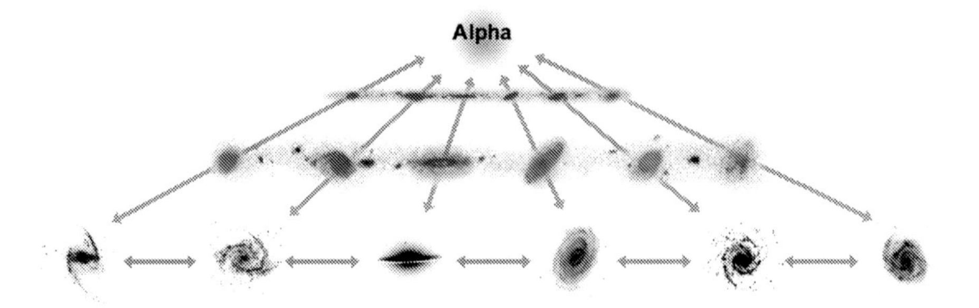

Figure 3.1: Arrows represent space expanding between the galaxies. Expansion in reverse becomes contraction and the collapsing of space can be followed back to the beginning of time at the Alpha extreme.

It matters not if time originated precisely from the Alpha extreme. Scientists today continue to debate over whether time traces backward all the way to Alpha. What is far more important is the role Alpha plays in our thinking and our ability to imagine. Alpha is not just a place where time may have begun. That issue is secondary to the significance of there being an extreme possibility such as Alpha. Alpha is a limit to what is ultimately possible. Alpha even represents an ultimate boundary to what is possible in the realm of all conceivable universes. Alpha is a limit even to what can be imagined.

Collapsing inward, the physical cosmos can shrink to a point, but once the volume of space reaches zero, once all space is vacuumed away, the collapse is complete, and physics finds itself at the outer edge of what is possible to be existent. Time may or may not have begun precisely from the absolute extremity of Alpha. But the marker of Alpha as a boundary defines a physical limitation to what is possible and in doing so Alpha plays a key role in envisioning timelessness.

We are actually very fortunate the cosmos is expanding as it highlights the fact that the extremity of the Alpha state is there. Seeing Alpha as an extreme is especially key in understanding the shape of the big picture. Even if the number of other worlds is infinite, Alpha creates a boundary within the infinite. We often hear the claim that "possibilities are endless", or "anything is possible", but the Alpha state exposes the fact that there is at least one ultimate limitation out there. There is a boundary in the world of all possibilities. We can even think of Alpha as a cornerstone in the foundation of reality itself, a footing that shapes what is imaginatively and physically possible. This is why Alpha deserves the title of Cosmic Absolute, a possibility beyond which no other possibilities exist.

Don't forget Omega

Once recognized as an edge to what is possible, Alpha can help to expose another equally important boundary, because Alpha is not alone. In fact, we will eventually discover boundaries in every direction of possibility. If we now look in the opposite direction, toward the future, the cosmos is ballooning outward due to cosmological expansion, so the volume of the known cosmos is becoming ever larger. Which means the density of the visible cosmos is steadily decreasing and the temperature of the cosmos is steadily dropping as light and heat waves are stretched and elongated by the ballooning of space. Such processes are very gradual and have considerable impact only after many billions of years, but if we run the clock forward in the same way that we turn the clock backward to find Alpha, the expanding cosmos eventually creates the opposite extreme of absolute zero.

What is absolute zero? Absolute zero is commonly known as the hypothetical temperature at which all motion ceases, a temperature equal to -459.67° degrees on the Fahrenheit scale, or -273.15° degrees on the Celsius scale. There is no temperature colder than a zero absolute temperature (sometimes called ZAT). If you are wondering why there can't be a continually colder temperature, the issue of motion is the easiest to understand. Temperature or heat is determined by the motions of atoms. If we could freeze matter to zero, all molecular motion would stop and be frozen in place, so the passage of time as measured by clocks would stand still. However, absolute zero is commonly misunderstood. Absolute zero is not merely a temperature.

What very few people realize even in science is that there is a single common zero for all measures in physics. The real absolute zero is far more extreme than just a coldest temperature. Absolute zero is a condition of the cosmos in our future, where mass, energy, density, gravity, and temperature all reach zero simultaneously. We sometimes casually refer to the extremities of absolute zero's properties, using words such as nothing, empty, cold, straight, or frozen. How can there be less than nothing? How can space be emptier of things than perfectly empty? How can anything be more flat than perfectly flat? How can a direction in space be more straight than perfectly straight? How can anything be colder than frozen still? These extremities of absolute zero are literally an extreme edge to reality, which is why absolute zero, like the Alpha state, is also a great cosmic absolute beyond which no other possibilities exist.

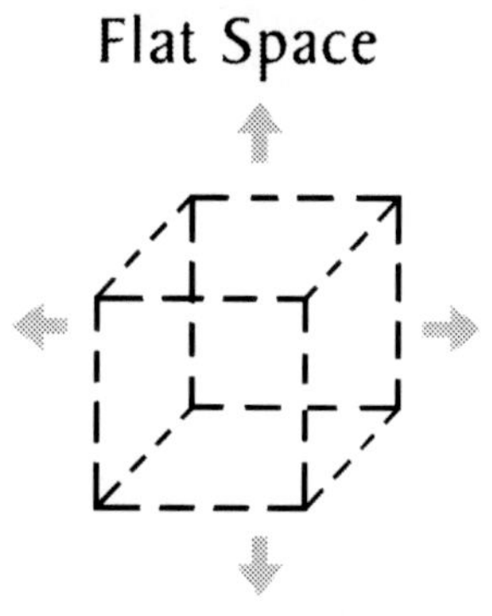

Initially, we can envision zero as we would imagine an empty space stretching out in all directions. The image of an invisible square, as shown here exemplifies a perfectly flat space, a space in which no objects exist. Any two parallel lines of the square shown would extend infinitely without ever converging together or diverging away from one another. In a perfectly flat space any two parallel lines remain parallel forever.

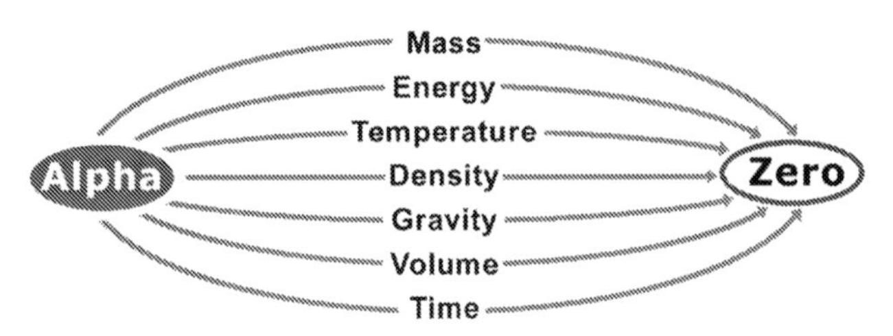

Figure 3.3: Mass, Energy, Temperature, and Density are all infinite at Alpha and are zero at Omega. The volume of space-time collapses at both ends, at Alpha and zero, while time is stopped at both ends. Gravity is turned around, considered a repulsive force during the big bang, so it is also zero on both ends of time.

The claim that ZAT is the point at which all molecular motion stops is helpful in one respect, as it highlights the fact that there is no passage of time at zero, but it also contributes to a thorough misunderstanding of absolute zero, a confusion that has made zero in the past seem to be physically impossible. Strictly speaking, matter should not be described as possibly being frozen at zero simply because it is impossible to make matter absolutely cold, which is a widely known fact in physics. Matter simply refuses to give up its energy and cease all residual motion. This actually makes a lot of sense if we think about it. The fact that matter cannot be cooled all the way to zero in some university laboratory is nature's way of saying that time for a group of atoms cannot be stopped while time is still occurring for those observing in the laboratory. However, although it is true that we cannot make atoms of matter stand completely still at this stage in the history of the cosmos, this does not mean the cosmos cannot cool to ZAT in the distant future.

One of the more interesting facts about a universal zero is that the only way that any of the physical parameters of the cosmos such as temperature or gravity can reach absolute zero is if all the parameters of space-time reach zero simultaneously. This fact is precisely what makes absolute zero both a cosmic absolute and the ultimate Omega State, Omega being the last letter in the Greek alphabet which means "the end". In a sense, there is only one way for the cosmos to reach zero in the future. The only way for the cosmos to become absolutely cold is if cosmological expansion stretches all the matter in the cosmos perfectly flat, at which point matter becomes indistinguishable from space.

The future scenario of the cosmos stretching and becoming perfectly flat has always been sort of ignored in science. In an obvious double standard, physicists have widely considered the possibility that in the past a fluctuation in a primordial vacuum somehow created a universe of matter, however in regards to the future reaching ZAT, the consensus has been that a matter universe cannot cool fully to zero. As scientists developed models of the future most have imagined only two general scenarios. It was believed the expanding universe would either stop cooling toward zero and collapse inward in a big crunch, and thus heat up again, or more likely, the universe would expand at an ever decreasing rate, moving ever nearer to, without ever reaching ZAT. We just didn't know what the future would be more like, fire or ice. But we did know for certain that the universe has always expanded and cooled toward zero and is moving ever nearer to zero as if magnetically attracted. Only recently did we discover how powerful the attraction of zero is.

The Big Rip Scenario

In 1998 NASA realized the expansion of the cosmos is accelerating. Then in March of 2003 the Dartmouth physicist Robert Caldwell, already known for his related theory of Quintessence, and two colleagues, presented to the scientific community what they called the Big Rip model of the future, which considers the scenario where the dark energy density, called phantom energy by Caldwell, increases with time. According to Caldwell this invisible phantom energy causes the expansion of the cosmos to literally rip apart all the galaxies, stars, and finally all atoms. In the Big Rip model all space is finally stretched perfectly flat, and the evolution of our cosmos ends distinctly in finite time at what Caldwell refers to as the ultimate singularity.

When the expansion of the cosmos was believed to be ever decreasing, it did not seem like time could ever reach the opposite extreme from which time began. So it's particularly interesting that we have now discovered that the expansion of the cosmos is accelerating, since the only physical process that can produce zero in the future is if an accelerating expansion stretches the final stages of the cosmos perfectly flat. Accelerated expansion is how the cosmos bridges the seemingly infinite gap between increasingly larger circles and the ultimate extreme of flat space where two lines can always be perfectly parallel.

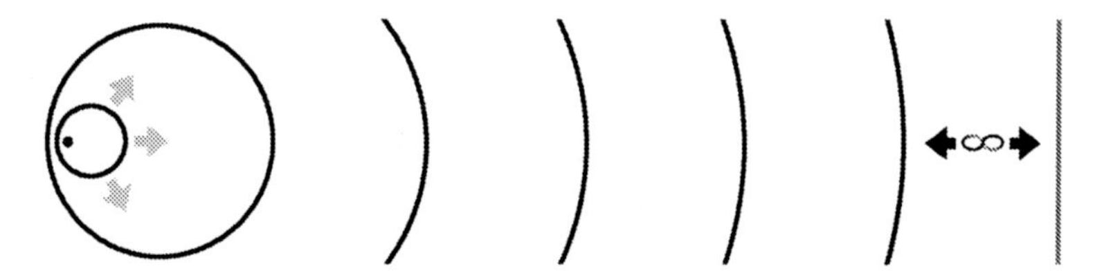

Figure 3.4: The curvature of an expanding circle moves ever nearer to the extreme of zero curvature. Similar to Zeno's paradox, it seems impossible for an expanding universe to become perfectly flat. Yet we so easily pull a curved string straight or straighten a curved rod. We can make a widening circle with our arms and imagine the circle growing ever larger but we can also stretch our arms out straight, to represent a perfectly straight line.

The Expanding and Accelerating Universe

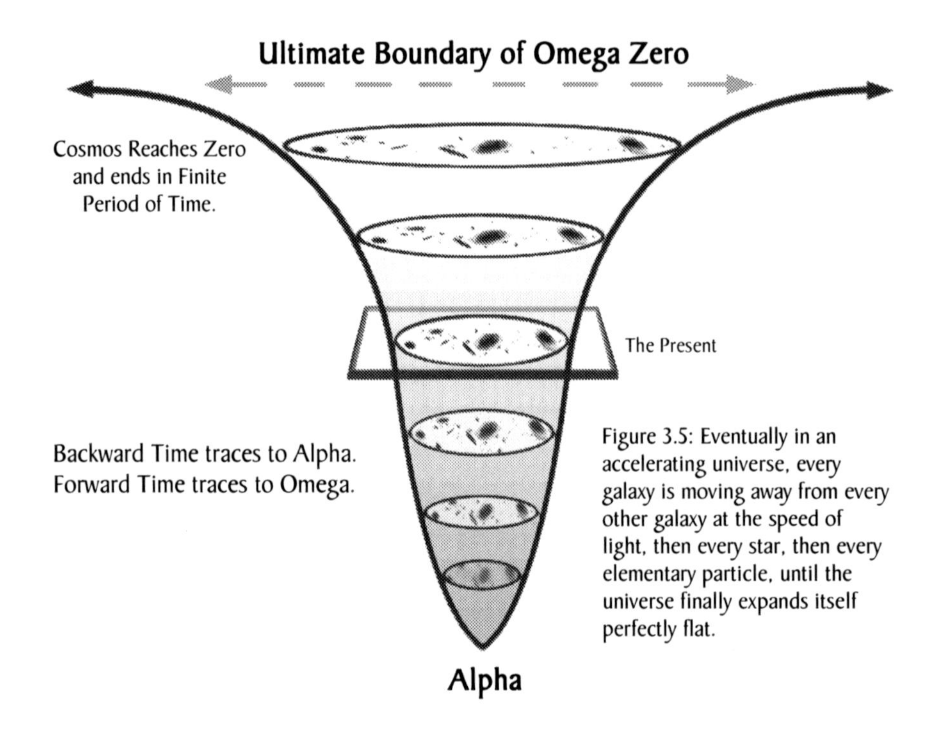

Figure 3.5: Eventually in an accelerating universe, every galaxy is moving away from every other galaxy at the speed of light, then every star, then every elementary particle, until the universe finally expands itself perfectly flat.

Outdated Expansion Scenarios

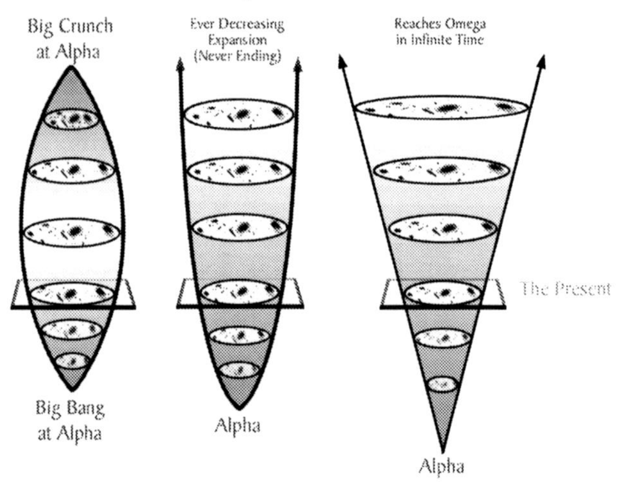

But what do we know about zero? Portraying zero in a logically consistent manner is rather tricky, in part because we make so many inaccurate assumptions about zero. The problem with referring to zero as an empty space, or using an invisible square to portray a real physical state of zero, is that a perfectly flat space is anything but empty. If it were empty we could put something into it. If it were empty it would be like a vessel that we could fill up with things, but the most basic rule about a perfectly flat space is that there is absolutely no way to introduce an object into it. Why? Because it is already full. Flat space is not empty. If flat space were empty we could travel through it, but it is completely impossible to travel through a flat space, in part because time stands still at a zero temperature, but more so the reason is that any matter such as the matter of our bodies requires spatial curvature. Any object introduced into a flat space would inevitably take away from what is actually a property of fullness and in doing so would take away from the perfect flatness of space. Objects require spatial curvature and curvature is always a reduction of perfect flatness.

Perhaps you noticed the unusual use of the word fullness. In physics, mass and spatial curvatures are inseparable. A major part of what Einstein discovered is that objects do not exist independently from space, nor does space exist independently from objects. We can imagine a flat space, but then imagining an object existing in and traveling through that flat space fails to consider the fact that objects or matter cannot exist without spatial curvature. The two are inseparable because they are one and the same thing. Spatial curvature is part of what an object is. Einstein described this by saying that space is the extension of mass, but one small step further is to say that mass is spatial curvature. Matter is nothing but curvatures in space. And all curvatures take away from the fullness of what we can only pretend is empty space. In reality empty space is the infinite whole, the fullness of everything combined together. It is the completed multiplicity of quantum mechanics, the superposition of all possible states and all possible universes combined into the ultimate singularity.

We think space needs to be empty because things move through it. Light passes through it. But light does not move through space from point A to B as we imagine of a thrown baseball. Light travels through space only as a probability, meaning it leaps from source to destination without ever having physically passed through space. This is true even of the thrown baseball. The particles that we see of a moving baseball are just a few which have assumed a single position in space long enough to collide with a photon. After the light bounces off the baseball both particle and photon vanish into a wave of probability, then the photon assumes a new position in your eye. The seemingly hard physical world literally bubbles up out of the fullness of flat space. If you think about this much, you realize the experienced physical world that has bubbled up out of the perfect void is really less than what we imagine to be empty space, and not more than empty space or nothingness.

Bose-Einstein Condensate

As space expands the cosmos is invariably becoming increasingly cold. Most everyone knows this, but there is also a hidden and very important underlying physical transformation occurring. At extreme cold temperatures far below water's freezing point, laboratory materials such as cesium gas become super conductive. At such temperatures, groups of oppositely charged particles magically arrange themselves into orderly columns and rows. Then at even colder temperatures, less than a millionth degree away from absolute zero, the individual particles actually unify into a single material. The many become one. This unified state of matter is called a condensate, which is a special form of matter first predicted by Albert Einstein and Satyendra Bose in 1924. This unique stage of matter that exists only at super cold temperatures near absolute zero was first created in a University of Colorado laboratory in 1995. Scientists weren't able to make time stop, but they were getting pretty close.

As the cosmos cools and expands, therein moving ever nearer to zero, literally all of the particles in the cosmos are moving toward becoming this single condensate. Condensates reflect a super orderliness near absolute zero where particles organize and smear together into a single unified medium. In other words, at the end of time all the many tiny particles become a single medium that is in perfect balance and is spread evenly throughout an area. Like ice cubes melting to become a liquid, like a liquid evaporating into a gas, near the end of time all that is left of the particles that now form stars and galaxies is an orderly cold and thinning gas that melts and evaporates into nothing but a low density space. This final form of matter need only be stretched a little further by cosmological expansion in order to push what remains of all the matter and energy of the cosmos to the extreme of perfect flatness. All the known matter in the cosmos will then be converted into pure space and time will have reached the end, the absolute zero of all physics, the great ZAT.

Two Boundaries in the Total Measure of All Possibilities

It might be surprising to discover how near we are to absolute zero presently. Nearly fourteen billion years of expansion has produced so much empty space between the galaxies, that the average temperature of the universe has been lowered to a minus -454.74° degrees, so on the Fahrenheit scale we are less than five degrees away from absolute zero. In Celsius the average temperature of the universe is -270.415°, which is less than three degrees away from zero. And finally, using the Kelvin scale which astronomers use since it is based on zero, the universe is only +2.735°K degrees above zero. The universe seems very warm living so near to a star, but out beyond the stars in deep space temperatures are very cold, that is, compared to where we started at Alpha.

Now suppose we take a step backward and consider how the cosmos has evolved from its beginning to present, which is one form of stepping outside of time. We know that time began from, or time began very near to, an infinitely hot Alpha, and then the cosmos expanded and cooled for billions of years, nearing zero even today. In being aware of the fact that Alpha and ZAT are the edges of what is physically possible in reality, we can now appreciate the revealing fact that time originates from one extreme of nature and travels all the way to the other extreme. The arrow below represents the direction time has taken since the big bang. The purposely simple image points out how the evolution of time of our cosmos spans across the whole spectrum of possibilities like a clothes line in between two poles. Imagine all the alternative directions that time could travel in. How relevant is it that time travels away from one cosmic extreme of an infinite heat and density all the way to the other extreme of absolute cold and zero density? Is there any discernable reason that this is the natural course of time for our cosmos?

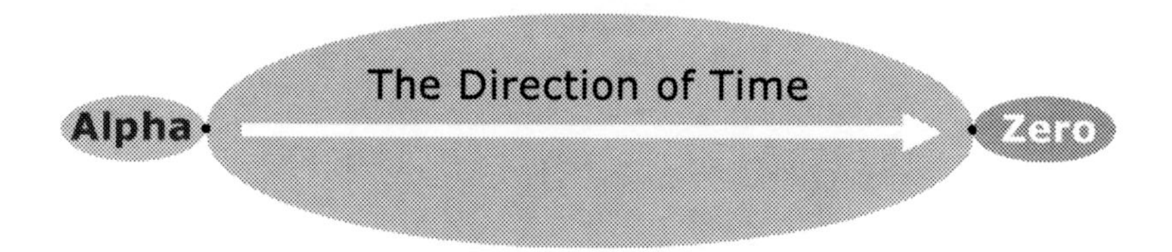

Figure 3.6: Many cosmologists of the last century spoke of the universe being "finite yet without boundary." Actually the universe is infinite but bounded by very definitive extremes, the Alpha in our past and the Omega in our future. Without the finite, the infinite would not be infinite, it would just be indefinite and therefore nonexistent or truly chaotic.

Years ago, discovering the expansion of the universe, taught us a great deal about the past, but only recently, due to the discovery of accelerating expansion, are we discovering the larger role that absolute zero plays in physics and cosmology. Science today is not merely coming to terms with the real and likely possibility that in many billions of years there is an abrupt edge to time at an Omega state. We are also beginning to focus on the physically real properties of zero. We are beginning to discover what zero actually is. The stage we are in now is quite similar to the period between 1910 and 1932 when Vesto Slipher began measuring the red-shifting of galaxies and later Edwin Hubble revealed how the universe contained many different galaxies all expanding away from one another. It was of course many years before the majority of scientists fully appreciated what the expansion of galaxies meant about the past, but that one piece of knowledge has led to virtually everything we presently understand about how the cosmos evolves in its early stages. What secrets and mysteries about the universe can be uncovered as we begin to better understand the zero in our future?

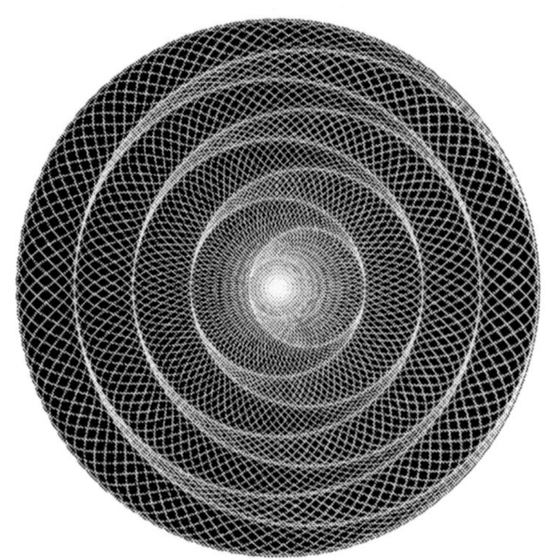

Fractal Art: Study © Kerry Mitchell

The probable is what usually happens.

Aristotle

~~~

Ultimately, the generation of probabilistic events seems to be an entirely mysterious aspect of reality. Asking about it is like asking about where the initial conditions come from in a deterministic theory, or like asking why there is something rather than nothing.

Matthew J. Donald

~~~

Time is a child playing dice; the kingly power is a child's.

Heraclitus

~~~

I believe that the vacuum, being the state in which all possible physical phenomena are present, in a virtual way, but still present, will win the record for the highest complexity.

Carlo Rubbia

~~~

Probability is the very guide of life.

Marcus Cicero

The real world, according to what we understand about physics, is described quantum-mechanically, which means, deep down, that everything has to be described in terms of probabilities. The "classical" world that we perceive, in which every object has a definite position and moves in a deterministic way, is really just the average of the different possibilities that the full quantum theory would predict.

Alan Guth

Chapter Four

Describing the Realm of All Possibilities

A Look at How Science Presently Sees the Big Picture

By the time we reach adulthood we are all at least vaguely aware of the range of possible events in our lives, beginning with the more probable and ending with the highly improbable to absolutely impossible. In addition to all the ordinary and predictable events, like the sun rising in the morning, there is each day also a chance of something extraordinary happening. We might meet the perfect friend or lover, or we might find ourselves in a car accident. We might win the lottery, or we might walk into a bank or store and find ourselves in the middle of a robbery. Our days are usually somewhat predictable, but in everyday life there is always a small chance that something exciting might happen.

Occasionally we hear the odds of winning the lottery, the chance of an earthquake, or a meteorite striking the earth. We hear of our chance of being in a car wreck compared to that of airplane crashes. The probabilities behind risks and opportunities flow and change depending upon where we are located, the time of day, the time of year, and the phase of the moon, and once we make note of it, it is surprising to realize how many of our actions and decisions are made based solely upon probabilities. We buy insurance due to the chance something bad might happen. We save for a rainy day. We wait to buy a CD until we have heard all the songs. Generally when we spend we try to invest wisely because of the risk of losing the value of our investment. Many pray and make self sacrifices secretly hoping to guide the future and avoid misfortune. Others eat healthy and exercise to ward off the threat of possible sickness and disease.

We constantly work to stabilize and control the possible events in our lives. Yet we can be wise enough to also intentionally open ourselves up to the chance of opportunities by adventuring into the unknown. We play with probabilities. We seek out risk. We challenge our fears. We rock climb, river raft, drive fast, strap boards to our feet and whisk down snowy mountains not only to strengthen our sense of power and control over the chance of danger, but for the thrill of directly challenging the possibility of injury and death.

So much of what we do is exploring what might emerge from the realm of possibilities. We read the newspaper to find out which of the more improbable events emerged from out of the whole of imaginable possibilities. Meteorologists forecast a probable weather. A sporting event is a play of probabilities, the stock market is a probability investment, a walk in the park is filled with probabilities of what we might experience or who we might see. There isn't much in life we couldn't gamble on because everything is uncertain to some degree, merely because there is so much that is equally possible.

Even our ability to know things involves a measure of certainty less than the unattainable one hundred percent certainty. Our only knowledge exists in what is probably true, not what is truly certain. What could we know with absolute certainty when we might wake up tomorrow to find that the person we think we are was only a complex dream of something else living in an entirely different universe. Our only comfort and security lies in the improbability of such things actually happening.

We live in an intricately shaped probabilistic world, in a well defined probabilistic society, and we are each a complex history that creates a unique suitcase of probabilities. Our brains are designed to calculate and navigate through oceans of shifting probabilities. We inherit specific probabilities from our parents and pass them on to our children. Growing up is very much about learning how to adapt effectively to the complex world of probabilities and the uncertainties it creates.

Without deeper consideration it seems as if probabilities change, because they adjust to our changing situations, but what is probable can be predetermined for any given situation. We can see this easily in card games or casino gambling. The probabilities in poker or roulette can be precisely calculated and never change based on time or place. Any gambling establishment knows the longer a person gambles the more their losses will exceed any fortunate gains, since the games are designed with more possibilities of loosing than of winning.

Of course there is also the impossible. We can't win the lottery if we don't buy a ticket. We can't be in two places at once. We can't walk through walls. We can't travel faster than the speed of light. We can't fly like superman. Most of us couldn't climb Mount Everest and would die trying. The impossible is real, which is why we commonly fight against it and challenge it.

On the positive side this same boundary keeps objects from materializing out of thin air in front of our cars and it keeps monsters from materializing under our beds at night. It is often said that anything is possible, but if that were really true, if there wasn't the strictly impossible, then there wouldn't be an ordinary or the comfortable, there wouldn't be the dependable or predictable or consistent. Without forces and laws anything could happen and would happen.

For some time now, scientists have known that just beyond the surface of the seemingly hard physical world, the individual particles that create atoms vanish into and then reemerge from an invisible realm of possibilities in a constant process of becoming and unbecoming. Beyond our vision, most of what we assume to be solidly real is really an invisible fluid of many options and alternative worlds. We call the process of moving in and out of the larger realm of possibilities *quantum mechanics*. All the particles of light that create a mental picture of the world in your mind have only just emerged from out of this fluid, only at the point of contact with your eyeball.

It would be helpful if we had a map of all possibilities. It would be helpful to be able to see the all bends and curves, the contours of potential, to know what lies up ahead, what lies around the corner, and to know the best path to take. It would be helpful to understand the shape of all possibilities well enough to answer profound but basic questions, such as, why does the possible realm place such limits on what happens in time? Why does it allow such wonder? Where are we going? What is certain in the future? Many people spend their lives trying to make such a map in their minds and many alter or oversimplify their sense of the real world in order to create such a map.

The American physicist John Wheeler referred to the whole of possibilities as Superspace, and the English theoretical physicist Julian Barbour named the same realm Platonia. Scientists in different fields generally refer to possibilities as state space, phase space, or configuration space, but the most commonly used name we have given the possible realm is 'Mother Nature'. Nature is the great mystical and mysterious guardian of the actual. She is the governor, the police woman, the facilitator of the possible realm. Mother Nature's forces regulate the physical universe in harmony with the underlying hidden possibilities. The only reason we experience a consistent and predictable natural world is because the possible realm is always there hidden behind time, existing so concretely and ageless.

Learning to see timelessness is first about learning to see the possible realm, which includes learning that there is a discernable shape to the whole of possibilities which can be modeled and understood. Of course we have to consider how modern science today models possibilities with the second law of thermodynamics, which sounds like a specific law about temperature, but it is actually a very rudimentary law meant to explain why the universe evolves and changes as it does. Today's version of the second law is widely recognized as making two related but different statements. First it states that the loss of usable energy, called entropy, always increases. There is always a deterioration of usable energy in the universe as time passes. That part of the second law will likely never be overturned. But the second law is unique among all the laws of science because it does not merely explain how this process occurs in nature based upon a bottom-up study of molecular behavior. It also provides a top-down explanation for why entropy always increases.

Boltzmann's Vision of Overall Possibilities

The physicist Ludwig Boltzmann was the first to imagine that the realm of possibilities has a shape and structure in 1868, as he further developed an understanding of nature that is known today as the second law of thermodynamics. Boltzmann was trying to understand the way that patterns evolve in nature, so he began to consider how an invisible world of possibilities might influence what is probable as the universe evolves and events unfold. He knew for example that gases disperse evenly throughout all available space. He knew that heat does not remain or collect in one area but rather spreads out, moving from warmer to colder bodies. He knew that although it is easy to break objects into smaller disorganized pieces, like a coffee cup or a glass vase, we never see the pieces organize themselves back together, at least not in forward time, as we would see of a broken vase if time were reversed. Why then is forward time different than backward time? Boltzmann concluded the reason is because there are fewer ordered possibilities than disordered possibilities.

Boltzmann discovered that he could model the world of possibilities when he realized there are more patterns of one particular kind than another. He considered patterns or states where things are organized and kept in a group, in comparison to patterns where things are broken up or dissipated throughout some larger area. Boltzmann realized there are more of the later type, there are more arrangements where things are spread out and disorganized, as opposed to the patterns where things are grouped together. The physicist Stephen Hawking in his book *A Brief History of Time* uses a puzzle inside a box to effectively explain the principles behind Boltzmann's thinking. Hawking writes:

> Consider the pieces of a jigsaw [puzzle] in a box. There is one, and is only one, arrangement in which the pieces make a complete picture. On the other hand, there are a very large number of arrangements in which the pieces are disordered and don't make a picture.
>
> Suppose the pieces of the jigsaw start off in a box in the ordered arrangement in which they form a picture. If you shake the box, the pieces will take up another arrangement. This will probably be a disordered arrangement in which the pieces don't form a proper picture, simply because there are so many more disordered arrangements.

Each time we shake the box we discover a new unique pattern where the puzzle is broken apart. In fact, in considering Hawking's explanation there would seem to be an endless number of disordered possibilities compared to a comparatively small group of ordered ones, while there is only one single most ordered state where the puzzle fits perfectly together.

Boltzmann concluded that since there are so many more disorganized and disordered patterns than ordered patterns, the flow of time is more likely to choose the disordered patterns over the patterns where things are increasingly organized into orderly groups. Essentially Boltzmann realized that a greater body of disordered possibilities will dominate a smaller group of ordered possibilities, causing time to have a preference for disorder. This he explained is why things tend to spread out in an environment, it is why things tend to rust and decay, and why heat spreads out causing things to cool rather than warm up, and it is why broken objects don't magically piece themselves back together.

We commonly play with this principle, this natural flow to the world, in various ways, most often in games such as pool or billiards. In pool we begin by grouping the balls together in the center of the table. We place all the balls (minus the cue ball) into various ordered shapes, in a group, then we break that order up by striking it hard with the cue ball, which spreads the balls out into one of a vast number of lesser organized patterns. Every time we rack and strike the balls they spread out into a new and unique pattern. So we move from one of the few possible arrangements where the balls are grouped to one of the many possible arrangements where the balls are spread out. Once we have created disorder, we then of course try to skillfully control and re-organize the arrangement by knocking the balls into the pockets.

What we never witness when playing pool is a case where after we strike the carefully racked group of balls with the cue ball, they bounce off the sides of the table and rebound back together into a carefully ordered pattern. They never group back together again, they never line up in perfect rows, even though we might imagine it as one of the possibilities of what could happen after a break. The probability of it actually happening is so small, it simply never happens. The same is true in card games. When we play cards, we hope we are dealt some measure of order. We want to gain cards of the same suit, such as a flush, or we want them arranged in consecutive order, such as a straight or a run. Unfortunately we always face the fact that the highly ordered patterns are the more unlikely results of a properly shuffled deck.

The actual number of unique patterns in a deck of cards is rather startling. With just three cards there are six different possible configurations for those cards to exist in. Add another card and you multiply six by four cards which is twenty four. Multiply twenty four by five cards and you have 120 combinations. With six cards you have 720 different patterns and any seven cards produces 5040 unique patterns. Now it begins to get staggering. Shuffling ten cards leads to 3,628,800 different possible patterns. And with eleven cards you multiply 3,628,800 by eleven, which is almost 40 million different configurations, and so on and so on, until you reach all 52 cards. The total number of unique patterns in only one deck of cards ends up being a bit more than 8 followed by 67 zero's.

80,000,000,000,000,000,000,000,000,000,000,000,000,000,000,000,000,000,000,000,000,000,000,000

Who would think one simple deck of cards could contain such a mind boggling number of possible configurations. And yet, now suppose you turn a shuffled deck of cards over only to find that each suit is separated and in perfect order, just like they are when you first take a new deck of cards out of its packaging. Of course if you spend your whole lifetime attempting to make this happen, shuffling a deck of cards and turning them over time and time again, it will never occur, at least not without the trickery of a magician, and yet within the myriad of all possibilities the most perfectly ordered pattern is one of the possible patterns for the cards to be in after being thoroughly shuffled. It is only one possibility among billions and billions of other possibilities, but it is a very special possibility, because that one pattern of cards is the highest state of order.

The Wedge Model

We can easily graphically represent the ultimate realm of all possible states according to Boltzmann's theory of fewer ordered states. In direct recognition of the second law, the physicist Julian Barbour portrays timelessness with a wedge shape originating from an Alpha state, as shown below. Having named his model of all possibilities "Platonia", in his book *The End of Time* Barbour writes, "Platonia is necessarily skew. It is easy to imagine that the cone 'funnels entanglement outwards', much as a trumpeter blows air from a bugle" (pg. 321).

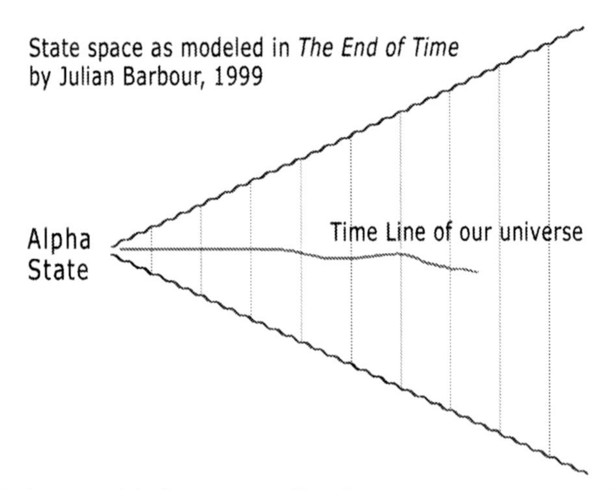

Figure 4.1: Barbour's model of state space reflects how many scientists imagine the realm of all possibilities based upon Boltzmann's second law of thermodynamics.

In this diagram, along a horizontal axis the number of ordered states decreases toward an ever fewer measure of highly ordered states, while in the opposite direction the measure of disordered states increases. I will be referring to this as *the wedge model.* This representation of the large-scale structure of possibilities has been portrayed albeit reservedly in science books as closing at the end of highest possible order which is often assumed to be where time begins, while in the other direction, the number of disordered possibilities grows, and the general assumption is usually of an endless and indefinite expansion of disordered states without end. Barbour writes:

> By sheer logical necessity, Platonia is profoundly asymmetric. Like Triangle Land, it is a lopsided continent with a special point Alpha corresponding to the configuration in which every particle is at the same place. From this singular point, the timeless landscape opens out, flower-like, to points that represent configurations of the Universe of arbitrary size and complexity.

Barbour imagines the possible realm in greater detail than anyone previously based upon the idea that the structure of timelessness can be understood by imagining all the possible triangles that can be formed by three particles. He explains his ideas generally in the passage below:

> Most strikingly, it [the possible realm] is lopsided with a most definite end and frontiers that are there by sheer logical necessity. For example, if you consider triangles as Nows, the land of these Nows comes to an absolute end in the degenerate triangle in which all three particles coincide. This point is so special I call it Alpha. Other frontiers, like ribs, are formed by the special triangles in which two particles coincide and the third is at some distance from them. Finally, another kind of frontier is formed by collinear configurations — all the three particles are on one line. The Platonia for triangles is like a pyramid with three faces. Its apex is Alpha. All the points on its faces correspond to collinear configurations, and the faces meet in the ribs formed by the triangles with two coincident vertices.

It is not hard to recognize that Barbour's ideas are working from the bottom up. He is essentially trying to break all physical reality down to the coordinates of three particles which is wonderfully fundamental, and triangles may be relevant as a way of understanding patterns, but then the difficulty is in showing how all possible triangulated particles might produce such a complex universe where we seem to experience time.

In his book Barbour describes Alpha as butting up against nothingness and from there extending outward without end, "for there is no limit to the richness of being". How the timeless existence of Alpha relates to a primordial nothingness toward the past isn't explained in his book, but it seems more reasonable to imagine that an eternal Platonia excludes the very existence of a nothingness of the type inferred, since Platonia has always been and always will be. On the other hand, if Barbour's nothing isn't considered to be nonexistence, and rather a physically real nothing, then it should be treated as just one of the many possible conditions in Platonia.

Using Barbour's ideas, one might also conclude that the whole of all possible triangles ultimately combine together to create a superspace that extends infinitely in all directions. If such a state were acknowledged, it would also need a proper location in the wedge model. But in being different than Alpha, where would we locate it in the wedge?

Although the model of timelessness explained here in this book is far more developed and more easily understood than Barbour's Platonia, in the following passage Barbour recognizes some basic principles we both share concerning timelessness:

> My conjecture is that some Platonia is the true arena of the universe and that its structure has a deep influence on whatever physics, classical or quantum, is played out in it. In particular, I believe the phenomenon that we call the Big Bang is not some violent explosion that took place in the distant past. It is simply the highly special place in Platonia that I call Alpha.

I originally began to envision the timeless universe in my youth, before I was influenced too severely by the materialistic and reductionistic tendencies that exist in science and particularly in physics. I made what I thought back then was an obvious conclusion. The direction of time is moving toward nothingness, rather than moving away from nothingness. This early conclusion allowed me to let go of the paradoxical necessity that physical existence somehow begins or evolves from a nothing, and it allowed me eventually to more easily consider other possibilities and reevaluate how we presently conceptualize nothing.

The Law is the Law

Boltzmann's version of the second law has become a true paradigm of modern science which physicists rarely scrutinize. Although there were early objections to the asymmetry of Boltzmann's vision in his own time, the most vocal advocate of the second law lived in the era of Einstein. The famous astrophysicist Arthur Eddington designated the second law as holding "the supreme position among the laws of nature". Eddington once remarked, "If someone points out to you that your pet theory of the universe is in disagreement with Maxwell's equations, then so much the worse for Maxwell's equations. And if your theory contradicts the facts, well, sometimes these experimentalists make mistakes. But if your theory is found to be against the Second Law of Thermodynamics, I can give you no hope; there is nothing for it but to collapse in deepest humiliation."

Albert Einstein himself remarked generally of thermodynamic laws, saying "[A theory] is more impressive the greater the simplicity of its premises, the more different are the kinds of things it relates, and the more extended its range of applicability. Therefore, the deep impression which classical thermodynamics made on me. It is the only physical theory of universal content which I am convinced, that within the framework of applicability of it basic concepts will never be overthrown." Seth Lloyd, a professor of engineering at MIT once remarked that "Nothing in life is certain except death, taxes and the second law of thermodynamics."

The basic concepts which Boltzmann proposed in the 1860's have remained relatively unchanged for over a hundred years. One of the main reasons scientists hold the second law in such high esteem is that it provides for science a rare ability to appreciate a reason "why" behind a law of nature, even if that reason is the uncomfortable preponderance of disorder or chaos. Another reason the second law is so prized is that the second law explains something as vital as time's arrow. It explains a fundamental difference between the past and future. The past is always more ordered than the present, and the future is always less ordered than the present, which makes the second law a perfect companion to our observations of an expanding universe as well as the big bang theory.

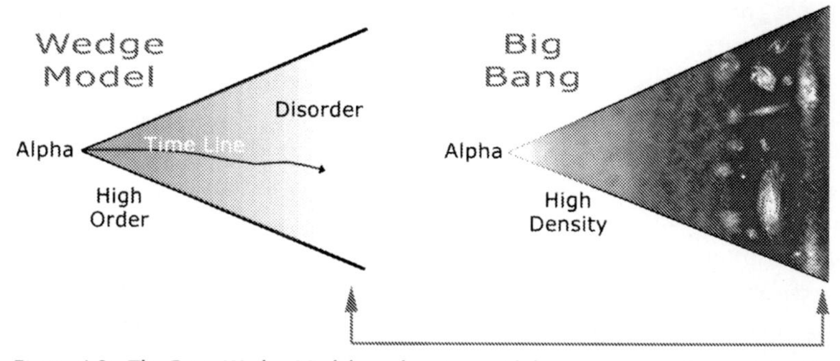

Figure 4.2: The Basic Wedge Model synchronizes with how many scientists envision the evolution of time based upon the big bang model.

Obviously, just looking at the wedge model one can see how the second law supports the big bang model, since as time is traced backward the universe must exist in increasingly higher states of density and order. Boltzmann could have predicted the big bang after 1868 purely from his basic concepts, long before Hubble discovered the expanding universe around 1930. Boltzmann surely would have made such a prediction except that he had no idea that space can contract or expand. He certainly never imagined that space can be removed from the universe, and so he never considered that the whole universe could physically collapse inward or be expanding outward.

Perhaps a young Einstein had an even better opportunity to predict the big bang by combining his new theory of relativity with Boltzmann's second law, since Einstein's own theory originally indicated that space should collapse due to gravity. In his day, Einstein and others believed that the stars and galaxies had existed in time forever, so Einstein invented a make believe force he called the cosmological constant which held the large-scale universe in a steady unchanging state. Later, after the expansion of the universe was discovered, Einstein considered this tweaking of his own theory to be his greatest blunder, because he realized he had invented a force to explain an assumption he was making about the universe, instead of accepting his theory at face value.

The Reason "Why" Behind Time

Now in taking a step backward, and looking at the second law and what we have discussed so far, we can appreciate how the second law suggests that probabilities are the fundamental force driving the evolution of time. This fundamental method of comparing types of possibilities (ordered and disordered) has provided us a reason for why events happen as they do. I am pointing this out because Boltzmann's general approach is the only major explanation of "why" the universe works as it does presently known to science. So far we are very adept at describing how things work, but we know very little about why the universe is the way it is. So the second law is unusually important compared to most other science because it provides a general reason for why time moves forward and it indicates the past is different than the future. In fact as a type of science, as a way of understanding the universe, the study of possibilities is actually a rather unexplored field in science. The second law is very general. For example it doesn't explain why there are forces active in time such as gravity and electromagnetism which work to order the universe. And Boltzmann's description of possibilities does not explain why we experience a systematic and cooperative natural world. In fact, although the abundance of disordered possibilities suggests a reason for the direction of time, it accomplishes little else.

As an important side note, in order for the second law to explain anything, our universe and the flow of time must somehow originate in a highly ordered state. In fact this is a tremendously important issue, and it is an unresolved problem with the second law. The key requirement underlying the second law is that a system has to originate in a condition of high order to then probabilistically evolve towards disorder. As Stephen Hawking also explains:

> Suppose a system [or a universe] starts out in one of the small number of ordered states. As time goes by, the system [or universe] will evolve according to the laws of science and it will change. At a later time, it is more probable that the system [or universe] will be in a disordered state than in an ordered one because there are more disordered states. Thus disorder will tend to increase in time if the system [or universe] obeys an initial condition of high order. (pg.146) [my comments]

Science doesn't yet know for certain why a universe ever managed to begin, or why it began so highly ordered, scientists just believe that once time begins the flow is naturally toward greater disorder. So a considerable stumbling block of Boltzmann's vision is that it doesn't explain how the universe started off pre-organized, which one might say contradicts the law itself. How could the universe ever have organized itself into the most ordered state in all of nature, or at least very near to it, if that state is logically and mathematically so improbable? From another angle, why would the universe pre-exist in a timeless ordered state if that order is so unstable and instantly decays once time begins?

Boltzmann himself found this issue difficult to explain, and the only answer he came up with he attributed to Dr. Schuetz, who was Boltzmann's assistant at the time. Long before the big bang theory, the two entertained the idea that if the universe was sufficiently old and sufficiently vast then given enough time the proper materials will accumulate in some region of space, producing a highly ordered state that would then decay. Of this Boltzmann wrote:

> If we assume the universe great enough, we can make the probability of one relatively small part being in any given state (however far from the state of thermal equilibrium), as great as we please. We can also make the probability great that, though the whole universe is in thermal equilibrium, our world is in its present state. It may be said that the world is so far from thermal equilibrium that we cannot imagine the improbability of such a state. But can we imagine, on the other side, how small a part of the whole universe this world is? Assuming the universe great enough, the probability that such a small part of it as our world should be in its present state, is no longer small.

Of course without any knowledge of the big bang theory Boltzmann only felt it necessary to explain the degree of order that the world existed in back in the pre-modern era when he lived. Today, in knowing the universe is increasingly denser in the past we have to explain how time began in a far more highly ordered state. Boltzmann's explanation would suggest the universe has existed long enough to defy all probability and evolve into the most ordered state of all, a state endlessly more improbable than shuffling and finding the deck of cards in perfect order. Pretty amazing considering the hands I get dealt in poker.

With all this said, after learning the ideas that support the second law, most people appreciate the logic but many feel like something isn't quite right, like something must be missing from the equation. Regardless of how perfectly logical the second law seems, most people recognize that in contrast to the uncomplicated order of an increasingly dense past there also exists a complex orderliness that has increased as the universe has evolved forward in time. How can the second law be true with so much orderliness to account for? Why is so much order maintained from one moment to the next throughout the course of time? In ways the universe is far more ordered than it was in the past. Atomic structure, galaxies and star systems, people, the human world, all appear to be of a more complex kind of order than the simple density of the big bang. If the second law explains the arrow of time, shouldn't the universe have simply decayed from the first moment? Most importantly, why are there systematic forces of nature that work against the trend toward disorder?

Actually there is something fundamentally wrong with the way that Boltzmann modeled all possible states. The second law is not a complete explanation of how the large-scale realm of all possibilities guides the flow of time. The basic approach that Boltzmann used was correct and ingenious. He was correct to focus on order. His way of considering how one ultimate group of possibilities competes with another ultimate group of possibilities is critically important as a method of understanding the probabilistic world we live in. But today we have a much broader picture of what is physically possible. We know the universe is expanding. We even know the expansion of the universe is accelerating. And today we know the direction of time is moving directly toward the extreme state of absolute zero. Where then do we locate absolute zero in the realm of possibilities?

Zero and the Wedge Model

The wedge model is a descriptive representation of how Boltzmann's second law describes all possibilities, and yet it is not widely used in science education mainly due to one reason. The logic seems flawless and irrefutable but when the second law is translated into a visual image, the ambiguity of portraying an endless extension of disordered possibilities becomes quite pronounced and is quite unsettling. Visually the model raises a lot of questions that otherwise are mute. Note the question mark in the diagram of the wedge model below. Does the measure of disordered possibilities continue indefinitely? Is there a boundary in the direction of increasing disorder? Most importantly, where is the extreme of absolute zero located within this gradient of order to disorder?

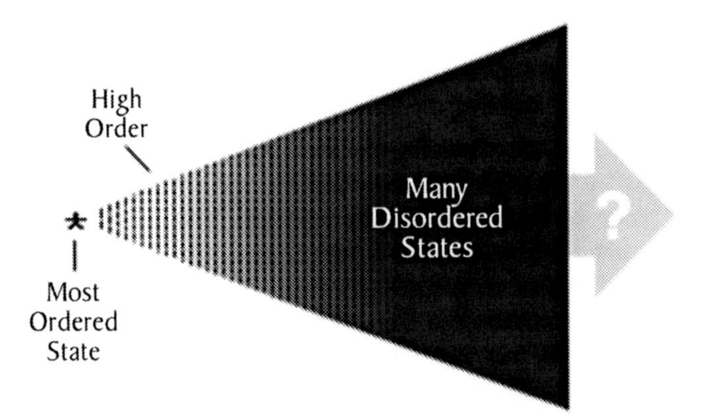

Figure 4.3: The classic Wedge Model of all possible states imagines reality is dominated by disorder and chaos. A small special region of order exists in contrast to an endless measure of disordered states. However, this conception of reality conveniently ignores the fact that absolute zero inevitably exists out there somewhere.

Does the expanding measure of disordered states really continue endlessly? The reader should already sense that the wedge does not extend indefinitely because opposite the Alpha state there is also an extreme of absolute zero. Hawking's puzzle in a box can be expanded to include the pattern where the box is empty. All the cards in a deck can be blank. When we play the various games of billiards we hit the balls into pockets around the table until there are none left on the slate. In addition to all the other patterns there is also the pattern where all objects or things are removed in some way, where parts are hidden or made formless. There is always the single extreme pattern where all things are combined together and everyplace is the same. With cosmological expansion accelerating it becomes quite noticeable that a wedge or a cone cannot represent all possible states because it leaves out a very important boundary.

Absolute zero isn't a myth. Science just didn't find it where we expected it to be. We expected to find it in the past immediately preceding the point where time begins, however, in actuality, when we move in the direction of the past we move toward the top end of physics, approaching infinite mass, infinite energy, infinite density, and infinite heat. There is no evidence to suggest that an ultimate zero ever existed in the direction of our past. There is no valid reason to suspect or believe some type of absolute nothing ever existed before Alpha. In reverse of what we expect, the nothingness of absolute zero is distinctly evident in the direction of our future, not the past. The absolute zero in our future is the bottom end of physics, at the complete other end of the scale. And now that scientists have discovered that the expansion of the universe is accelerating we are being forced to acknowledge zero as an extreme where the universe is stretched perfectly flat. Which means the universe isn't merely moving toward increasing disorder, as the second law suggests. The extreme of zero distinctly appears to exist beyond disorder, on the other side of what we presently imagine as a great bulk of disordered states, so there are extremes at both ends of time that bound the realm of all possibilities.

Large-Scale Realm of All Possibilities

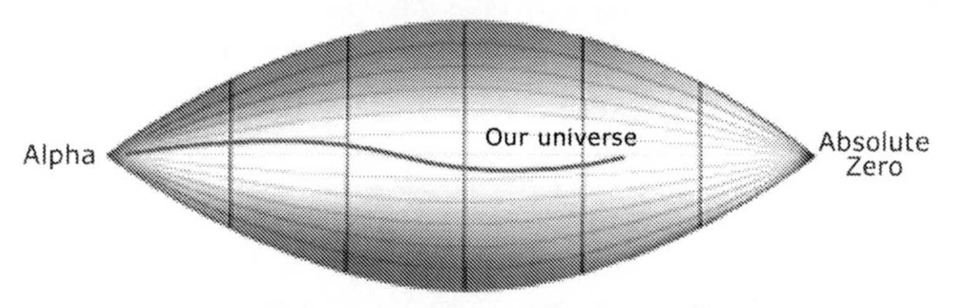

Figure 4.4: Cosmic Absolutes: Even a totally Infinite Universe that includes all possibilities is bounded by two extremes of possibility, beyond which no other possibilities exist. With an Alpha boundary and a Zero boundary, we can roughly draw a picture of the large-scale realm of possibilities. Why is our single universe moving toward zero and even accelerating in that direction, seemingly being pulled toward absolute zero?

For odd reasons, our envisioning of all possible states has been missing a state of absolute zero, as well as the inevitable decreasing number of states in the gradation of states approaching the single state of zero. It isn't like scientists haven't known that an absolute zero is out there somewhere in the space of all possible states, so why has the natural boundary of zero been excluded and not acknowledged when scientists have envisioned all possible states? Also, why don't we expect the universe will eventually become a perfectly flat space — considering we live in an expanding universe? There are various reasons why modern science in the past has failed to integrate an absolute zero state into the realm of all possibilities, one reason being how loudly the second law was defended by Sir Arthur Eddington and others. But more so the reason was because Einstein himself rejected the notion of empty space existing independently of material objects, which made the very idea of empty space extremely unpopular for many years. Before we discovered expansion accelerating in 1998, the expansion rate was thought to be slowing down at an ever decreasing rate, and time in an ever cooling universe was theorized as not having an end, so it seemed okay only a few years ago to imagine that the progression of disordered states in the wedge continues without end.

And of course in the big crunch scenario, which until recently most physicists favored, any scientific concern about whether the quantity of disordered possibilities are actually endless is somewhat irrelevant because the universe stops expanding and returns to Alpha. Had the big crunch scenario turned out to be correct it would have kept zero out of the picture altogether. And since no one ever imagined the universe would simply disappear back into an original nothingness after the big crunch, we might never have rediscovered zero. But since discovering accelerating expansion the big crunch theory is old news, while the complete absence of zero in the future, as portrayed by the wedge model, is suddenly very pronounced. The extremities of zero have certainly become an issue as scientists freshly consider the range of possible futures.

Unfortunately today discussing ZAT as if it is a real physical condition is still controversial, even frowned upon. For example, when the Dartmouth physicist Robert Caldwell presented the Big Rip scenario in a scientific journal article he invariably described time as ending at the ultimate singularity, but even this mention was very reserved for a paper describing the fact that time may end specifically due to the extremities of absolute zero and flat space.

Undeniably, the notion of empty space is difficult to imagine or reconcile conceptually. We tend to look at zero from our place within a world of things, so we see zero as zero things. Zero has often been semantically confused with nonexistence, even though the real zero, the common point of absolute zero in physics, is one of the most fundamental axioms in all of science. Why has zero been overlooked? It has just been a blind spot in science, inevitable perhaps and reasonably so, but we are slowly awakening to the reality of a zero future.

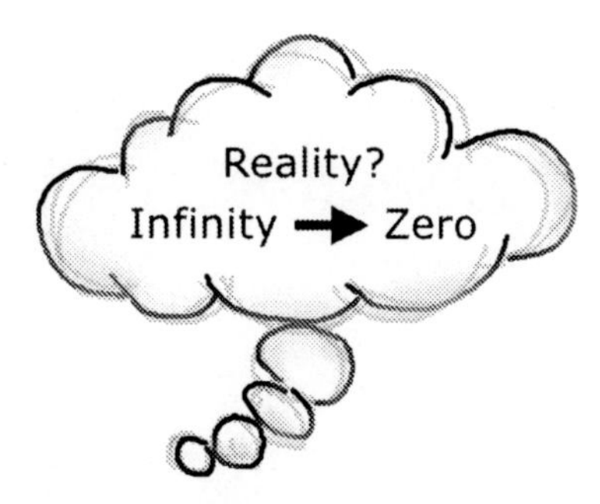

What is Reality really like out there?

Once we include zero in the set of all possible states, a plethora of new questions arise that demand to be answered. How should we conceptually describe absolute zero? What happens to the existence of our universe if time ends at Omega zero? Why would time travel all the way to absolute zero if that state is the extreme edge of what is possible? How do we reconcile absolute zero in reference to order and disorder? Should zero be described as the most disordered state, or is zero something all together different in terms of order and disorder? These are just some of the questions that must now be answered.

What follows in the next chapter is so basic and fundamental that it seems like something that should have been discussed years ago in the time of Galileo and Newton, or perhaps even further back in history by ancient philosophers. However, it wasn't exposed or discovered back then plainly enough to be passed on to us, so here we are discovering something very basic about the universe amidst so much other information. It is not easy to relearn fundamentals, but when we begin to wake up to the reality of zero and confront its nature and its role in time, what is explained in this next chapter is what we discover.

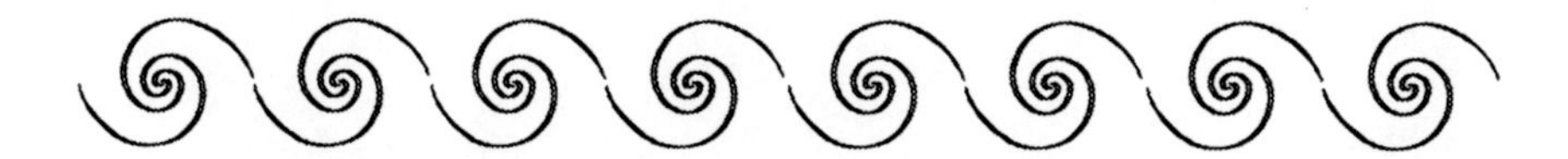

A Brief Forward to the Two Kinds of Order

So far on this planet we can't seem to grasp a full understanding of the Universe. We don't even understand how or why anything exists. Is the Universe comprehensible? Consider all that we have accomplished already. And yet something is wrong, we even know something is wrong, but we can't see the problem to fix it. What is this problem we can't see?

We have recently entered a new millennium and interestingly there were several major new discoveries in science that will undoubtedly mark the present as the beginning of a great time of discovery. We are also approaching a remarkable paradigm shift as we learn to understand physical reality in a new light, a shift not only in physics and cosmological science, but more so in the general way that we humans view the universe. This paradigm shift marks the beginning of a great age of reason, where humankind begins to fully understand the whole.

What is the problem we can't see? Presently the most fundamentally important concept we use to understand the universe is faulty, that being our understanding of order. Presently we think in terms of order and disorder. And that is the problem, because there are actually two kinds of order in nature, rather than simply order and disorder. And once we recognize the two orders, once we fix the problem in how we see the world, suddenly the entire flow of time and the world of human events appear as an interplay of two contrasting orders, while the idea of general disorder or chaos gradually loses its meaning. Two orders is as important as Einstein's Relativity or Quantum Theory. What follows opens us up to the true Universe to create a far deeper comprehension of time and ourselves.

Fractal Art: Bubbles © Kerry Mitchell

While empty space appears to be devoid of properties, to a modern particle physicist empty space, also called the vacuum, is an enormously complicated substance.

Alan Guth

~~~

Nothing is so powerful as an idea whose time has come.

Victor Hugo

~~~

Something deeply hidden had to be behind things.

Albert Einstein

~~~

We think the world apart. What would it be like to think the world together?

Parker Palmer
~~~

The problem of finding the one thing that lies behind all things in the universe is called the problem of the one and the many. Basically stated, the problem of the one and the many begins from the assumption that the universe is one thing. Because it is one thing, there must be one, unifying aspect behind everything. This aspect could be material, such as water, or air, or atoms. It could be an idea, such as number, or "mind." It could be divine, such as the Christian concept of God or the Chinese concept of Shang-ti, the "Lord on High." The problem, of course, is figuring out what that one, unifying idea is.

Richard Hooker

Chapter Five

Caught Between Two Kinds of Order

A System of Understanding the Order in the Universe

In the same way that the presence of a universe seems like a miracle, as if there should instead be nothing at all, so also are we perplexed at the order that is such an elementary part of the universe in which we live. There should instead be chaos, it seems, rather than the one particular universe observed, for we naturally consider the infinity of less consistent, less trustworthy, and more chaotic, worlds that might exist instead. Yet suppose for a moment a rarely considered idea, that this universe is not actually unordinary or improbable, and instead we ourselves are making some critical mistake in how we see the world, a mistake which if we could somehow see beyond, suddenly the order in the universe would seem exactly how things should be.

Very simply stated, at present most everyone, even a scientist, believes order is properly defined with a single concept. In most dictionaries order is defined as "a comprehensible arrangement among the separate elements of a group." Disorder is commonly defined as chaos, clutter, confusion, derangement, disarray. If we translate our sense of order into an image, we would draw an axis, with greater order in one direction and disorder in the opposite direction. Thus if the order of a pattern increases its disorder must decrease.

❄ ← Increasing Order / Disorder →

Although the following is more developed than our existing vague definitions of a general order and disorder, the basics of what I am about to explain aren't very complex or difficult to envision. We all know a great deal more about this subject than we realize because of our immersion in nature and because of our participation in the ordered flow of time. As it turns out, we are surrounded by two kinds of order which oppose one another and yet work together to create all the complex and diverse patterns we experience. One order comes to us from the past, while the other order comes from the future.

Grouping Order

The type of order we are all most accustomed to recognizing is *Grouping Order* which can be understood as any class, or similar kind of thing grouped together, and located in a specific area or separate place usually apart from another group. Grouping order is the precursor of things and responsible for the definition we know as the finite world. It is very common and very easy to recognize. By nature, the identity of like things grouped together becomes more pronounced. Things stand out more as a group.

For example, when we go to the store, there are groups and sub-groups of different products, each grouped separately from one another. At any grocery store the apples, the oranges, the bananas, each of the vegetables, are grouped together separately. Products are also grouped into larger groups, the meat section, the bakery section, the dairy section. If all the fruit was displayed mixed together with all the vegetables only the largest individual items would stand out. Yet separated, each group is very defined and pronounced. In fact we find grouping in every store, every business, every city. Likewise, stores and businesses are grouped apart from residential areas.

At home we keep the socks, tee-shirts, and underclothes each in their own place in the dresser drawer. The dishes are kept in the cupboards, and the canned food is kept together in another cupboard. Books are all grouped together in a library, where they are organized into sub-groups by subject or title. When we communicate with others, when we convey ideas in writing, we tend to discuss one topic at a time, and we prioritize our subjects. There are places where we congregate to do things, where we shop, where we eat, where we pray, where we play. Everywhere we humans inhabit we group things together as opposed to the chaos of individual items being randomly located throughout a room or a space.

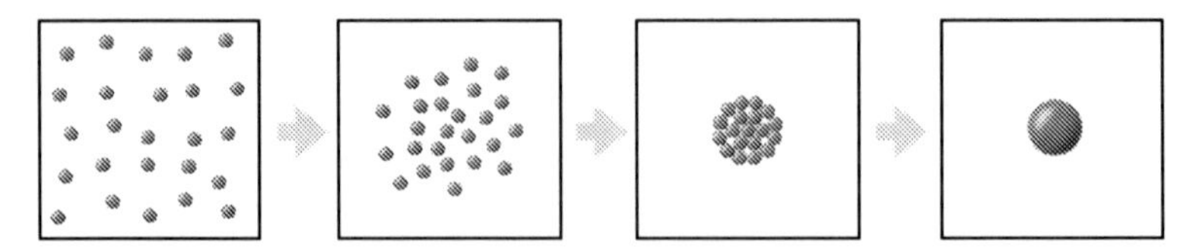

Figure 5.3: Generally, in the direction of Grouping Order objects are more dense within a frame of reference, in extreme creating a more pronounced single object, as when stacked checkers become a king.

In science the classic example of order given is a concentration of gas particles contained within a flask, as opposed to the gas being spread throughout a room, which is said to be disordered. But we could recognize a concentration of gas particles more specifically to be grouping order.

A prototype idea to represent grouping order is the game of checkers or chess. To begin the game we divide the pieces apart and place pieces of one color on one side and the other color on the opposite side. With the pieces previously mixed together randomly inside a box we would say that they were disordered or mixed irregularly, until we separated them by color into two distinct well organized groups.

Symmetry Order

Grouping order is not the only kind of order we find in our surroundings. We actually exist in nature between two different kinds of order. The other type of order is best referred to as Symmetry Order which if I simplify its definition to extreme is an even and regular pattern or arrangement in which all different types of things are combined together and distributed evenly throughout the whole frame of reference. Where grouping order divides and separates things into many groups, symmetry order mixes and combines things together ever more evenly.

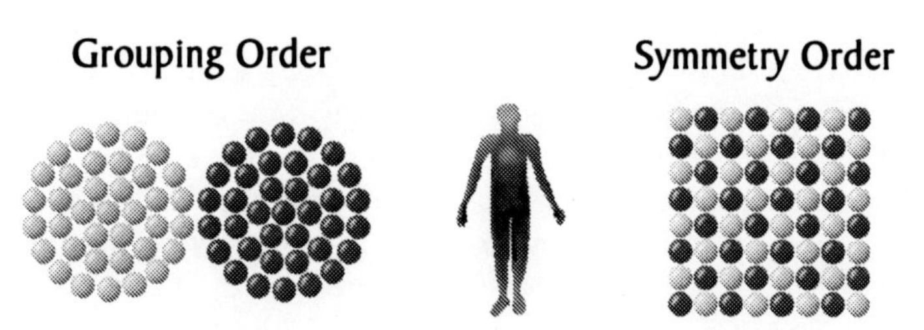

Figure 5.4: Grouping Order moves toward division and separation while Symmetry Order moves toward balance, integration, and unity.

The prototype idea for symmetry order is also the game of checkers or chess, that is, if we set the pieces aside and focus our attention on the board on which the games are played. A simple checkerboard is obviously ordered but we normally might not reflect upon it as a special type of order, however, consider how the checkerboard is ordered in an opposite way to that of the game pieces separated into two groups. The colored squares of a checkerboard are mixed together evenly, white, black, white black, while the next row alternates black, white, black, white. Where the game pieces begin perfectly grouped apart, the squares of a checkerboard lattice are mixed together evenly, making the two patterns opposites.

Each square in the checkerboard lattice is a measure of grouping order. So we can increase the symmetry order of a checkerboard by dividing up each square into smaller, finer squares and then mix them back together again into a finer checkerboard lattice, which decreases grouping order and increases symmetry order. If we divide up each square and continue to mix the squares and colors evenly, increasing the measure of balance, the black and white squares visually merge like two colors of paint to finally become a neutral gray. Once the pattern is made uniform, we have reached an extreme. There are no further possibilities in the direction of symmetry, because we have produced a oneness. Many things or many colors have unified into a balanced whole. Of course a simple pattern of gray doesn't appear to be ordered at all, which has been the stumbling block in the past. It is why we don't commonly recognize that there are two opposing directions of order in nature, because while grouping order produces an obvious order, symmetry order produces a hidden type of order.

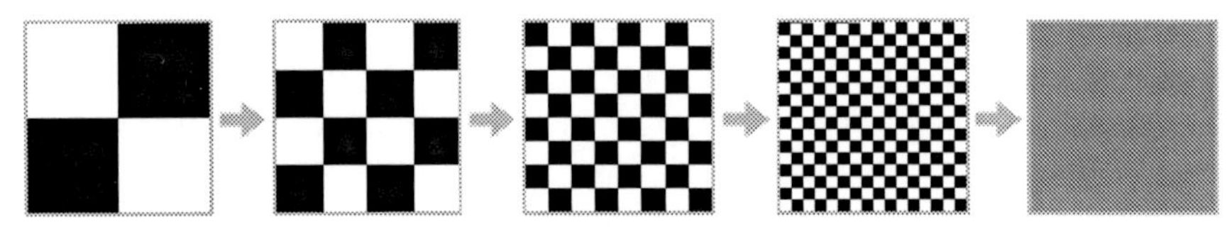

Figure 5.5: As Grouping Order decreases another form of order, Symmetry Order increases towards an extreme of perfect balance and uniformity.

In this example, the uniform gray color is only one type of order so it looks plain to us. Keep in mind that the word symmetry refers to the measure of sameness in a pattern. Nothing is more visually symmetrical than everything painted the same color. Even though a checkerboard pattern seems to be more ordered than a simple gray color, the single uniform gray is hiding its order. It is hiding its fine structure and its fullness beneath its uniformity the same way an uncut pie doesn't show all the pieces it can be cut into.

Suppose we take two cans of paint, red and yellow, and we pour them together into a larger bucket, and then stir them together. Just a moment ago we had two separate and distinct colors, but now we only have one. We try adding another color to the stir, blue, and the distinction of that color disappears also. With paint the product of many assorted colors is brown. Brown soil is the most common case of many individual things, a variety of chemicals and minerals, combining together to create an integrated less pronounced medium we often derogatorily call dirt. The idea already mentioned of taking everything in the frig and cooking it into a thick paste applies here also. There is nothing much exciting about dirt or paste because we don't appreciate mediums and balances. In being indistinct they seem hardly more than nothing at all. It is only when we begin to notice how many even distributions and symmetries we observe around us that we begin to appreciate the great power of symmetry order.

There isn't merely one kind of order where things form distinct powerful groups, there is also a direction where previously distinct separate objects or colors unify into a single form. Increasing symmetry order ultimately causes a shift in the nature of the pattern, since in the final stage of balance, what before were distinct and separate things end up as one, not one group, rather one overall form that takes over an area or a volume.

Most everyone knows that if we shine white light through a glass prism we find that the content of white light includes an entire rainbow of other colors. Where did all these colors come from? The prism merely exposes how the color is somehow embedded in the fullness of white light. Typically this is not how we think. We tend to think from the bottom-up rather than from the top-down. We refer to certain colors as primary colors and imagine for example that yellow and blue combine together to create green, while yellow and red create orange, and red and blue create purple. We further mix colors to create the tertiary colors, or a third level of colors all produced by the primary colors. We imagine the primary colors are fundamental and create all other colors, and in this same way of thinking, scientists today imagine that atoms are fundamental and create the universe, not the other way around. Yet we all know there is the top-down perspective where all colors are born out of the true primary color that is white. If we stay in the top-down perspective then orange is a primary color that divides into yellow and red, or a primary green divides into blue and yellow, or a primary purple divides to create red and blue. The top-down is a whole other way of looking at things where a seeming formless whole divides apart into lesser colors and lesser shapes. The same general principle can be applied to the space that surrounds us. Matter isn't more than space, matter is of space.

We actually see evidences of the whole are all around us. Lesser measures of symmetry order exist everywhere. Where ever you are as you read this, you can easily observe something uniform, or something spaced or mixed evenly in your environment. Every mass of material is grouping order, but all the balances and uniformities in the world are of symmetry order.

Figure 5.6: Recognizing the two orders leads us to specify any measure of balance as a separate component of a pattern distinctly separate from the order of grouping. Any evenness, sameness, or balance in a pattern is an expression of symmetry order.

It can be surprising to notice at first how opposite grouping order and symmetry order are to one another. In the image below, we see the extreme results of each direction of order, with the checkerboard lattice caught in the middle between both orders. On the left the contrast and pronunciation of the pattern has increased to an extreme of white and black. On the right side the individuality of black and white squares of the checkerboard has dissolved into a neutral gray. All definitive form has been given over to the uniformity of a single color. On the right the many squares become a single whole.

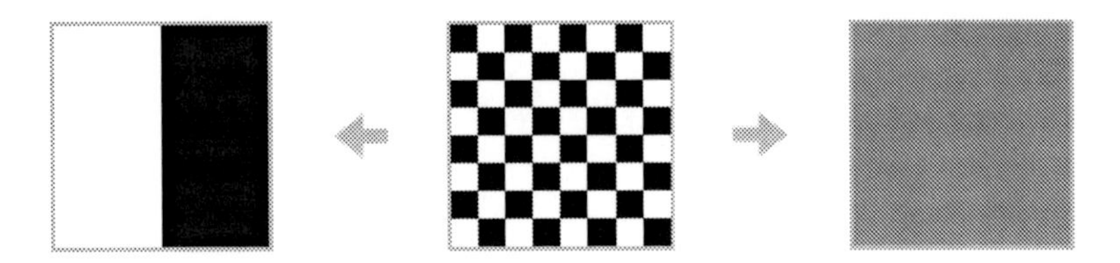

Figure 5.7: Above, in one direction of order away from the checkered pattern the parts form two pure groups of opposition and difference, while in the other direction the parts merge into one color. Below, the same principle but presented in terms of matter and in three dimensions. Toward grouping order matter unifies into increasingly dense objects such as stars while in the direction of symmetry order, equal positive and negative particles annihilate (unify and hide) to leave behind the singular uniformity we know as space.

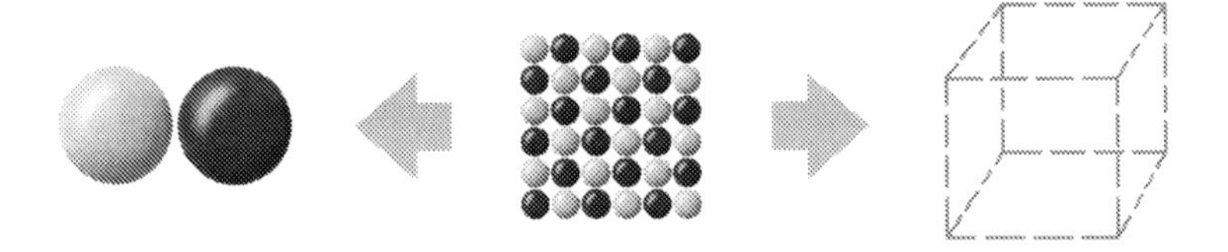

When we separate and group any two likenesses apart, be it apples from oranges, large coins from small coins, men from women, matter from anti-matter, or any imaginable classification, we invariably create an imbalance. In terms of matter, when all the positive particles in the cosmos are grouped apart from all the negative particles, we have reached the perfect extreme of imbalance where there are no other possibilities in the direction of further grouping order. This is important because it essentially means that what is ultimately possible is bounded by a grouping order extreme. Just as there is an Alpha boundary in science, there is a boundary to what is possible of grouping order.

There also exists a symmetry order extreme. When we combine things together evenly the distinction and qualities of separate things also combine together to create the balanced sum and whole of all parts. The squares of a checkerboard lattice become gray, making everyplace within the dimensions of the board identical. Many colors become one color. Matter and anti-matter when combined together, even the infinity of all possible universes combined together, become the singularity of space. Pushed to extreme, symmetry order naturally produces sameness and oneness, even though nothing of what existed before is actually lost to the unity. Each quality forms the whole because each quality was always a fragment or an abstraction taken from the whole.

The process of definitive objects and colors merging into a singular form is terribly under-appreciated and completely misunderstood at this time in history. We are constantly experiencing similar uniformities in our surroundings where many parts are forming a whole, but if we don't see the parts we often wrongly evaluate the whole to be a 'nothingness', when it is really an 'everything'. Consequently in regards to understanding order, we are in a sense only seeing one side of the coin. We are only appreciating grouping order. We are only respecting the direction of order where things group into a larger object (which is actually always an imbalance). When we see a plain white canvas hanging on a wall, instead of thinking it represents nothing, we can see an integration of all colors hiding behind that oneness. We should know it as an 'everything' color, since otherwise we are being perceptually lazy.

Finding Two Kinds of Order in Nature

In recognizing the order of balance and symmetry apart from the order of grouping, we can now begin to identify the two orders separately in virtually every imaginable scenario. We can view complex images and patterns and identify the two orders separately. We can look out at the universe and see the two orders working together. We can also learn to see the two orders as two very basic processes occurring everywhere in the dynamic flow of time. One order exists in extreme in our past. The other order exists in extreme in our future. In one direction of change, toward the past, matter separates into elementary parts, while in the other direction, toward the future, the universe integrates those parts into a whole. These two fundamental directions of change are clearly opposite of one another, which is what necessarily defines two very different types of order.

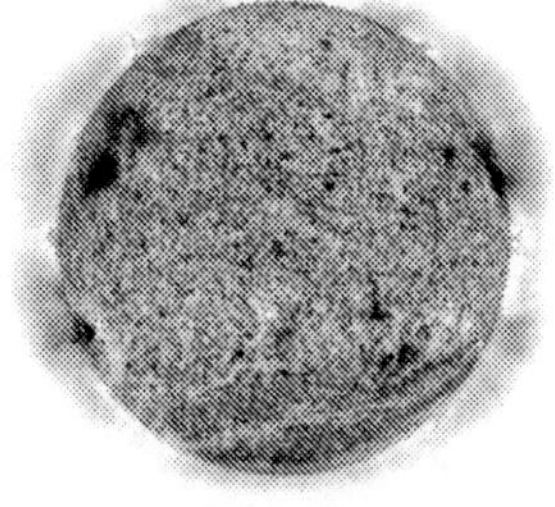

The great hierarchy of grouping order begins with the massive group of atoms we call a star, the most noticeable feature of the universe. Then our nearby star, the sun and its nearby planets and asteroids, all held together by gravity, form the group we call a solar system. Then at another level, all the stars and their planets are gravitationally grouped into galaxies, which often contain billions and trillion of stars. Beyond even the galaxies, there are still larger groups, there are thousands of galaxies bound together by gravity called clusters and superclusters. As we shall see later, gravity, the force that produces all these groups is the forceful influence of grouping order.

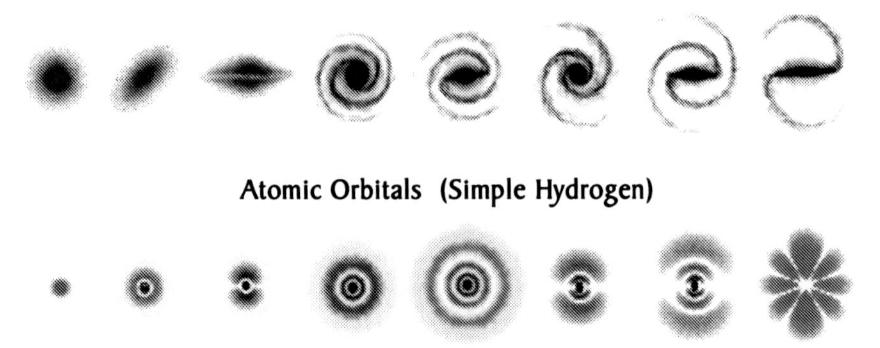

Figure 5.8: We see in galaxies and atoms the range of patterns from simple grouping order on the right side to the complexities of two kinds of order working together on the left. We can easily learn to identify and distinguish between grouping and symmetry in all patterns in nature, from the macro-world of galaxies to the micro-world of atoms.

The Earth is a complex study of grouping order, from its iron core to its diverse crust. The divide between land and ocean, as well as the layers of the Earth's atmosphere, are obvious examples of grouping order. All of our natural resources, coal, natural gases, petroleum, minerals such as copper, silver, and gold, require like atoms to form large pure groups. Of course the entire table of elements is a classification arisen from grouping order, and there are family groups of elements such as metalloids and noble gases which have similar properties. Generally there are three basic groups; pure gases, solids, and liquids.

All the increasingly distinct groups we know of exist in stark contrast to another universe we might imagine completely void of the value of grouping, where all the elements are spread chaotically throughout space so that there would be no stars or planets, just an unruly cosmic soup of atoms. Further still, we can imagine even the absence of sub-atomic particles where the universe is just smooth fluidic plasma, a dense nothingness spread evenly across the entire volume of an infinite space.

While groups of elements and solar masses give the universe its definition and bring about order as we know it, the universe also combines and fuses different materials together producing various patterns of increased symmetry order. The oceans, the soil, and the atmosphere of the Earth, each contain a varied mixture of unique materials. Rock, glass, wood, soil, plastics, and metals such as bronze and steel are all admixtures of atomic materials. The mysteriously systematic and ordered universe is not simply chemical elements existing in pure form, the ninety three pure elements that occur naturally on Earth also combine together creating a nearly endless array of cooperative coordinated materials. Where like atoms form distinct and pronounced groups, unlike atoms bond to produce thousands of different molecules, compounds, and solutions in our environment.

In the larger world, as in the small, there are rhythms, cycles, orbits and oscillations, which all reflect measures of balance. Patterns are always coordinated, repetitious, and consistent around some point of balance. And pattern balances exist everywhere. We don't simply group things, we also arrange things evenly, we space things out rhythmically, we try to balance our lives, and we believe it all evens out in the end. Balances are everywhere in nature. Across the vast reaches of space there is an isotropic distribution of galaxies as far as our telescopes can see. We don't look out at the overall universe and see all the galaxies grouped in one direction and nothing but dark voids in other directions. Nor do we see distant inexplicable regions where the laws of physics shaping our region of the universe don't seem to apply. Even the governing of reality is evenly distributed.

The reason all these distinctions can be pointed out is because there is not simply order in the direction of our past and disorder in the direction of our future. Rather like an ultimate chess game, time as we know it is literally the one type of order transforming into the opposite type of order. That process forces the two types of order to compete and cooperate together in creating the entire range of simple to complex patterns we observe in nature. The two orders work together to create the complexity of our experience.

We live in the great divide between the extremes of two kinds of order, so we can't help but know the natures of these two types of order because we encounter them everywhere we turn. In imagining dividing up the game pieces of checkers we can recognize the nature of grouping order. Words such as opposition, distinctiveness, definition, pronunciation, duality, and conflict, all describe the nature of grouping order. All result of dividing the whole into its constituent parts. Grouping order turns small differences into great differences. Then in imagining many colors becoming one or the way the checkered lattice transformed into the complete uniformity of gray we can recognize the nature of symmetry order. Words such as unity, sameness, similarity, integration, combination, harmony, and oneness describe the nature of symmetry order. We shall eventually discover quite vividly how these two basic natures give shape to and rule the universe.

Grouping and Symmetry as Separate Components of a Pattern

The key to distinguishing between grouping order and symmetry order involves recognizing properties of grouping and symmetry as entirely separate components of a pattern. Once the two orders are identified and understood it becomes apparent that the entire variety of simple to complex patterns in nature are forced to utilize two different types of order, this being true even of what we consider to be disorder.

To begin to recognize how the two orders cooperate to create orderliness and complexity, we shall start off with a few very basic patterns in which the two orders are both separate in one way and combined in another. Each pattern below utilizes both grouping and symmetry order in a unique way. Notice how the arrows show these simple patterns of order are actually two types of order working together.

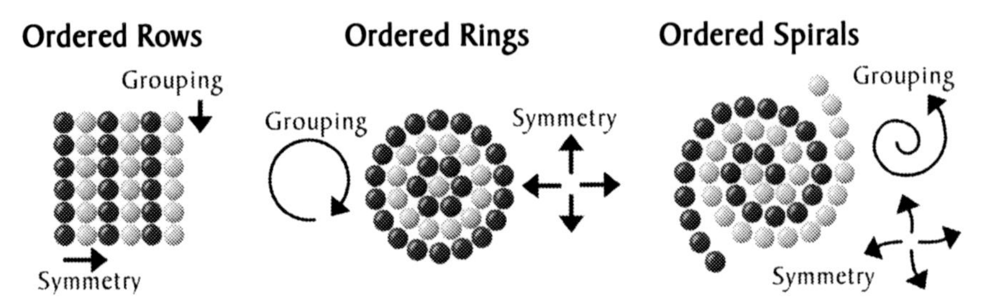

Figure 5.9: It is not difficult to learn to recognize how any pattern is composed of both grouping and symmetry order properties. In the first example above, grouping is necessary to create the vertical rows, and mixing creates the horizontal symmetry. In the ringed pattern, which is rows in a closed circular pattern, the grouping of a common material creates each ring, while the layers create oscillating rings. In similar fashion a spiral is an elegant and attractive mix of rows and rings.

In each pattern shown above, one of the two orders is dominating in one direction while the other dominates in another direction, shown by the arrows. The exclusivity and opposition of the two orders requires that they compete with one another, since increasing one order decreases the other. One order wins the battle in some particular way, while the other wins the battle in another way. Intense competition makes it appear like the opposing orders are also working together, even cooperating with the goal of creating complex orderliness.

Common examples of evenly spaced rows include trees in a forest, buildings in a city, crops planted in fields, fences and posts, telephone poles, city streets, and the sentences on this page. Examples of rings or concentric shapes in nature include anything rounded or layered, such as the majority of fruits and vegetables, as well as most plants and flowers, and also most embryonic eggs and seeds. Man-made circular shapes include rings, buttons, cones, donuts, wheels, most carnival rides, and of course hula hoops. In the larger cosmos, internally we find the concentric layers of the Earth versus its core. Such layers exist in all planets and stars. Externally there are the layers and densities of the Earth's atmosphere, planetary rings such as those around Saturn, and stellar asteroid belts in solar systems.

Figure 4.10 Orderly Rows

Even spacing of rows and lines is an elementary way grouping and symmetry combine.

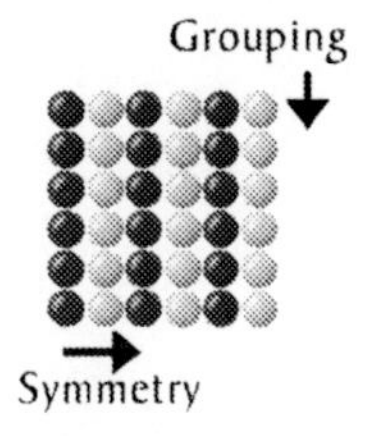

The volume of a tree is grouped in the line of the tree while the trees are generally spread out evenly through a forested area. A mass of buildings is pronounced grouping order, as is each separate building, yet in some measure they are spread out evenly throughout the city. Like pebbles of sand, apple trees are collected in groups called rows but only symmetry creates the even spacing of an orchard and ripples on a beach.

Photo Credits. Trees: © Mike Levin. Buildings: © Max Bian

Mushroom: © Piotr Pieranski, Sand: © Josh Wickham

The design of flags must naturally utilize simple measures of grouping and symmetry.

Figure 4.11 Rings and Layers
Both grouping and symmetry are easily recognized in round circular patterns and concentric layers commonly found in nature. Note that rings are curved rows.

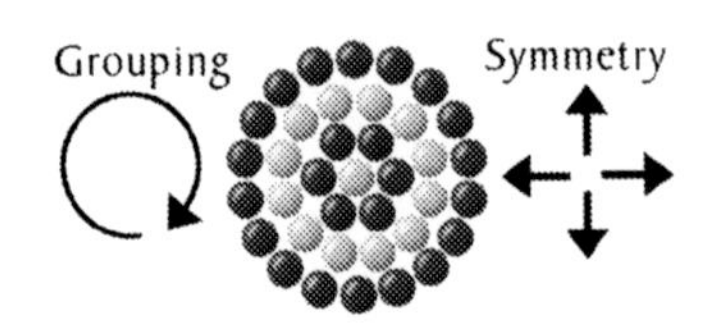

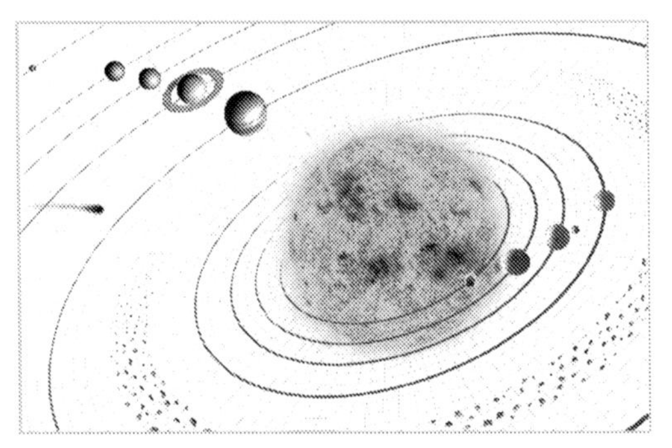

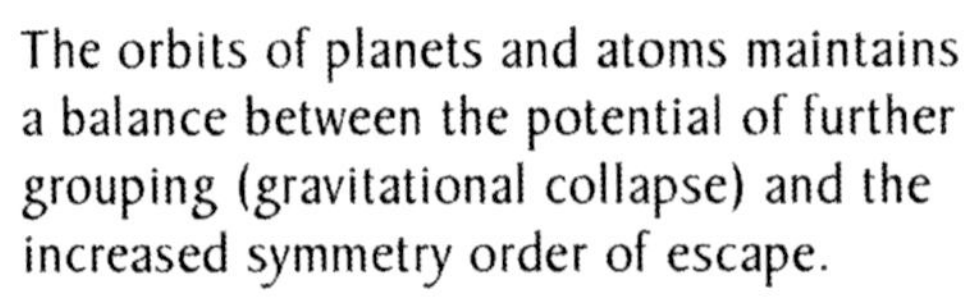

The orbits of planets and atoms maintains a balance between the potential of further grouping (gravitational collapse) and the increased symmetry order of escape.

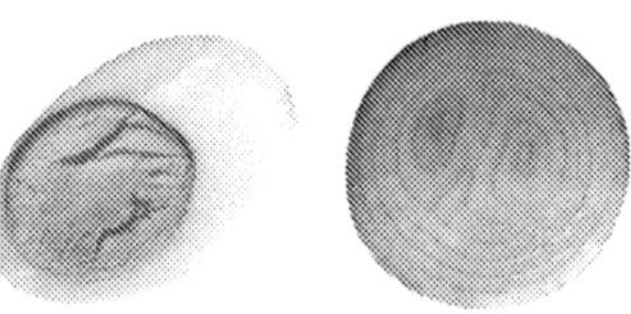

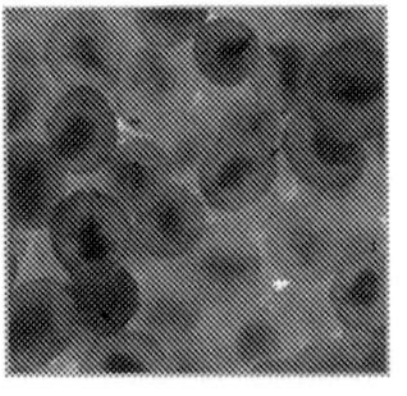

Above and left: Just a few of the nearly endless number of biological examples where we see two orders at work producing concentric layers, such as fruits and vegetables, eggs and cells, flowers and plants.

Grouping Transforming into Symmetry

Presently we are surprised about the existence of order in the universe when we should expect order and see it both as inevitable and natural. In fact, all there is, is order. There are only two directions of possible change, either things separate into groups to form greater definition or distinction, or things mix together toward an indistinct uniform whole. Either patterns move toward separation or move toward integration. There simply isn't any other way in which to physically arrange things.

Surely the most beautiful and eloquent method by which the two orders combine, in a very simple and identifiable way, is in the shape of a spiral. Spiral shapes represent a third stage of increasing complexity and systemization beyond simpler rows and rings. The spiral is the prototype example of how the two orders cooperate in pattern composition, leading to greater complexity. A spiral clearly exposes the natural opposition of the two orders, with the decay of grouping giving way to the symmetry of the spiral.

The great spiraling galaxies portray the most basic struggle in nature of grouping order competing with symmetry order. In the mass of the galaxy we see grouping. And in the spiral we see symmetry. Why spiral galaxies form is not yet completely understood in astrophysics. Just as dense gases are turned into stars by gravity, all galaxies originate as more rounded elliptical galaxies. With two orders in mind we can appreciate how spiral galaxies utilize rotational spin and centrifugal force to counter further gravitational collapse. The rotation of a galactic body as a whole producing the spiral is necessary to keep the body from collapsing toward the center of mass. In this way the rotating spiral reflects the dynamic balance that exists between the order of the past and the order of the future. The formation of galaxies reveals a natural tension, a sort of competition in the universe, between further grouping and increasing symmetry.

Common examples of spirals include galaxies, storms or hurricanes, the rotations of planets and meteorites around a star, whirlpools, seashells, a vast variety of plants and flowers such as Cauliflower and Romanesque, cacti, pine cones, roses, and fingerprints. Man-made spirals include fans, windmills, saw blades, drills, springs, propellers and turbines, not to forget old style lollipops and licorice. Any search on the internet for spirals will reveal an almost endless array of elegant spiral shapes and fractals, and the beauty we see in spirals, our fascination and appreciation of them, arises because of how each distinctly displays a simple cooperation between grouping and symmetry.

As to the question of why there is all this order in the world around us, the answer is really that we should expect it to be here. What we think of as disorder exists only as a temporary stage in the transition of one kind of order becoming the other. There really is no such thing as general disorder. All patterns necessarily consist of two kinds of order, even disorder.

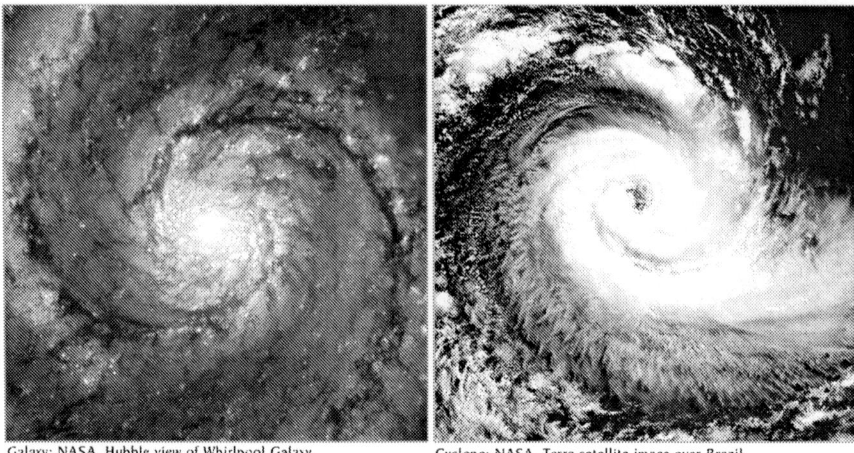

Galaxy: NASA, Hubble view of Whirlpool Galaxy.

Cyclone: NASA, Terra satellite image over Brazil

Figure 4.14 Spirals: The rotation of a spiral is the most common pattern in nature where we see grouping order and symmetry order competing against one another.

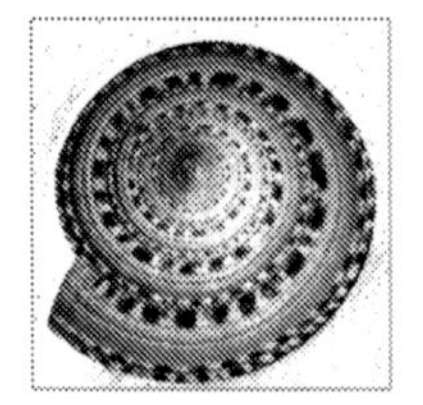

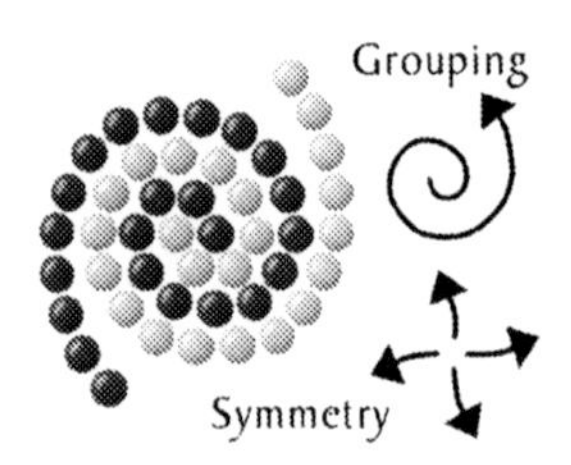

A variety of common spirals and fractals. The grouping and symmetry in each pattern can be recognized separately. The grouping in the pattern makes each spiral distinct and stand out while the spiral expresses an unraveling symmetry.

Photo Credits. Rose: ©John Evans, Romanesco: ©John Walker, www.fourmilab.ch
Wave: © Paul Topp, Fractal Art: ©Kerry Mitchell www.fractalus.com

All patterns combine grouping and symmetry in some regular or irregular fashion. Since disorder includes in its composition constituent parts of both grouping and symmetry, recognizing the two orders challenges the very existence of a generally disordered pattern.

The Two Opposing Directions of Order

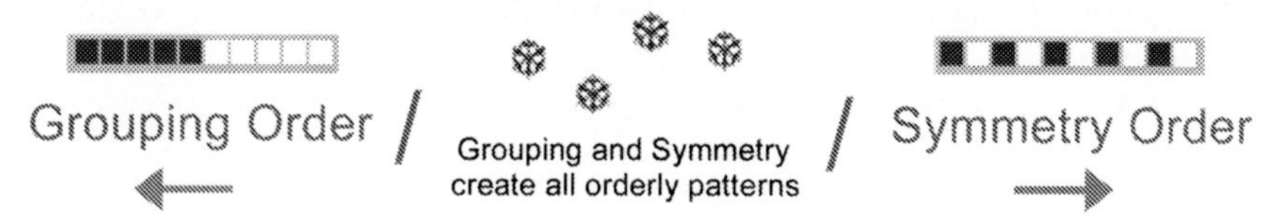

Imagine playing a game of checkers half way through and then stopping. If we study the positions of the pieces against the backdrop of the checkerboard they invariably exist in one of many possible irregular patterns, in which we can find areas that the pieces are grouped, as well as a measure that the pieces are evenly distributed. In what direction is the flow of the game evolving? Imagine having no knowledge of how the game came to be in its present state. What has transpired in the past? What will happen in the future? If we only know of the first half of the game we end up with a distorted view of the transformation that is occurring, but as the pieces are removed the game is moving toward the symmetry of the checkerboard. The same is true of the universe in general, it is only when we begin to study the transition of patterns from start to finish that we begin to genuinely appreciate the forward direction of the grand cosmic flow of time.

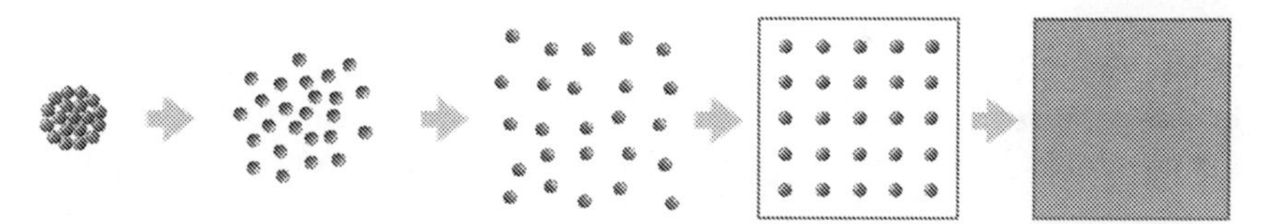

Figure 5.13: From order to disorder to order. The typical way that we expect changes to occur involves order (grouping) becoming disorder as shown in the first three stages, yet in that same direction we see that a state of even spacing (symmetry order) exists beyond disorder, as shown in the fourth stage, then finally perfect balance, where parts unify with the reference frame, i.e., space.

What we think of as general disorder is actually just a temporary stage within the transition from grouping order to symmetry order. In the first three stages above we recognize a common portrayal of order becoming disorder. A single dense object breaks down and spreads randomly within the volume of the square. In the first three stages it appears the order of the single object has disintegrated into disorder, however, in those same steps the pattern is potentially moving toward balance. If we imagine the transition as moving toward balance, the next stage will be to move toward the pattern where the objects are perfectly balanced within the space of the square, as in the fourth stage. Beyond the fourth pattern the objects can only disintegrate or dilute toward the smooth extreme, where the objects integrate with the reference frame and so disappear.

All patterns exist trapped between the two extremes of order, and the only difference between ordered patterns and what we see as disorder is a measure of tension between the two orders which varies in intensity. If the strength or intensity of both orders is high, then the pattern exhibits a high measure of both grouping and symmetry, which we recognize as orderliness, as shown below:

Figure 5.15: In this very simple transition between grouping and symmetry the tension between the two orders is high, so there is no irregularity. Each stage begins grouped, then breaks up, then spreads evenly. Once the objects are spread evenly they are forced to undergo a phase transition by dividing apart, this being necessary if the pattern evolution is to continue toward symmetry.

The transition shown above displays a very high tension between the two orders which decreases the measure of irregularity, making each stage in the transition from one order to the other order appear orderly. At each stage in this transition above the influence of each order is intense, therein creating a lattice. Lattice structures are extremely basic to nature and exist at the atomic level in most of the chemical and material compositions of our environment.

If the intensity between orders is low, then the pattern is freer to exhibit irregularity. Below the two transitions are compared. In the orderly extreme series we see the influence of each order extending throughout the transition, as if the invisible influence from the order on each side is reaching all the way across to the other side. With lower tension the influence of each order is weak.

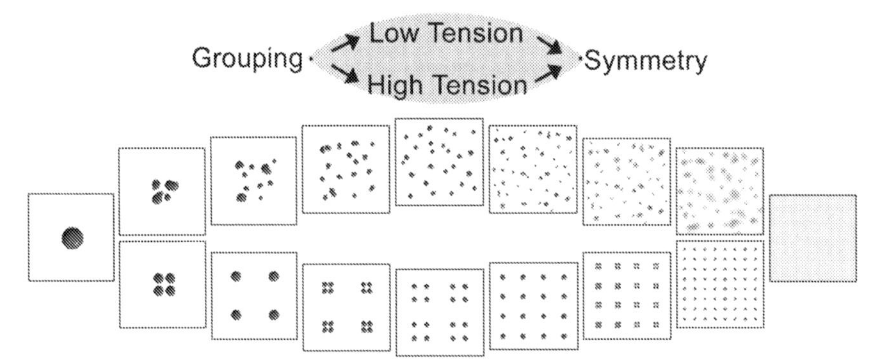

Figure 5.16: The library of possible patterns in between the two extremes of grouping and symmetry portray high and low measures of competition. Along the upper span of patterns we see the disorderly extreme and along the lower span of patterns we see an orderly extreme. It is intermediate tensions that allow complexity and diversity to exist.

In the formations of all patterns of possibility, the grouping order of our past reaches into the future, and the symmetry order in our future reaches into its past. Both orders inevitably reach across time, from beginning to end, from end to beginning, contributing to each stage in the overall transition of all possible patterns. We exist caught in the world in between these two great powers, controlled by them, given shape by them.

The Intensity of Competition between Two Orders

In an orchard the trees are spread extremely even, and we quickly appreciate that evenness as order, while in a natural grove the trees are less even where the competition between both orders is at a natural level, and so less intense to that of an orchard. Can we say that an orchard is more ordered than a forest? That is not an easy question to answer. Technically no pattern is more generally ordered than another, as that would require a general measure of disorder. If the order of one is the disorder of the other, then as one is decreased, the other increases, keeping the measure of order constant. In a forest there are areas where the trees are grouped closer together than in other areas, and some of the trees are larger than others, reflecting the probable degree of irregularity that we expect of nature. In an orchard, due to human intervention, the trees are more likely to be the same age and size, and are evenly spaced. Most human-made things are less irregular, since we can intelligently select and produce patterns where the tension between the two orders is high.

Orchard Rows © Daniel West

Trees © Gevin Giorbran

Orderly Fields: © Teun Van Den

Perhaps where we see the highest measure of competition between both orders is in atomic and chemical structure. The distinction between grouping and symmetry is plainly visible in the patterns of atomic orbitals which often form square lattice structures. Crystals and crystalline solids are formed by a highly symmetric lattice built from square or hexagonal blocks. Crystalline materials in which the intensity between grouping and symmetry is extremely high include diamonds, quartz, salt, aluminum, gold, platinum, mica, graphite, nylon, polyester, polypropylene, sucrose, glucose, and fluorides. Many crystallines contribute to the rigidity of plastics. Generally in nature it is typically true that a round structure indicates high grouping order while square or lattice structures indicate high symmetry order.

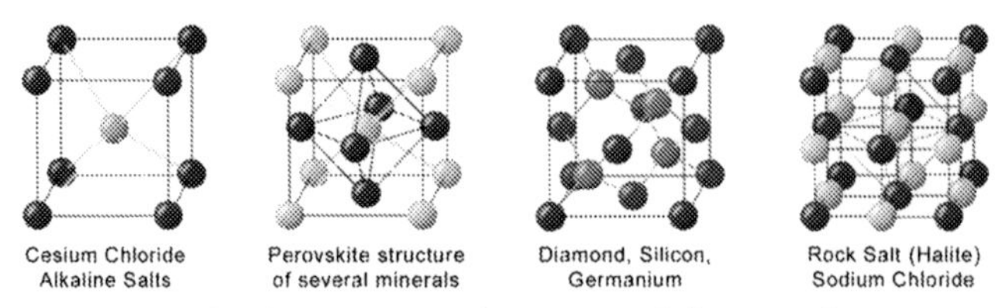

Figure 5.18: The Electromagnetic force creates balances and symmetries in the chemical world, with many materials held together in square lattice configurations called cells.

What is pattern cooperation? A pattern exhibits cooperation when the parts of that pattern are positioned in alignment, or are held in position, in accordance with other parts of the pattern. The two kinds of order are each essentially two different methods of cooperation. Cooperation between parts is most clearly evident when either grouping order or symmetry order is dominating. The cooperation of grouping order is when particles gravitate together into a star. The cooperation of symmetry order is when atomic particles are electromagnetically held in a perfect checkerboard type of lattice.

When are the two orders competing with one another? A pattern exhibits competition when the cooperation of both orders is high simultaneously. A helpful example of two sides cooperating and then competing against one another might be the more systematic American style of football. A great deal of planning and training is necessary in order for the group of eleven players on the field to function as a single coordinated unit. The individual players cooperate with one another in order to break down the cooperation and coordination (dynamic positioning) of the other team, so they compete against each other by cooperating with one another. Consequently, in human terms, football is an exhibition of extremely high orderliness, at least most of the games anyway.

Likewise, as the two kinds of order create patterns, one kind of cooperation, grouping order, competes with the other type of cooperation, symmetry order, with varying degrees of intensity. When the cooperation of grouping order is high and symmetry order cooperation is low the result is a single dense group. When cooperation is low for grouping and high for symmetry the result is a smooth and uniform substance, such as Bose-Einstein condensate. When cooperation is high for both orders we distinctly witness both the grouping of matter and symmetry, and so the result is an intricate spiral or a lattice of dense pronounced particles. When cooperation for both orders is low the result is irregularity and apparent disorder.

So in the combination of grouping and symmetry there are two general possibilities, there can be orderliness, and there can be disorderliness, but there cannot be a general lack of order. All of the more orderly patterns in nature are produced by the two orders intensely cooperating as they compete with one another, while disorderliness, or what we think of as general disorder, occurs when the cooperation of each individual order is low. Since it isn't possible for a pattern to exhibit a lack of both grouping and symmetry, we shouldn't imagine a weakly cooperating pattern to be generally disordered; rather it is merely an irregular combination of the two orders, which makes the presence of both orders less apparent.

Figure 5.19: We often see the two orders weakly cooperating in the patterns of clouds, even though we can still identify areas of grouping and uniformity, but occasionally even the clouds exhibit higher orderliness, in cloud patterns sometimes called 'cloud streets'.

Above: Ireland Coast © Jon Sullivan, www.hpphoto.org; Left: © Manthy Maragoudaki; Right: Cloud Streets © Viren Shaw, www.viren.org

We could and probably should also create here a distinction between disorder and chaos. What we imagine to be disorder exists when the intensity between grouping and symmetry is weak. However, I like to define chaos as an even greater level of disorder where a pattern seems to exhibit an effort toward avoiding either order. Having now recognized the complementarity of the two orders, we can appreciate the fact that extreme chaos is a difficult task. A chaotic pattern seems to avoid being designated as either grouping or symmetry by being extremely irregular, first in a smaller micro-frame (of reference), which is some 'focused upon' fragment of a pattern, and then also on an expanding and increasingly larger macro-scale.

Consider for example the randomness of numbers. A random number generator on your computer that selects a number between 1 and 9 will immediately create, when turned on, an irregular pattern such as 78192850. The selection of each number is highly random and unpredictable, however, a block of 1000 of such numbers will add up and sum very close to a value of 5000, which reflects the average value of numbers between 1 and 9, i.e., 5 multiplied by 1000. The immediate production of numbers is random but the macro-scale value of larger groups of random numbers is increasingly more predictable. A thousand groups of 1000 numbers will sum toward an average of 5,000,000, and so on and so on. Such large groups will never sum toward an average of 1,000,000 (where every number generated is by pure chance a 1). In fact, for the number generator to be random on a larger-scale it would have to be specially programmed to be less random on a finer scale. The generator would need to be more apt to produce the same number as the first number selected, so that after the first random choice of each block of 1000 it then creates a large group of the single number selected, i.e., grouping order. Then in the next block it can produce a randomly selected new group, which pushes the unpredictability one level up. Extreme irregularity or chaos is actually extremely difficult to create. What is immediately random is not random on a larger scale and only what is immediately predictable can be extremely irregular on the larger scale.

Chaos on a macro-scale is highly improbable because there is always symmetry on the largest macro-scale, such as the isotropic distribution of galaxies throughout the cosmos, and ultimately Omega, which is why any large-scale system or summation tends to reach equilibrium and balance out, therein resulting in symmetry order. And so with the two orders exposed we can begin to see the reason why the universe we experience is orderly and not chaotic. In more accurately considering what is ultimately possible we see that chaos is actually very improbable and requires an increasing measure of work to create.

Two Directions in the Possible Realm

We typically imagine there is a terrible unpredictable chaos at all levels out there somewhere, opposite to the diversity we know that exists in the balance between freedom and the forces of nature. Instead there is control in both directions. There are two different types of cooperation in the two fundamental directions of possibility that exist for us. There is the control of grouping order in our past. And there is the control of symmetry and balance in our future. As our own universe continues to evolve toward zero a more strict and simple orderliness will eventually remove the freedom and diversity that exists here in time. Stars and galaxies eventually disintegrate as the universe cools and the influence of zero increases. Finally the waveforms of particles unify into a single material, and near the end of time the condensate predicted by Einstein and Bose will take over and fill the entire universe. Fortunately, in confronting this boring sameness and control, we at least should be completely consoled by knowing of the enfolded order that exists behind the surface of increasing symmetry and sameness. It is wise to never forget the supporting background of timelessness.

For those concerned about our immediate future, there is no reason to believe complex orderliness is presently on the decrease, or that sameness is set to take over soon. Both orders are still very active and present in our cosmos. We ourselves, our bodies, our perceptive minds, are examples of the complexity and super cooperation that results from the two orders competing against one another. There are likely levels of super-complexity destined in our future such as what is often portrayed in science fiction, not merely in our own region of the universe, but a super complex future may be perfectly in synch with the natural evolution of patterns. Some kind of complex future is inevitable, whether we humans are here to enjoy it or not. Nanotechnology and computers probably offer the greatest potential for increasing complexity in the near future on Earth, that is, if we manage to lessen our individual impact and decrease the growing populations that are so furiously using up the Earth's natural resources. In any case, the most important lesson to be learned from two orders is that the general evolution of the universe is not simply moving from order to disorder, as is now believed to be the case.

In the same way the north and south poles allow us to map the unbounded surface of the Earth, there are two great poles in the whole of reality that bound the whole of all possibilities. In a view of reality from the top-down there are the two great boundaries of Alpha and Omega, the extremes of grouping and symmetry. If there were no cosmic absolutes at all then finite things might still find definition in relation to other finite things, but we would have no way of making sense of the whole of reality. Fortunately we can make sense of the big picture. In that the two great absolutes exist there is also a spectrum of patterns between them, and fortunately time worlds that travel between them. That is of course where we find ourselves, in a sampling of the diversity of nature, in between the two ordered extremes.

Presently we don't collectively see the big picture, it hasn't quite come into focus yet. Today we see only a distorted and incomplete fragment of the greater whole, because we view the world mainly from only one side of nature, one side of knowing and understanding, somewhat similar to how we view the Earth relative to the North Pole. We imagine the North Pole as the top or up side of the Earth and then we imagine everything else relative to that single pole. We don't as readily make note of the other pole, the South Pole, as if we don't like the fact that the surface area of the Earth ends down there.

As a sort of a joke or a twist to make people think, there is a man in the land down-under the equator known as the "Wizard of New Zealand" who sells maps of the Earth which are upside down so that the southern hemisphere is considered the top. Of course there isn't really a top side to the Earth, so there is nothing inaccurate about such a map. There is just our tendency to make reference of ourselves or the world against one chosen place. And so in terms of the general evolution of the universe we tend to refer to the beginning of time and Alpha.

Figure 5.20; An alien Planet Landscape with New Zealand on top of the world. Photo: NASA

Just as there is a north and South Pole to the Earth, there is both a top and a bottom to our existence. The beginning of time is our north pole. We identify with the past. We assume the past created us and hardly imagine the possibility that the future might influence us, or create us, because in our minds the future hasn't happened yet, or that is what we think! The possibilities are endless, or so we think! What we haven't yet realized at this point in the history of discovering the Universe is that there is a distinct south pole of reality. There is a south pole in the realm of all possibilities, and there is a south pole for the duration of time we measure with clocks. There is a south pole for the expansion of the universe. There is even a south pole for disorder, or what we think of as disorder.

In terms of how the universe will transform in the deep time of cosmic evolution we hardly notice what takes place at the extreme south pole of time. We hardly consider the end result of moving continually toward balance and equilibrium. So we have failed to notice there is a single extreme state toward the end of time, in comparison to the single extreme of the big bang at the beginning of time. We more often take for granted the fact that balance finally blends everything together into an invisible whole, blinded perhaps because of all the interesting diversity that we experience in the present. Never the less, all such diversity is within the whole. The things of the world are real, but they are really all fragments of something much greater.

Whilst the world is thus dual, so is every one of its parts. The entire system of things gets represented in every particle. There is somewhat that resembles the ebb and flow of the sea, day and night, man and woman, in a single needle of the pine, in a kernel of corn, in each individual of every animal tribe. The reaction so grand in the elements is repeated within these small boundaries...

Ralph Waldo Emerson

All are but parts of one stupendous whole.

Alexander Pope

You cannot conceive of the many without the one.

Plato

Sunflowers: © Charles Beck

Part Two

The Governing Dynamics of Symmetry

To see the world of timelessness and appreciate the infinite whole it is necessary to make a significant mental shift away from how we ordinarily see the everyday world. In the classic view of reality we quite reasonably conclude that a definite world exists, many things exist, so what we think of as an absolute or ultimate "nothingness" does not instead exist. It follows logically that a true and absolute "nothingness" cannot exist simultaneously in some parallel dimension in complete contradiction to the existence of our something world.

We do however in our classic view require a lesser form of "nothingness" to divide apart material objects or things. Perhaps the universe began as one thing. But now the world is made up of many things. We see the world as many separate things and we have grown comfortable in thinking it's okay for a *nothingness* to divide things apart, even though its limited presence here in a something world is extraordinarily paradoxical. As we see the world today, we need nothingness to be defined separate from everything else, to break away from the mold. Our diverse world is necessarily a holy place, meaning it must have lots of gaping holes in it for there to be individual things. If there isn't a "nothingness" to make things separate from one another, then the whole world would be just one seamless thing.

In appreciating only objects as reality we see the world abstractly, and quite inaccurately as well. We see things, we count things, and we multiply and divide things, so when we imagine all the things are taken away we are suddenly lost. Our method of modeling reality fails without things. We commonly say, "there is nothing in here", or "nothing over there", but the nothingness we are always pointing to and moving around within is really more than we are, not less. We can only pretend space is a nothing. In truth, all that we know as reality is a fragment embedded within a vast and measureless whole. We ourselves always and forever exist inside that whole, sort of like pieces of a whole pie. The world is really just one thing.

We can cut a pie into many different pieces, and consider each piece of the pie to be a separate thing. We can even slice the pie into an infinite number of unique pieces, but no matter how we cut up the whole pie it is still an undivided oneness. We can divide up a continent into countries but the continent remains one land. We don't see this great whole oneness mainly because we never concern ourselves with the whole. We identify with things. We see all these fragments, our selves, the sun and sky, a planet, a solar system, a galaxy, a field of galaxies. If we did see everything at the same time, we would say, "oh, that is just 'nothing' (no-things)". But ultimately that single whole is what we are, a nothing that is everything, and I say this as factual science.

We all know there is a timeless realm for imagination, for mathematics, for the meaning of words and ideas. We never imagine such things begin or end, nor do we imagine they exist in a place and not some other place. The meanings of ideas are everywhere and always whether someone thinks with them or not. Likewise, physical things are everywhere and always. All the form, all the definition of the solid physical world, is created from all the different ways the whole can be divided apart. We are not more than nothing, we are less than everything. This space we swim through does not really separate us from anything else. In fact it connects us. It is the oneness of everything else our world is not.

Ultimately the great whole that any recognition of the two kinds of order exposes is everywhere. The whole exists in every part. It doesn't have a particular location. It is not out there somewhere or elsewhere. The great illusion is not our perception of time, it is our inability to perceive all of time all about us, or as the American physicist John Wheeler put it, "time is what keeps everything from happening all at once". A unified whole has long been expected and two orders finally lays a solid foundation for a grand mental picture of the world as one thing. Instead of seeing existence as mostly empty and fragile, it is possible to see definitive form as a tiny fragment of what exists all around us, just beyond the surface of form in the enfolded timeless background.

What subject could be more interesting than timelessness? The timeless realm is essentially God's realm, meaning that timelessness is not just configurations and mathematical laws but rather the whole space of ideas and a great integration of everything living that exists in time. In the permanent stasis of timelessness we see all worlds and all lives simultaneously. We can appreciate even scientifically that the universe itself is alive, at very least through each one of us. Even timelessness must therefore be considered alive which illuminates the definition of life in an unexpected way.

Mankind likes to think in terms of extreme opposites. It is given to formulating its beliefs in terms of Either-Ors, between which it recognizes no intermediate possibilities. When forced to recognize that the extremes cannot be acted upon, it is still inclined to hold that they are all right in theory but that when it comes to practical matters circumstances compel us to compromise.
John Dewey

~~~

Our mind is capable of passing beyond the dividing line we have drawn for it. Beyond the pairs of opposites of which the world consists, new insights begin.

Herman Hesse

~~~

The One has never known measure and stands outside of number, and so is under no limit either in regard to anything external or internal; for any such determination would bring something of the dual into it.

Plotinus

~~~

In our life there is a single color, as on an artist's palette, which provides the meaning of life and art. It is the color of love.

Marc Chagall

~~~

If one feels the need of something grand, something infinite, something that makes one feel aware of God, one need not go far to find it.

Vincent Van Gogh

The Landscape becomes reflective, human and thinks itself though me. I make it an object, let it project itself and endure within my painting....I become the subjective consciousness of the landscape, and my painting becomes its objective consciousness.

Cezanne

Chapter Six

Natural Order

Two Orders in Art, Music, Architecture, and Poetry

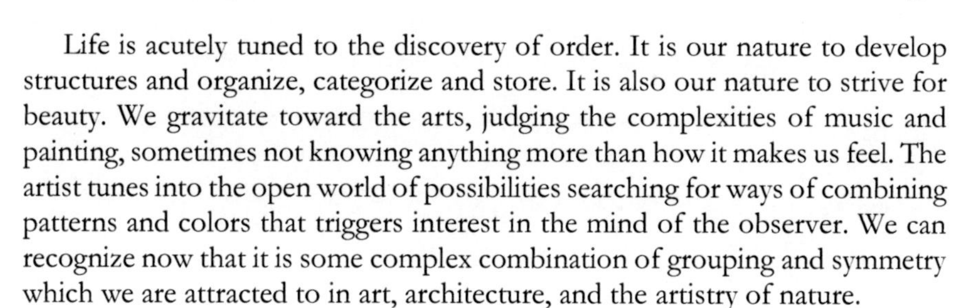

Life is acutely tuned to the discovery of order. It is our nature to develop structures and organize, categorize and store. It is also our nature to strive for beauty. We gravitate toward the arts, judging the complexities of music and painting, sometimes not knowing anything more than how it makes us feel. The artist tunes into the open world of possibilities searching for ways of combining patterns and colors that triggers interest in the mind of the observer. We can recognize now that it is some complex combination of grouping and symmetry which we are attracted to in art, architecture, and the artistry of nature.

In the wide range of human-made visual art there are compositions more toward the nature of grouping order that emphasize the order of definition and form, by being realistic, distinct, and bold, from artists such as Michelangelo, Picasso, or Salvador Dali. And there are compositions in which distinctiveness and boldness are traded for showing the connectedness and commonality between things, where objects flow together, where a unity is felt, as found in paintings from Vincent Van Gogh or Winslow Homer.

Figure 6.1: Comparing the bold pronounced work of Rembrandt with the flowing low-contrast unity of Van Gogh.

Works toward the nature of symmetry order are often low contrast with blended colors and wide brush strokes. Generally speaking, the overall expression of a painting can either be more uniform and flowing or it can be more variegated and distinct. Paint colors can blend smoothly or they can stay pure and sharp in contrast. Each of the different mediums in art, such as watercolor or inks, express either the unity of the world or the distinction of things. Watercolor paintings overlap and flow together while ink pen drawings are contrastive with sharp lines and distinct shapes. This range of possibilities we are comparing is the two directions of order governing both materials and composition.

There are artists who in exploring composition have learned to recognize and utilize two orders at an intuitional level years ahead of science. Some art work seems to capture the rules of the unseen underlying orders, especially visible in the woodcut prints of Charles Beck. In Beck's art we see a rigid combination of grouping and symmetry. Generally in Beck's work, distinct subjects are spread almost perfectly even in the painting, like rhythm in music, sometimes with the only asymmetric element being a trace of human activity taking place among the combining of grouping and symmetry. In his art Beck often seems to convey an intuited awareness of our place in nature between the two orders.

© Charles Beck Maplelag © Charles Beck Poplars

Figure 6.2: Woodcut Prints of Charles Beck visually reveal the two orders cooperating.

In Beck's prints, objects or things are portrayed evenly and symmetrically and although the evenness of the trees and the buildings might be increased to an even greater extreme, they would then appear so orderly that they would not convey the message. They would not relate in a mysterious way to similar patterns we see in nature. In nature we commonly see things combined together evenly although the evenness is less rigid and distinct than what artists often portray. Artists often add-in or choose a point of view that frames greater symmetry, since we all enjoy symmetry. Beck's landscape compositions above compared to those below communicate that we are constantly witnessing balance and symmetry in nature, it is just a less rigorous and exacting symmetry.

Figure 6.3: Similar scenes in nature show a less tense combination of two orders although the two orders are cooperating in every real life setting.

Two Orders in Music and Poetry

Near the turn of the century the art philosopher George Lansing Raymond struggled to understand the common underlying structure of order as it exists in art, music, and poetry as equally as it exists in nature. In *The Genesis of Art Form* Raymond writes:

> When the child first observes the world, every-thing is a maze; but, anon, out of this maze, objects emerge which he contrasts with other objects and distinguishes from them. After a little, he sees that two or three of these objects, thus distinguished, are alike; and pursuing a process of comparison he is able, by himself or with the help of others, to unite and to classify them, and to give to each class a name. . . . All his knowledge, and not only this, but his understanding and application of the laws of botany, mineralogy, psychology, or theology will depend on the degree in which he learns to separate from others, and thus to unite and classify and name certain plants, rocks, mental activities, or religious dogmas. Why should not the same principle apply in the arts? It undoubtedly does...
>
> (Classification) enables one to conceive of many different things—birds or beasts, larks or geese, dogs or sheep, as the case may be—as one. Classification is, therefore, an effort in the direction of unity. It is hardly necessary to add that the same is true of art-composition. Its object is to unite many different features in a single form. Unity being the aim of classification, it is evident that the most natural way of attaining this aim is that of putting, so far as possible, like with like...

Grouping and symmetry provides a simple way of understanding the common order of all patterns in nature including art and music. In separating grouping and balance into two components the complexities of art and the harmonies of music are perhaps more fundamentally comprehensible than with any other method of description. The first level of grouping order in music is just a single note breaking silence, like a particle in space, or a star in the sky. Then in the rhythm of music we find the evenness and balance of symmetry order. Often in a song several different notes are played simultaneously with one instrument to create a chord, which is somewhat like cooking various vegetables into a soup. A careful selection of certain like notes create harmonious chords, just as many different instruments create songs.

In the same way that we can define a frame of reference in space, we can define a frame of reference in time, which in music is called a measure. A musical example we can pick apart which most everyone is familiar with would be the first notes of Beethoven's fifth symphony. Three sudden notes are followed by a drawn out fourth note. A series of pronounced individual notes played close together as opposed to further apart in time is of course intense grouping order, although the even rhythm is symmetry order. A steady rhythm is perhaps the most important ingredient of music. Rhythm is what distinguishes music apart from all other sound. But we also find symmetry in a lasting note duration, which is like a smooth dense space in which matter is spread evenly.

The complex ordering of many musical instruments can be interesting, pleasant, saddening, spirited and exciting, able to compliment and enhance every human emotion, perhaps suggesting a connection. Musical instruments create either more pronounced blunt sounds akin to grouping order, as with percussion instruments such as drums. Other instruments play sustained notes and chords more evenly across a measure of time, such as a violin or a flute, representing symmetry order. Music genres like rock and roll or rap that are pronounced with a strong beat are more of the nature of grouping order, while the smooth and drawn out rhythms sometimes produced by a symphony are more toward the nature of symmetry order. Regardless, a beautiful musical composition resonates with both the distinction of notes and instruments, and the limited harmonizing of those notes and instruments.

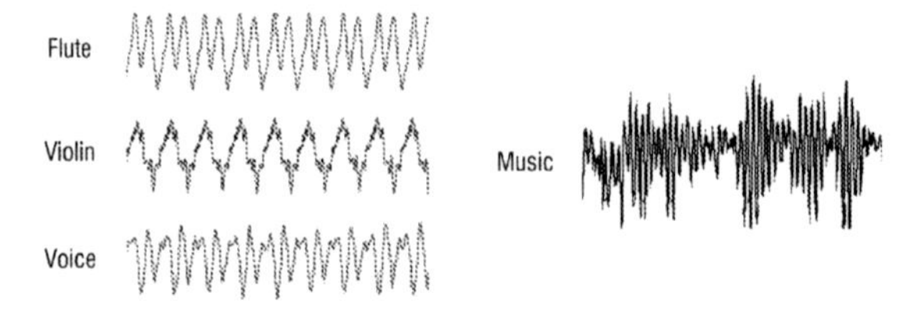

Raymond, who was a professor of aesthetics at Princeton, also considered how two very basic but very different elements contribute to the beauty of art and music. Raymond initially focuses on the role of "likeness".

> Applying [likeness] to art-composition, and looking, first, at music, we find that the chief characteristic of its form is a series of phrases of like lengths, divided into like numbers of measures, all sounded in like time, through the use of notes that move upward or downward in the scale at like intervals, with like recurrences of melody and harmony. So with poetry. The chief characteristics of its form are lines of like lengths, divided into like numbers of feet, each uttered in like time, to which are sometimes added alliteration, resonance, and rhyme, produced by the recurrence of like sounds in either consonants, vowels, or both. In painting, sculpture, and architecture, no matter of what "style, " the same is true. The most superficial inspection of any product of these arts, if it be of established reputation, will convince one that it is composed in the main by putting together forms that are alike in such things as color, shape, size, posture, and proportion…

The likenesses mentioned of lyrics and poetry are most apparent in very simple poems such as, "Mary had a little lamb, its fleece was white as snow. And everywhere that Mary went, the lamb was sure to go." But specifically what does Raymond mean by *likeness*. How many different ways can things be alike? Are there perhaps opposing directions that things can be alike? For example, each letter in the alphabet is different, yet some letters form a group called vowels, which are alike in ways but different than the group of consonants. And yet all are alike in simply being letters, so letters are simultaneously different and alike. Where letters are expressions of distinct form and thus grouping order they are different, and where letters are a general type of forms they are alike, and as a whole create a uniformity. It can be a surprise to realize that all the letters in this book, or every book that has ever been written in any language, exist within ■ any tiny black square.

It is said that someone asked Michelangelo how he had created such a perfect image of David out of stone, to which he replied that David was there in the stone all along, and just needed a little help in getting out. In being so focused upon likenesses that establish the differences we notice between things, which make classification possible, we easily forget that everything is of a single existence and thus is ultimately just one thing. Ultimately all difference is an illusion. All things are ultimately alike.

Is the beauty of music and poetry and art determined by a mix of these two opposite directions of likeness? Having considered likeness, Raymond then focuses on variety and contrast:

> ...the mind is confronted by that which classification is intended to overcome, by that which is the opposite of unity—namely, variety. If there were none of this in nature, all things would appear to be alike, and classification would be unnecessary. As a fact, however, no two things are alike in all regards; and the mind must content itself with putting together those that are alike in some regards. This is the same as to say that classification involves, occasionally, putting the like with the unlike and necessitates contrast as well as comparison. A similar fact is observable in products of art. One of the most charming effects in music and poetry is that it is produced when more or less unlikeness is blended with the likeness in rhythm, tone, and movement which, a moment ago, was said to constitute the chief element of artistic form. In painting and sculpture one of the most invariable characteristics of that which is inartistic is a lack of sufficient diversity, colors too similar, outlines too uniform.

In how likeness in one way makes things different and classifiable, and in another way makes all things the same, we can see distinctly why there are two different and incompatible kinds of order in nature. In one direction of being alike things are different and unlike other things. Birds of a feather flock together. This classification creates contrast and unlikeness. Unlikeness is even required of diversity. In another direction of being alike things are more general or similar to other things. All birds are mammals. All mammals are animals. All animals are life forms. All forms are matter. All matter is part of one great existence. In combining like with like we create the distinctions and definitive form of grouping order, yet in combining that same unlikeness we somehow further the unity of symmetry order. We bring all things together into a whole.

We are all attracted to symmetries and balance in music and art, yet unlikeness and imbalances are a critical ingredient of the aesthetics of all art and music. Too much symmetry, too much harmony, too much balance, and we end up with silence. No matter how intense, a synchronized positive wave and a negative wave turn each other into silence. And to us, silence, like the white canvas, is boring.

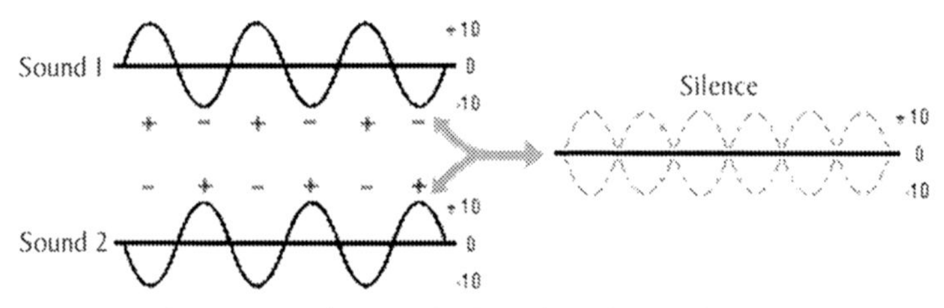

Figure 6.6: When two sounds are perfectly in phase they combine into silence.

The more the notes and chords are drawn out, the more the music is indistinct and unified over its measure of time, and here again the extreme moves toward a single sound that cannot be heard. Sound is fundamentally an oscillation. Without oscillation, sound is silence. Sound, like both energy and matter, is an oscillation of waves, and when we stretch that wave flat it becomes silence. Flat sound, flat energy, flat matter, is like the color white, our senses perceive it as a nothing even though all possible sounds combined together have the same consequence.

All music is created from limited measures of imbalance and balance. In music, in art, in poetry, we want to experience imbalances and balances eloquently woven together. We want the imbalances of grouped likenesses combined with the sameness and balance of symmetry. We want to see and hear the complexities possible in the myriad of ways of combining grouping and symmetry. We are attuned to those complex combinations possible of both orders. It isn't simply order out of chaos that gives art and music its pleasurable qualities, it is the cooperation of each order in strict competition with one another.

Two Orders Competing in Architecture

Is there a difference between the beauty we see in nature and the beauty we see in man-made creations. To focus more clearly on what we perceive as beauty we can turn to the less temporary art of architecture. The difference between common patterns in nature apart from those patterns that we commonly see in architecture that are constructed intelligently by humans is understandable as a variable of the tension between the two orders. A carefully planned intelligent design can increase this tension of competition, so that the competing two orders appear to be cooperating.

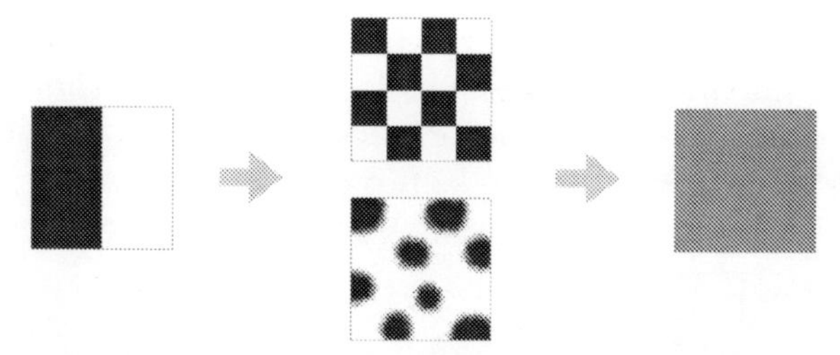

Figure 6.7: In the transition from grouping to symmetry order nature does group things and yet spread them evenly but not as rigidly symmetrical as is possible if the pattern is intelligently designed (checkered pattern).

Opposite to cooperation there is randomness. Normally the definition of randomness means to have no specific pattern or objective, and randomness creates disorder, chaos, and disorganization, but actually randomness is a measure of freedom that exists only between the two kinds of order. The ordinary patterns we experience in nature are always created only by two orders, yet most patterns in nature are less intensely symmetrical or less intensely grouped than man-made patterns, so there is a measure of randomness or freedom in the pattern, which is still the two orders working together, but not as intensely or synchronically as is possible. In comparison, man-made creations can fuse the two orders together in a way that eliminates randomness.

Photo credits: Left and right © Piotr Pieranski, Middle: © Quentin Rowe, Below: Gravitational lensing in Abell Cluster

Figure 6.8: Beach rocks, waves, and field plants, exemplify ordered patterns less rigidly trapped between intense grouping and symmetry. Below, a small sampling of distant galaxies (HST Abell Cluster), and the microwave background radiation measured by the WMAP Satellite revealing regional lumpiness in the early universe before galaxies had formed (images from NASA).

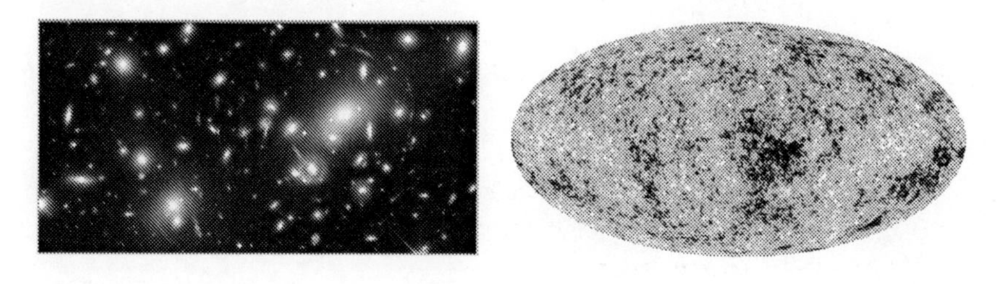

There is an increased symmetry of patterns at low temperatures, first seen in things such as window frost or a snow flake, but in warmer climates we commonly observe a measure of freedom or irregularity, which is essentially a measurable weakness in the influence of the two orders, in contrast to a man-made fractal which displays an extremely intense combination of two orders.

Snowflake © Ken Libbrecht

Figure 6.9: The uniformity of water vapor in clouds and in the air and symmetry in the chemical structure of water translates into the complex symmetry of a snowflake.

Usually we draw a line between patterns in nature and patterns that are man-made, as if man isn't natural, but the only real difference is the combined intensity of each order. Intense competition creates a level of cooperation in what can be called an *order game*, or some way in which the two orders are intelligently made to work together. Human beings play all sorts of order games, where the two orders compete against one another, most plainly in sports, card and board games (especially checkers), but also in social situations, business, politics, and conflicts such as war.

If we analyze the images below as an exercise of finding cooperation between two orders, in the first photo, clouds of fog float through a recognizably even distribution of trees in a forest. A cloud of fog is a group of particles, although there is an obvious symmetry in the distribution. Also this photo displays an "S" curve to the fog, well known to be an attractive pattern in art and photography because it has symmetry. To the right, grooves carve up the earth evenly and seedling plants sprout along each row evenly, forming the group of each row, then left to right we see the symmetry of the collective rows. Mountains in this satellite image of the Himalayas each represents grouping order, particularly Everest, but from afar we see the collective uniformity of the mountains.

© Mike Levin mikelevin.com © Frank Poulsen www.fp-engin.dk Photo: NASA - Himalayan Mountains

Figure 6.10: Various Patterns in which grouping and symmetry can be easily identified.

It doesn't take long before a person is able to recognize grouping and symmetry in all patterns, both natural and man-made. We most commonly find an intense competition between grouping and symmetry in man-made designs of governmental and religious architecture. The pronounced grouping of materials contrasting the even distributions of windows and columns are easily recognized in these buildings, as well as the general bilateral symmetries. All these symmetries are components of a balance showing little sign of the innate conflict with the grouping order nature of the buildings themselves. Yet take away the symmetries and we would have a simple pile of matter. Obviously it is pronounced masses ornamented with symmetries that we find to be beautiful in architecture, likewise true of sculpture, music, art and nature.

United States Capitol: ©Digital Vision

Angkor Wat, Cambodia: ©Stacy Rushton

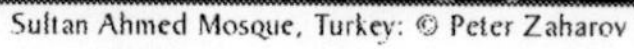

Sultan Ahmed Mosque, Turkey: © Peter Zaharov

Saint Basil's Cathedral, Russia: ©Yaroslav Ushakov

Figure 6.11: These great landmarks show the human preference for both orders combined together intensely competing and yet cooperating.

Of course we forget sometimes that literally everything that exists is of the natural world. All that human beings create is part of the universe and so nature. The only imaginable distinction between nature and the human world is that we humans can intelligently make the two orders cooperate in ways that nature cannot accomplish without our help. This does not mean however that there isn't an equal measure of cooperation occurring generally in nature.

Photo Credits: Peacock © Becky Rech, Larvae © Michal Fafrowicz, Fish © Gareth Peters, Leaf © Joanie Cahill, Passion Flower © Ingela Nordlund, Aloe © Chris Carter

We might even ask, when we consider all patterns, is there a greater measure of cooperation in human-made patterns over natural patterns? Note that most human-made patterns are not as rigidly perfect as the architectural examples shown, while many plants and animals rival our creations with exquisite orderliness. There are likely consistent measures of chaos and cooperation at all levels everywhere in the universe, even at the human level, almost as if we accomplish exactly what nature has designed into us.

Summary

The distinction between the two kinds of order, between dividing things apart and mixing things together, initially seems too simple to be of any great importance. Certainly, the important simple principles have all been discovered long ago. It is after all something a child could understand. How can something so simple dramatically change how a person sees the complex world? Indeed the distinction between the two orders is simple, but our existing definition of order is even simpler than what is being explained. Consequently the concepts of order and disorder only vaguely describe the patterns we experience, which actually makes the universe seem more complex and perplexing than it is.

After learning to recognize two orders virtually everywhere we look, we can try to return to the basic concepts of order and disorder, except an increase in symmetry order now means that grouping order is lost, and an increase in grouping order requires that symmetry order is decreased. This completely contradicts the commonly held belief shared by most that order decays into disorder and it contradicts the belief in science that there is high order in the direction of our past which is decaying in the direction of our future. The exclusivity of two orders necessarily replaces the commonplace concepts of order and disorder, showing that neither concept can be generally applied to nature. In essence, we have to start over.

A World of Water © Michel Meynsbrughen

There are waves on the vacuum sea corresponding to every conceivable quantum, even those we have not yet discovered. All of physics - everything we hope to know - is waiting in the vacuum to be discovered. Everything that ever existed or can exist is already potentially there in the nothingness of space.

Heinz Pagels

~~~

Space is a collection of all places

Liebniz

~~~

Everything happens at the same time with nothing in between.

Paul Hebig

~~~

In a certain sense, everything is everywhere at all times. For every location involves an aspect of itself in every other location. Thus every spatio-temporal standpoint mirrors the world.

Ralph Waldo Emerson *Compensation*

~~~

God is day and night, winter and summer, war and peace, fullness and hunger.

Heraclitus

~~~

Ideas come from space. This may seem impossible and hard to believe but it's true.

Thomas Edison
~~~

The existing scientific concepts cover always only a very limited part of reality, and the other part that has not yet been understood is infinite. Whenever we proceed from the known into the unknown we may hope to understand, but we may have to learn at the same time a new meaning of the word 'understanding'.

Werner Heisenberg

Chapter Seven

Enfolded Symmetry

The People of Symmetry Order

The two directions of order are probably the most important features of nature that we will ever understand. There are endless applications of two orders in every field, in physics, biology, geology, politics, and psychology, just naming the more obvious fields. The theory of two orders generally describes all change, all patterns, all definitive form, so what if anything could be excluded. As we learn to perceive the cosmos as one type of order transforming into another, doors of comprehension will open beyond our wildest dreams.

A surprising number of others in the past have managed similar comprehensions of the two orders, including Henri Bergson and William Yeats in the early 1900's, but the most recent and most notable is the American physicist David Bohm who introduced concepts of Implicate Order and Explicate Order. A student and friend of Albert Einstein, many of the general ideas about order being explained were introduced by Bohm to modern science in the 1960's. In describing Implicate Order Bohm writes:

> This order is not to be understood solely in terms of a regular arrangement of objects (e.g., in rows) or as a regular arrangement of events (e.g. in a series). Rather, a total order is contained in some implicit sense, in each region of space and time. Now the word 'implicit' is based on the verb 'to implicate'. This means 'to fold inward' (as multiplication means 'folding many times'). So we may be led to explore the notion that in some sense each region contains a total structure 'enfolded' within it.

In order to understand how the whole can exist in every part, Bohm became interested in the mechanics of holographic photography. Using lasers, a holographic image is recorded evenly across the photographic film. Consequently any region of the film contains information about the whole image, so any small region of the film can recreate the image, although in poorer resolution than the entire film produces.

Bohm believed all matter is unfolded out of what he eventually described as a holomovement, which meant that matter could also enfold and so return into the holomovement. Bohm considered quantum mechanics to be a process of unfolding and enfolding. He imagined the universe as an infinite sea of space and energy out of which matter could be unfolded, which he called explicating, and enfolded which he called implicating, which, in Bohm's words, "together are a flowing, undivided wholeness. Every part of the universe is related to every other part but in different degrees." In an interview published in Omni magazine conducted by the physicist F. David Peat and John Briggs, Bohm explained his concept of enfoldment:

> "Everybody has seen an image of enfoldment: You fold up a sheet of paper, turn it into a small packet, make cuts in it, and then unfold it into a pattern. The parts that were close in the cuts unfold to be far away. This is like what happens in a hologram. Enfoldment is really very common in our experience. All the light in this room comes in so that the entire room is in effect folded into each part. If your eye looks, the light will be then unfolded by your eye and brain. As you look through a telescope or a camera, the whole universe of space and time is enfolded into each part, and that is unfolded to the eye. With an old-fashioned television set that's not adjusted properly, the image enfolds into the screen and then can be unfolded by adjustment."

The process of unfolding and enfolding suggests a whole exists primarily as a base, and although Bohm described the whole as being dynamic and in constant motion in most of his writings, in his later years he began to describe time as occurring within timelessness. In hindsight it appears that Bohm didn't focus enough on the timeless nature of patterns. He didn't recognize consciously enough that ordinary patterns reveal the implicate order, although in Bohm's analogies you find the same type of examples that I use to explain the transition from grouping to symmetry order. One of Bohm's favorite analogies spawned from a scientific program he'd seen on a television show. The program featured a small scientific wonder, where an insoluble drop of dark ink in one process disappears uniformly into a glycerin, then in the opposite process the drop reappears.

> About the time I was looking into these questions, a BBC science program showed a device that illustrates these things very well. It consists of two concentric glass cylinders. Between them is a viscous fluid, such as glycerin. If a drop of insoluble ink is placed in the glycerin and the outer cylinder is turned slowly, the drop of dye will be drawn out into a thread. Eventually the thread gets so diffused it cannot be seen. At that moment there seems to be no order present at all. Yet if you slowly turn the cylinder backward, the glycerin draws back into its original form, and suddenly the ink drop is visible again. The ink had been enfolded into the glycerin, and it was unfolded again by the reverse turning.

Of course we can accomplish this same transformation less dramatically with any container filled with water and oil, and simply shake the container until the two liquids mix together evenly. Then at rest electrostatic cohesion and gravity will rather quickly re-separate the liquids into two pure groups. What these analogies of mixing and separation attempt to convey is that material form can integrate and disappear into a whole. This suggests an equivalency between matter and that invisible background, a background which we know simply to be space. All the empty space in the universe, which we naturally assume to be less than the physical matter we are able to interact with, is actually more full of content than the surface of form we see due to light waves. What we imagine to be empty space contains the whole of everything. It is a considerably different way of looking at the world, but the message is that matter is constantly unfolding out of and refolding into a larger balanced whole. Bohm writes:

> Classical physics says that reality is actually little particles that separate the world into its independent elements. Now I'm proposing the reverse, that the fundamental reality is the enfoldment and unfoldment, and these particles are abstractions from that. We could picture the electron not as a particle that exists continuously but as something coming in and going out and then coming in again. If these various condensations are close together, they approximate a track. The electron itself can never be separated from the whole of space, which is its ground.

In the book, *The Holographic Universe* the writer Michael Talbot integrated Bohm's vision of two kinds of order with the work of Karl Pribram, a famed neurophysiologist who also became interested in holographic photography as a means of explaining the mysteries of human memory. Talbot also suggests the whole universe is a hologram. He writes:

> Just as every portion of the hologram contains the image of the whole, every portion of the universe enfolds the whole. This means that if we knew how to access it we could find the Andromeda galaxy in the thumbnail of your left hand. We could also find Cleopatra meeting Caesar for the first time, for in principle the whole past and implications for the whole future are also enfolded in each small region of space and time. Every cell in our body enfolds the entire cosmos. So does every leaf, every raindrop, every dust mote...

Unfortunately, the mainstream of other scientists didn't catch on to Bohm's vision of two kinds of order, and he struggled with despair and depression at not being able to convince the scientific community of the scientific value of an enfolded order, a struggle I inherited. Without knowing I was dramatically influenced, I encountered Bohm's concept of implicate order many years before I discovered and learned to present my own system of seeing the same thing, the same principles, the same process in nature. Then rediscovering and studying Bohm's work in greater detail, I now realize my practical grouping order and symmetry order concepts are simply an extension of Bohm's concepts. I now believe the only reason Bohm's visionary accomplishments were kept from having greater impact is that he was led more to challenge how quantum mechanics was being interpreted in his day rather than led to explore how the terms order and disorder are defined, as well as the second law's usage of the terms order and disorder. Also I think he tried to convince the wrong audience. Judging from my own experiences I am sure it would have taken years for even a fully convincing challenge to sink in to the sometimes dull collective mind of science, which grounded in the present dry paradigm tends to shy away from profound concepts altogether.

One of the most important lessons I convey perhaps uniquely from Bohm, is that symmetry order is an entirely separate component of the patterns we find in nature, so that we can see the unfolding and enfolding process as a governing system, visible not only within the general evolution of time, but also present within each individual, and within humanity as a whole. I expect anyone who reads this book will eventually be able to observe any pattern in their world and recognize the two components of order not only contributing to the appearance of order, but also the appearance of what we think of as disorder.

The Absence of One Order Creates the Other

Both the Irish poet William Yeats, who received the Nobel Prize for literature in 1923, and the French philosopher Henri Bergson who received a Nobel in literature in 1927, wrote of two fundamental orders at work in the evolutionary process of the universe. Although the nature of the orders was not described as clearly as Bohm managed with his implicate and explicate concepts, both men recognized the exclusive quality of the two orders more clearly than Bohm. In his book entitled *Creative Evolution* Bergson writes:

> The two orders are not organized into a linear hierarchy or a graduated spectrum in which one is on the top, the other beneath it, and absolute 'disorder' constituting a third alternative, at the very bottom of the hierarchy - 'one nature in graded powers', to use Plotinus' words. Instead, the 'absence of one of the two orders, consists in the presence of the other'.

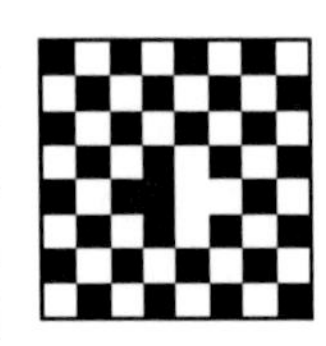

Obviously Bergson somehow recognized how any decrease in the measure of one type of order, increases the order of the other type. This can be recognized if we consider a breakdown in the purity of each order. Suppose we switch two oppositely colored squares on a checkerboard. The inconsistency in the symmetrical arrangement stands out like a beacon. The rhythmic order of symmetrical pattern has been broken. The symmetry order is clearly diminished. And yet the fusing of three colored squares into one larger block is an increase in a type of order. Grouping order has increased. The greater measure of symmetry order isn't suddenly destroyed, we are only one step away from an unbroken perfect lattice, it is merely diminished.

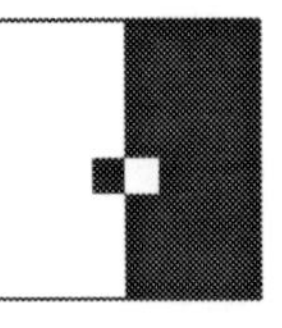

Alternatively, at the other end of order, if we group together the game pieces by colors on each side of the board, and mistakenly displace one colored piece with another, the diminished purity of the grouping order stands out vividly here also. The purity of each of the two groups is contaminated, yet note here how this is a first move toward increased symmetry order. There isn't actually an increase in any general form of disorder. Each of these first steps away from an unbroken high order is both an increase of an order and a decrease of an order. Hence the order of one type is the disorder of the other.

Anywhere that a substance gravitates together, grouping order has increased while symmetry order has decreased. Anyplace where things spread more evenly throughout any frame of reference, symmetry order has increased while grouping order has decreased. Generally, one type of order cannot increase without decreasing the other. Each order has its own direction. Each order is unique. And all there is throughout the universe is order.

We observe order in the universe and are amiss at why it exists over disorder. We typically imagine that our corner of the universe must be a tiny island of order within a greater chaos. The term *order within chaos* is common. However, once we learn to appreciate the two kinds of order, existentially speaking, it seems more apt to say that *chaos exists within order.* If the disorder of symmetry is inversely the order of grouping, and if every pattern consists of a combination of two orders, then all there can be is order. In the poetic words of the photographer Catherine Ames, "the order and chaos are one."

The Gyres

After achieving his Nobel in 1923 the poet William Yeats came to portray the transformation from one order to the other with what he called gyres. These images relate to the double wedge shape of pattern space, but they more accurately portray the measure or intensity of one order compared to the other throughout the entire span of what is possible. They represent how the two orders combine and so become two parts of the same description.

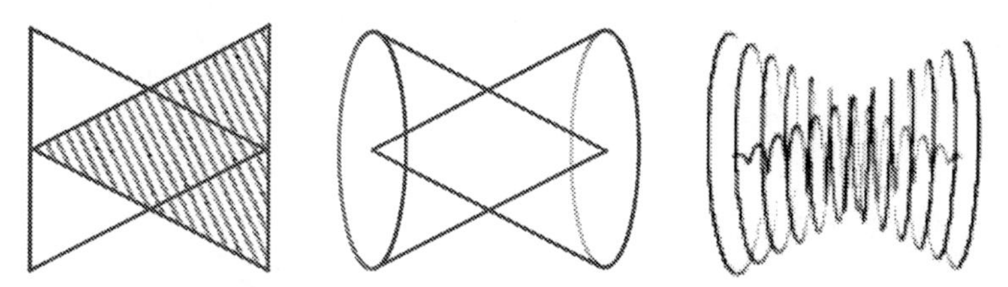

Figure 7.3: Different Gyres described by Yeats.

The two cones effectively represent the integration and inseparability of grouping and symmetry orders. It is said that Yeats discovered the images through the mystical experience of his wife Georgie Hyde-Lee, who demonstrated a gift for automatic writing. In various ways Yeats diagrammed two cones or wedges embedded together. Yeats notes that the information his wife acquires mystically seems to come from a common dream they and a few others shared. Noteworthy, it is well known that Yeats studied the writings of Emanuel Swedenborg, a Swedish scientist in his time turned religious leader, who wrote, "All physical reality, the universe as a whole, every solar system, every atom, is a double cone; where there are 'two poles one opposite the other, these two poles have the form of cones."

The second law is meant to describe the flow of patterns and processes occurring in nature. It attempts to describe the fundamental way that the physical universe changes as time passes, stating that the overall entropy of a system; entropy being the measure of spent energy, always increases with time, which is true, there is no way to get around that law of nature, but the second law also makes a statement which erroneously assumes a connection between a loss of usable energy and an increase in disorder. And we find now that this part of the second law cannot be true, primarily because there is no such thing as general disorder. Symmetry order is the disorder of grouping order, and grouping order is the disorder of symmetry order. If the order of one is the disorder of the other, then there is no room for a general disorder. All there is in nature is ordered patterns of one type or the other, and combinations thereof.

We envision order and chaos as being opposed, and chaos theorists speak of an order at the edge of chaos, but neither accurately represents order as it exists in nature. The order that is so visible throughout our universe springs from combinations of imbalance and balance. Whether we speak of the even distribution of galaxies, or gaseous particles that disperse evenly throughout available space, we find what is happening is that time is evolving away from the most extreme state of imbalance to the most extreme state of balance. When one learns to see this transition, it suddenly is visible in everything from red hot flowing materials that solidify into rock or steel, to water vapor which crystallize into a snowflake.

My own early diagram, meaning to show the opposition and exclusivity of two orders was simply of a square split in half by opposite corners, and likewise was meant to show that the extreme of one order type is the disorder of the other.

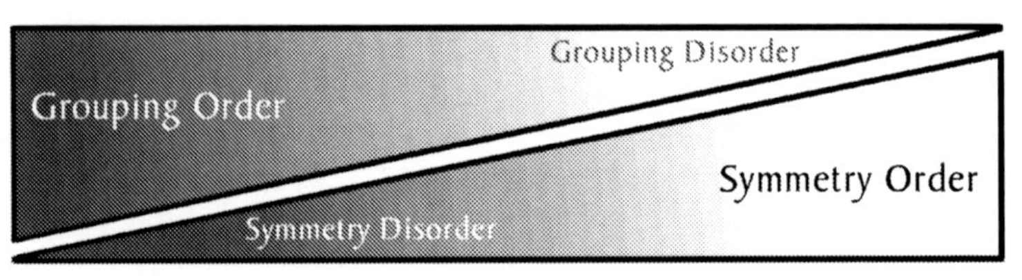

Figure 7.4: Order gradients indicating the absence of a general disorder. The order of one type is the disorder of the other.

Symmetry order is responsible for all the qualities of, and preferences for, balance in the world, from spiral galaxies to the dissipation of gases, even perhaps to the human yearning for peace. In regards to the Universe as a whole, the ultimate singularity, and the ultimate example of perfect symmetry or sameness, is plain old empty space. The single property of the whole universe, from which everything is made, and which everything is transforming back into, is just space, although in learning this, it is so important to understand that things are less than, not more than space. The entire world of material things is a fragment of the whole. Everything we know as real is less than the space it exists in, and not more than empty space or nothing. That low opinion we have of nothingness is all wrong. What we think is nothing is perfect symmetry. Presently we have it all backwards.

Physicists and cosmologists have long questioned whether a state of perfect symmetry ever existed in the past, we just haven't ever considered it as a possible future. We have been convinced instead, because we observe a measure of randomness that the order of the universe is simply winding down. But if anything it is winding up! Everything is enfolding together. The final state of zero which the universe has been evolving toward since the very dawn of time is simply the native state of the Universe. It is the timeless whole. It is truly everything forever. And we aren't really becoming, we are already there. The universe we know, the past, the future, and the infinity of other universes, all exist simultaneously. We are inside that whole. We are a part of the native state of zero, part of the eternal present.

The electrons in a carbon atom in the human brain are connected to the subatomic particles that comprise every salmon that swims, every heart that beats, and every star that shimmers in the sky. Everything interpenetrates everything, and although human nature may seek to categorize and pigeonhole and subdivide the various phenomena of the universe, all apportionments are of necessity artificial and all of nature is ultimately a seamless web.

Michael Talbot

~~~

Frequently consider the connection of all things in the Universe. ... Reflect upon the multitude of bodily and mental events taking place in the same brief time, simultaneously in every one of us and so you will not be surprised that many more events, or rather all things that come to pass, exist simultaneously in the one and entire unity, which we call the Universe. ... We should not say 'I am an Athenian' or 'I am a Roman' but 'I am a Citizen of the Universe'.

Marcus Aurelius
~~~

According to the movement of reason, plurality or multitude is opposed to unity. Hence, it is not a unity of this sort which properly applies to God, but the unity to which neither otherness nor plurality nor multiplicity is opposed. This unity is the maximum name enfolding all things in its simplicity of unity, and this is the name which is ineffable and above all understanding.

Nicholas of Cusa

Chapter Eight

Beautiful Diversity

Unifying the One and the Many

For every north there is a south, for every up a down, for every forward a backward. For every action, there is an equal and opposite reaction. Every school kid learns how to pair together opposites such as hot and cold, smooth and rough, short and tall. On the playground we learn what goes up must come down. And as we grow older there are lessons to be learned about pleasure and pain, strength and weakness, love and hate. In stranded moments throughout life we contemplate opposites of good and bad, wrong and right, darkness and light. The philosopher and poet Ralph Waldo Emerson wrote, "Every sweet has its sour; every evil its good." But does absolutely everything have an opposite? And if so, and it is also true what they say, that opposites attract, then what happens when all the opposites meet in the middle?

In between opposites there is always a middle ground, and yet the middle can be awfully difficult to describe. Between thin and tall there is the average height. Between heavy and light there is the average weight. We usually have to use neutral words and phrases such as medium, average, the most common, the norm, to define the middle ground between opposite attributes, because oddly enough there aren't special words that identify the middle ground. What word defines the middle between strong and weak, hot and cold, sharp and dull, hard and soft, or easy and difficult? The middle ground is almost always nameless and yet we can easily recognize it exists between each opposite.

There are opposing directions in politics, the left wing and the right wing, but of course the balance between liberal and conservative is simply called the middle of the road? Why aren't those 'middle of the roaders' allowed a special name of their own, or a political party of their own, like everyone else? And why aren't there more of these people. Where do they hang out? I can't remember ever meeting any of them. Could it be that when people reach the middle of the road they just disappear without a trace? Or is it that they just don't speak up, or don't have an opinion? Why are the rest of us so clearly on one side or the other? There are left wingers and right wingers in politics, religion, justice, education, and even art. Why is everyone so polarized into camps, or sides, or groups?

In the study of human personality, Carl Jung, and later Isabel Myers and Katherine Briggs identified the four temperaments that define the essence of each person's personality. A multiple choice test identifies a person as more thinking or more feeling, more sensory oriented or more intuitive, more introverted or more extroverted, and finally in the last divide it tells if a person is more spontaneous, flexible, and free to flow with the ups and downs of life, or oppositely if they are more inclined to be planned, rigid, structured, and organized. The four divides effectively define sixteen basic personality types. It can be quite surprising to find how accurate one's own personality and behavior is described in respect to being one or the other of each of these temperaments. Yet I have wondered, with billions of people on this planet, isn't there one person out there who is right in the middle, who isn't any one type more than the other. Really there must at least be many thousands, but why hasn't their personality been identified as the seventeenth type? What is their personality like? Do they have one? And how come I hear somebody telling these individuals (if I can call them that) to "take a stand", "be somebody", "make your mark!" Why are we expected to be off center? What is so terrible about being in the middle?

All opinions, all traits, all characteristics, all forms, have a middle ground, but we don't name the middle ground apparently because we think it's too plain and boring to be given a name. We usually act as if there is nothing in the middle, as if when two sides blend together they cancel out or disappear. But just because the middle ground is always less distinct and pronounced than the definitive extremes on either side, why do we go and think the middle ground is a formless nothing.

This elusive middle ground between opposites is itself a physical part of reality, although it depends upon how we look at it as to whether we define it as a combination or a cancellation. It can be seen as inclusive, as the combined sum of opposite properties, or it can be seen as exclusive; the negation or cancellation of opposite properties. It can be seen as the potential to be either or it can be seen as a nonexistent neither. And since we are each defined by our own particular imbalances, it often depends on our own temperament as to whether we see the middle ground as the whole, or as a void.

The great egos, the loud and obnoxious, the pronounced types, of course see the middle ground as boring, empty, and repulsive. The practical, the conservative, the sensory oriented skeptic, tends to see the middle ground as irrelevant because it doesn't, or it doesn't seem to, have identifiable qualities. What isn't physically definitive doesn't exist. The classical physicists of the past century, in concert with mathematical logic, have strictly seen the merging of opposites as a cancellation. Someone with a practical and physical personality prefers to define reality as limited to physical things and measurable properties. On the other hand, the more intuitive, the progressive, the insightful, the more spiritual types, tend to sense the middle ground as a whole containing all opposing sides. They tend to depreciate the physical and see the balance between opposites as a unity, as two sides of the same coin. The middle ground is seen as a foundation or axis, from which form springs outward. For some, the combination of all opposites forms a single unified whole, a common oneness. This is the central core of many philosophies and religions in the east.

A common belief in Hinduism, Taoism, and Buddhism asserts the unity and interrelatedness of all things. Brahman in Hinduism and for the Yogi is the unchanging and infinite background of all physical being. It is the sum totality of all. Likewise, in Chinese Taoism the word Tien or Tao refers to the ultimate sum of all. Everything exists in relation to the Tao and everything is a part of the Tao, even though the Tao is one thing. Therefore nothing can exist or have meaning apart from the Tao. Out of the Tao comes the Yin and Yang, the two opposing forces or natures. In Buddhism the Dharmakaya is the experience of a timeless unity devoid of all physical characteristics, which is said to be true reality. The Heart Sutra of Buddhism states "...Form Does not Differ From the Void, And the Void Does Not Differ From Form. Form is Void and Void is Form..."

Oneness was also a common message of many great philosophers, including Xenophanes, Heraclitus, Parmenides, Plato, Plotinus, and Giordano Bruno. Xenophanes appears to have influenced a long line of other philosophers with his belief in an infinite and eternal Universe that is unable to change. He undoubtedly influenced the development of religion as he described the infinite whole as an omniscient God that sees all, thinks all, and hears all, "one god greatest among gods and men". A few years later Heraclitus called the unity of opposites "Logos". And following Xenophanes, the logically minded Parmenides described being as innate and without any opposite of non-being, since non-being cannot exist, and he also argued that being is ultimately timeless and unchanging. In that belief Parmenides treated time, form, distinction, and all duality, as illusion.

Plato also considered the visible world to be an illusion, one that produces weakly assumed beliefs in the illusion. In the allegory of the cave he suggests the world we experience is like the shadows of another much deeper reality. Plotinus, like the Buddhists saw the great Oneness as beyond all attributes, including even being and non-being. More recently, the Italian Philosopher Giordano Bruno wrote: "Everywhere is one soul, one spirit of the world, wholly in the whole and in every part of it, as we find in our lesser world also. This soul...produces all things everywhere; so that for the generations of some even time is not required..."

In modern times, Ralph Waldo Emerson in believing that opposite halves inevitably produce a whole writes:

> POLARITY, or action and reaction, we meet in every part of nature; in darkness and light; in heat and cold; in the ebb and flow of waters; in male and female; in the inspiration and expiration of plants and animals; in the undulations of fluid and of sound; in the centrifugal and centripetal gravity; in electricity, galvanism, and chemical affinity. Super-induce magnetism at one end of a needle, the opposite magnetism takes place at the other end. If the south attracts, the north repels. To empty here, you must condense there. An inevitable dualism bisects nature, so that each thing is a half, and suggests another thing to make it whole; as spirit, matter; man, woman; subjective, objective; in, out; upper, under, motion, rest; yea, nay.

It is extremely difficult to contemplatively turn a switch within oneself and suddenly see the world in an entirely different way. But if we really take a careful look at the way all opposites are bound by a neutral center, and simultaneously consider the likelihood that beyond our personal experience of time literally everything exists timelessly, meaning that all the opposites exist simultaneously...and then we try to imagine what the universe would be like if we could glimpse that whole if only for a brief moment, the vision we would see could be interpreted to be nothing at all. In seeing everything we wouldn't see anything but an endless single uniformity, i.e., empty space. In perceiving the whole, the only property or characteristic to observe is the serene surface of balance and uniformity. So even though the great oneness is full, it is our nature to be blind to the endless dichotomies that exist beneath the surface we emerge from.

Scientists are already aware that the seeming empty space around us, or between the stars and galaxies, is not ever actually empty. Underneath the quiet surface of space, the microscopic background is bubbling with virtual particle pairs which emerge from nowhere, always both a positive and a negative paired together, so that they immediately attract and collide. An equal positive meets an equal negative, and so the sudden finite creations are instantly destroyed. For an instant the particles are small expressions of grouping order until they merge together again and the balance of symmetry order returns. This keeps the surface of our surrounding ocean calm.

Of course we are not accustomed to recognizing the simplicity of perfect uniformity and sameness as being properties of an underlying order. It is a challenge to genuinely appreciate our surroundings as being so full, containing an infinite number of other worlds. We easily recognize the balance and symmetry in the pattern of a snowflake as being highly ordered. Yet the same qualities of symmetry and balance pushed to ever greater extreme leads to an indistinct sameness. The funny thing about our experience of physical reality is that we are so tuned into appreciating the organizational properties of grouping order, or the pattern complexity of both orders, that the uniformity of symmetry order doesn't seem like anything of any importance.

Ugly Symmetry, Beautiful Diversity

As mentioned, the simple reason that we don't commonly recognize symmetry order distinct from grouping order is that the blending together of objects, the dilution of form into its surrounding space, in extreme, produces a bland uniformity. Uniformity in appearance is plain and boring. It is the world painted one color. So when oneness is compared to the diverse complexities of human experience, grouping order steals the show.

In addition to those who have discovered two types of order, there is one person who has consciously recognized symmetry as a distinct and separate component in nature which is increasing as the universe evolves. The chemist Shu-Kun Lin, creator of *Molecular Diversity Preservation International* and founder of the journal *Entropy* has recognized the connection between increasing entropy and symmetry. In Lin's similarity principle, higher symmetry correlates with increasing entropy. Lin boldly states that "the universe evolves towards a maximum symmetry." Thus Lin understands as I have explained here that symmetry increases with time, and likewise has concluded inversely that the universe is ever more asymmetrical in the past.

Lin does not yet refer to symmetry as a form of order. Instead Lin has recognized the ugliness of high symmetry in relation to the diversity of lesser symmetry and asymmetry, which adds an interesting perspective to our discussion. To explain the ugliness of symmetry and the beauty of diversity Lin writes:

> Children understand the beauty of diversity. On the walls of a child's classroom or bedroom a visitor may find a lot of paintings, drawings and even scrawls created by the kids. If you do not put some colorful drawings there, the children will create their own. The innocent children want to destroy the ugly and boring symmetry surrounding them. If your kids destroy symmetry on the walls, they are doing well to be creative. Diversity is beautiful. Symmetry is not. Coffee with sugar and cream is an interesting drink because of its diversity in taste (sweet and bitter) and color (white and dark). The United States is considered a healthy country because of its tolerance to all kinds of diversity (racial diversity, cultural diversity, religious diversity, etc.). Without such appreciation of diversity, this country would become much less colorful and less beautiful. If everyone behaves the same, looks the same, and there is a lot of symmetry, the world would be truly ugly. I see democracy as a sort of social diversity.

Lin very interestingly focuses on the surface uniformity of symmetry, and he uses the term ugly symmetry to describe the plain character and decreasing information of high symmetry. As Lin describes it, "Symmetry is in principle generally ugly because it is associated with information loss or entropy increase." which is a quite reasonable conclusion when we consider what patterns of symmetry present to us as compared to the vast diversity of asymmetry existent in the human world of experience.

In contrast to the focus throughout this book on the enfolded nature of symmetry order, Lin's focus is on how we as humans relate to extreme symmetry. In order to convey an important new way of looking at the world, Lin is pointing out that extreme symmetry is ugly compared to the beautiful diversity we experience. Human beings spend their lives working to maintain and possibly improve our physical stability as a distinct thing in the world. Most people don't describe themselves as an object in a world of objects, and yet most do see themselves as a separate being apart from other things. We go to great lengths to maintain the objects that we identify with. Most people shrink at the idea of the universe becoming empty of things. So then how should we interpret this increase in symmetry taking place in the universe? What good is it if it threatens our continued existence?

Lin being a scientist also relates the plain (ugly) uniformity of symmetry to information loss. In what is called Information Theory, entropy is considered very generally to be a loss of any information (which happens to be related to the present paradigm where one sees increasing entropy as increasing disorder). Information can be thought of in this way as all physical form, and the loss of information is like a segment of writing or a map on a white piece of paper fading with time. As the ink fades the information is lost, and finally all that remains is the white paper, which conveys no information, as least from our perspective in a world of finite objects and things. Lin writes:

> The Greek word symmetry means the "sameness measure". It is therefore closely related to distinguishability or similarity. Symmetric structure is stable but not necessarily beautiful. All spontaneous processes lead to the highest symmetry which is the equilibrium or a state of "death". Life is beautiful but full of asymmetry.

Lin's perspective is certainly valid and extremely helpful. His vision lends another perspective to the mix, and reminds us to appreciate the beauty of asymmetry. For science his theory constitutes a major step forward toward recognizing the role of increasing symmetry in the evolution of time. It could be said that in all of science, Lin's recognition of increasing symmetry is the most advanced view of the universe from the bottom-up perspective, since he recognizes the actual direction of time is toward symmetry. However, once we have learned to see symmetry as a separate component of a pattern, the next step is to recognize symmetry as a distinct type of order.

When we make the mistake of merely identifying with material things, and see reality as a product of material things, we are thereby assuming that imbalances and asymmetry are all that exist and are all that constitutes physical reality, in which case we run into paradoxes, where for example, existence and order seem unlikely or impossible. Once one recognizes the enfolded order behind the surface uniformity of high symmetry, that recognition, that step forward, leads to a universal perspective in which all facets of the universe start to make sense. We then learn why the universe is so ordered. We recognize the actual difference between the past and future, and we find a meaningfulness to time. We recognize the impetus of time, the drive toward balance. Those mysteries can't be solved until we acknowledge the enfolded implicate order as a timeless base, which is not actually beneath but above the surface of things.

Of course subjectively one's instinct abhors the notion of loss of form. The thought of formlessness is almost horrifying to our identification with form, even if the complex forms around us are becoming the oneness we otherwise idealize. The uniformity of symmetry from the vantage point of an intricately diverse world in which life is manifest is not merely ugly but dangerous, a threat to our very being. We are conditioned to see oneness as a threatening emptiness because we identify so intimately with things. We see ourselves as a thing. We collect things, we sell things, we ingest things, we expel things. The science of physics is all about the broken symmetry of things. Physics doesn't study wholeness or uniformity, it studies thingness. Basic mathematics counts things. Science is all about describing what we consider physically real, and the underpinning of physicality is grouping order. Grouping order is an entire worldview. I cannot stress this point enough. The backward direction toward grouping order is one of two basic forces of nature. So we shouldn't be surprised to find an entire view of the world derived from grouping order. That view has been the dominant view in the past, but it is not the only view.

Ugly and Beautiful Extremes

Diversity is altruistically beautiful, and nothing exposes the diversity of human beings more than the myriad of ways that we group together. People group together when they associate with one religion, or a sports team, or a club, or a hobby, or an interest like art or science. There are baseball fans separate from auto racing fans, painters separate from sculptors, physicists separate from biologists. People group as a city, as a state, as a country. We group by fortune, by health, by skills, by developmental stages, and by education. We group by class, by race, by wealth, by age, by fitness, and by sex. And yet without question it is our diversity and measures of distinctiveness that also divide us apart from one another. Groups, identities, classifications, define boundaries and often times not only do they divide up the world, they lead to opposition and conflict. They lead to teasing, to harassment, to terrorism. Inequalities are arguably the leading cause of hate, crime, and war. So although diversity is unquestionably beautiful, diversity also has an ugly side, in that such differences often bring out the very worst in people, certainly so in those who have, as well as in those who don't.

Alternatively, at times, in ways, we also break down and cross barriers, we tolerate or completely see beyond individual or group differences, and come together as a whole. We come together as states, or regions, or cities, to form a larger government. We come together as countries that share the same ocean, or continent or region, or the same planet, making laws to protect and safeguard common interests. We invest in one another. We care for one another. At times we forget or forgive and unite as people, as life, as a whole. So isn't it when we harmonize with one another, when there is cohesiveness and consensus, when we recognize our relatedness, isn't that when we are most beautiful? Isn't it diversity or difference that divides and sets us against one another, which defines when we are most ugly?

Actually, sometimes our differences bring us together in wonderful harmony or in play, or in exploration, without regular conflict, as in the way opposites attract. And sometimes sameness divides us apart, such as a parent and child who are alike, perhaps in being stubborn or opinionated. Children as they grow up almost always avoid being too much like their parents, and try to develop their own identity. Often times the detraction or repulsion away is caused from not wanting to make the same mistakes, or in not wanting to become the perceived negative side of another person or group. Still in other ways people sometimes unify together because a powerful or forceful person or group dominates or destroys the distinctiveness of an individual or another group. The will of one person or group imposes on the will of another, imprinting their own character, their ideals, their religion, their culture, onto another person, group, or another land. Self identity on one side is degraded or lost. Countries are invaded. Cultures and belief systems are suppressed. In extreme, entire cultural systems are attacked and horribly massacred.

The loss of culture not only occurs by destructive force and by coercion, but it also can result of positive forces. Sometimes a counterproductive dark side of a culture is changed by persuasion or by setting an example, in an appeal to reason, or through kindness and caring, or all of the above through regular interaction. Sometimes the deeply embedded ignorance and neuroses in a culture are overcome as wealth, health, education, and technology is developed, or transferred from one group to another, or traded between groups, but notably sometimes this forward step happens at great cost to the uniqueness of that culture. The flip side of progressive influences is often the loss of a deeper less rational connection to nature, to the past, to history, and to spirit.

When we look at the big picture, imagine all the historical events of nations, all the different sports and festivities of various countries, the unique foods, the architecture, the belief systems, all this culture and diversity appears unquestionably beautiful, yet differences do create conflicts that are sometimes horrifyingly ugly. Gulfs of difference are often the cause of great evil and destruction. Moving in the direction of becoming increasingly different, we are ever more inclined to polarize into some kind of conflict. And we do tend to polarize into two groups divided by some boundary. Such is the nature of grouping order, while the great beauty of diversity occurs somewhere in the middle between the ugliness of extreme grouping and the surface ugliness of extreme sameness. It is when the most disparate tightly knit groups break apart into smaller groups, the branching outward, the disintegration of stark difference, that we attain the highest measures of beautiful diversity.

The disintegration of stark difference into variety and diversity is a disintegration of groups that inevitably leads toward an increase in sameness and symmetry. In the step from extreme polarization to greater diversity, people become less starkly different and so invariably more alike. So increasing symmetry is an important contributing part of diversity. Both grouping and symmetry share a role in diversity and its beauty. They are in fact the two defining forces. Diversity is an intermediary domain, a stage in the transformation, of one order becoming the other.

Dr. Lin isn't actually meaning to suggest that symmetry in all measure or forms of expression is ugly, but rather, he is pointing to the same fact I am pointing to here, that a blend of two orders creates the wide range of interesting diversity. As an example he points to the small mole on Cindy Crawford's left cheek. Cindy Crawford is of course one of the most widely known supermodels of her era, and the mole has been noted by many to be a feature that actually enhances her otherwise highly symmetrical attractiveness.

Diversity is beautiful and it could be argued that sameness is ugly, however, the direction away from diversity toward grouping also results in ugliness. That direction away from symmetry toward ever more extreme grouping is also toward a less than ideal condition for life. In the past the universe is increasingly dense and hot. In the same way that the increasing symmetry order in our future is related to loss of energy and extreme cold, to the point of freezing time itself, the grouping order in our past relates directly to brutal heat and density. We all know that time viewed in reverse portrays the universe melting into a dense plasma. Most people would rather be warm than hot, but everyone would rather die from cold than from heat. The conditions of the past are also unpleasant to life and destructive to diversity. Further still, the very nature of grouping order being increasingly divisive hints at a dark side to that extreme that we have only begun to expose here. In summary, hardly anyone would argue with the idea that we should appreciate the now, appreciate what we have, take care of it, and don't let it slip away. From our subjective perspective, in this illusion of being separate things, nothing else is as wonderful as the great stage of diversity between extremes.

It sounds incredibly idealistic at this time in history, but the trick of course to harmonizing and living together peacefully, to us all becoming the same from some other reason than a forced conformity that destroys diversity, is for each person to become diverse, meaning everyone must develop all the many sides and potentialities of themselves, as opposed to being imbalanced or one sided, so that we all appreciate the fullness of one another. The same is true of groups, of communities, of countries. The people of the Earth will only live in blissful harmony when all societies have developed their fullness. It sounds so simple and fully impossible I know. However, what seems to be the answer, the thing that increases the measure that communities manage this kind of individual and collective fullness is simply quality education, matched with wide experiences in youth, and contemplation. As long as we develop fullness our darker side is overwhelmed by the wisdom and understanding inherent in balance.

Either/Or

Fundamentally there are just two distinct directions of change or transformation that keep surfacing here. In one direction groups of people can become more powerful and dominant, in contrast to other groups. In the other direction there is a dilution of distinct groups toward the function of a single whole. In a marriage one person can become increasingly more dominant and define the character of both, or the two can become increasingly more accepting of one another, each developing the traits of the other which they lack, and consequently grow individually as well as grow together as a couple. The same dichotomy is true of the largest groups of people. The two directions would apply even if an alien civilization from another star system were to make contact with humanity.

We might find alien visitors to be either backwardly aggressive warriors bent on galactic domination, who are selfish and brutal, or an advanced cyborg race in a constant state of being at one with each other. There are also two ways that people would react to alien visitors. People either resist change out of preference for the way things are or have been. Or they support and enforce change out of a preference for a perceived future goal of growth and progress. Nothing highlights how powerfully people might resist change than the idea of aliens landing and wanting to become friends. Individual people or groups can only react to such changes with a backward pull or a forward push. The backward pull is what I like to call human gravity, since it relates to the way gravity attempts to recreate the dense past. Backward people want to keep things as they are or even recreate the past. Forward people want to change everything and make the world better than what has been in the past. Of course cosmically speaking, the real force of gravity settles for maintaining the past against what would otherwise (without gravity) become an overly rapid expansion and growth into the future.

Most importantly, considering Lin's perspective in contrast to that of Bohm's, we can identify two different ways of perceiving increasing symmetry, one as a bottom-up view, one a top-down view. When we see the outer surface of extreme symmetry becoming increasingly plain and boring we are seeing reality from a grouping order perspective. From that place we see the expanding universe becoming flat and empty, and we see order decreasing (denying the increasing symmetry all together). From that place we see the universe purely as finite and materialistic. The world is arisen above nothingness. Matter is prized and worshiped. Pronunciation is honored. "Rise up and be counted." Such views arise out of the spirit of grouping order, or Bohm's explicate order. The religion of this perspective could be paraphrased as "only imbalances are real".

Ordinarily we view reality from a perspective derived from grouping order. Much if not all of modern physics is based upon the axioms of grouping order. But there is another perspective of reality, another mode of perception, which we can shift into mentally. There is the dramatically different top-down view. The universe and the integrity of form is defined altogether differently from the unique perspective of symmetry order. Rather than viewing the substantive world as magically arisen above nothing, there is a valid alternative, a perspective where all we know is less than everything.

The symmetry order perspective senses an inner content to the world. It perceives all the possibilities within the uniform surface of space. As Bohm recognized, "We may be led to explore the notion that in some sense each region contains a total structure 'enfolded' within it". From this perspective the Universe has a timeless and eternal nature. Things are not seen as separable from the whole, and so space and matter become one. In this unity the deeper meaningful, metaphysical, or spiritual aspects of reality are sensed existing beyond, yet integrated with, the surface of appearances.

It is never the less important to keep in mind how the top-down view comes along with its own religion where the whole is considered to be all that matters, such as when we see God as only what matters and see the finite human world as unimportant in comparison. From that place material things are unimportant. Achievement is unimportant. Only being is important. Only perfection is important. Such an extreme view can be just as one sided and mindless as the grouping order perspective.

Even as the two perspectives seem to contain an inner self consistency that repudiates the other view, as if there is an enormous gap in the transition between sides, when the two views are respected as the two sides of reality then we understand reality. Only then can we move to experience the diverse world properly. Without the bottom-up there is no top-down. Without the top-down there is no bottom-up. As some Eastern Sages have long explained, when we understand the true nature of reality there is no exclusion, no dividing boundary between thingness and wholeness, between nothing and everything. Silent meditation has proven to be an extraordinary tool for moving a person toward a peaceful and balanced state of mind in which the fullness and unity of the world can be known and appreciated.

Photo: Curious Crab © Nico Smit

. . . part of metaphysics moves, consciously or not, around the question of knowing why anything exists - why matter, or spirit, or God, rather than nothing at all? But the question presupposes that reality fills a void, that underneath Being lies nothingness, that de jure there should be nothing, that we must therefore explain why there is de facto something.

Henry Bergson

~~~

There is still left a single story of a way. Along this way there are signs exceedingly many, that being is uncreated, and also imperishable, whole, unique, immovable, and complete. Nothing was not once nor will it ever be, since being is now altogether.

Parmenides

~~~

Besides learning to see, there is another art to be learned, not to see what is not.

Maria Mitchell

~~~

The idea of nothing has bugged people for centuries, especially in the Western world. We have a saying in Latin, Ex nihilo nuhil fit, which means "out of nothing comes nothing." It has occurred to me that this is a fallacy of tremendous proportions. It lies at the root of all our common sense, not only in the West, but in many parts of the East as well. It manifests in a kind of terror of nothing, a put-down on nothing, and a put-down on everything associated with nothing, such as sleep, passivity, rest, and even the feminine principles. But to me nothing -- the negative, the empty -- is exceedingly powerful. I would say, on the contrary, you can't have something without nothing. Imagine nothing but space, going on and on, with nothing in it forever. But there you are imagining it, and you are something in it.

Alan Watts
~~~

I am well aware that it is not easy to elucidate in Latin verse the obscure discoveries of the Greeks. The poverty of our language and the novelty of the theme compel me often to coin new words for the purpose. But your merit and the joy I hope to derive from our delightful friendship encourage me to face any task however hard. This it is that leads me to stay awake through the quiet of the night, studying how by choice of words and the poet's art I can display before your mind a clear light by which you can gaze into the heart of hidden things. The dread and darkness of the mind cannot be dispelled by the sunbeams, the shining shafts of day, but only by an understanding of the outward form and inner workings of nature. In tackling this theme our starting point will be this principle:

Nothing can be made out of nothing.

Titus Lucretius Carus 96 B.C.

Chapter Nine

Something From Nothing?

Why would anything simply exist? How could existence be the default state of reality? It seems too good to be true, and yet, here we are. So why then does nothing seem more probable, more simplistic, and more natural than a universe of many things? The physicist Max Tegmark points out that nothingness would have zero information content, whereas a something universe contains information. For this reason, a nothingness seems to require no cause or explanation where in contrast a world of things being physical, being definitive, being diverse in character and quality, requires an explanation or reason for existing. "The fact is, nothing could be simpler than nothing — so why is there something instead?" remarks the astronomer David Darling.

Like everywhere else, in science *nothing* is a general term with a vague definition, and the word is always used generally and perhaps presumptuously in ambiguous applications. The word *nothing* isn't considered a scientific term, yet like most of us, physicists and mathematicians do commonly make conceptual relationships between the word nothing and other concepts such as zero, the empty set, a vacuum, and empty space. It is for that very reason scientists are astonished about how these same phenomenon behave in contradiction to our expectations of nothing. Indications of a vacuum or empty space having a hidden content or producing things such as virtual particles, is expressed as one of the great curiosities of physics and nature. Many times scientists, due to this issue, have remarked that apparently you CAN get something from nothing.

There's a Gary Larson cartoon that expresses in ingenious Larson fashion the absurdity of entertaining the notion that something can come from nothing. It portrays two professors talking at a chalkboard, one exclaiming to the other a breakthrough in his equation. "Yes, yes, I know that, Sydney ... Everybody knows that! ... But look: Four wrongs squared, minus two wrongs to the fourth power, divided by this formula, do make a right." the professor states with astonishment. In my opinion, Larson is teasing science particularly for accepting the notion that the universe might have begun from nothing. I think the cartoon conveys how the deductive logic of science and math can sometimes miss out on what the intuitional mind knows to reject. If two wrongs are genuinely wrong they don't ever make a right, any more than the existence of the Universe can suddenly pop out of nonexistence.

Real nothings, and things that look like nothing, make up a large part of our everyday life. We can see through air and water to the point that both can be invisible...but then the wind blows and the rain falls. Any physicist will tell you there are all sorts of surprises hidden within empty space. We can logically and reasonably relate the word nothing with empty space, with uniformity, with a white canvas, all of which is fair and accurate, until we step over the line and pollute the reality of existent phenomena with a comparison to nonexistence. One has meaning while the other drains all inference of meaning. When we view something from nothing as some kind of miracle, behaving as if we gained 'something' physically existent from nonexistence, then we have crossed the line, because the nothing that we 'get something from' in nature is always being. And being is innately full. The creations we observe in nature are always the unfolding of an already enfolded timeless whole.

The Roman philosopher Titus Lucretius presented his own version of this argument, he recognized that space can never end, for what would happen, Lucretius asked, if we throw a dart at the outer edge. "Wherever you may place the ultimate limit of things, I will ask you: 'Well then, what does happen to the dart?' The universe has nothing outside to limit it", wrote Lucretius, recognizing then what we still believe today scientifically, that space or the universe cannot become thin or simply end, beyond which there is nothing. Later Einstein showed that space is flexible, and is curved by massive objects, and we now call the curvature of space-time gravity. It followed from Einstein's theories that space might be curved into some sort of loop, such as a figure eight, making the universe finite. But as mentioned, in the last few years scientists have determined that Lucretius was correct about the infiniteness of space, since the known universe shows no sign of being curved into any kind of repeating loop.

"Nothing comes from nothing", Lucretius said, and I remember discovering this paradox in boyhood, because I walked around for several weeks fully convinced that the universe could not possibly exist. As my own existence persisted, I finally relented to the idea that something had somehow cheated its way past the original nothing. And a lot of people make that erroneous conclusion. Literally everyone recognizes the simple logic that something cannot come from nothing, and yet in the face of our own existence and the unwavering presence of the universe, and in the absence of any other explanation, we conclude that by some fluke chance the impossible must have happened somehow. It's actually a terribly damaging form of surrender, not merely because it is a faulty explanation, but because it plants a seed in our mind that our internal logic is out of synch with the Universe. It even places into question a sensible reality. Consider how damaging it is to place into question our ability to reason, and how damaging it is to not see the Universe or reality as ultimately sensible.

"Nothing" isn't a science word but the principle that existence is neither created nor destroyed is universally true throughout science. Everyone knows the objects of everyday experience do not simply pop into existence. When we look out with telescopes there are no distant edges to the universe beyond which lies the domain of nonexistence. Matter and energy change form, they can even transform into space. But all is just the reshaping of something else, the reshaping of what already exists. The First Law of Thermodynamics, which is perhaps the most fundamental law of physics, states that energy is neither created nor destroyed. There are all sorts of conservation laws and equivalency principles in physics and mathematics. What exists on one side of the equal sign is the same as what exists on the other side. All sides exist balanced around equality. It appears now that the universe begins as all energy and no space, and ends as all space and no energy, so either the first law is wrong or there is a fundamental equivalency between matter-energy and space. Space isn't a form of nonexistence. Matter, energy, and space, are all part of the same thing. They are all part of the same existence.

There is always an intricate logic hidden behind our expectations. In the case of whether a universe should exist versus nothing at all, the existence of a universe even seems to violate Ockham's razor, which holds that the simplest answer is usually the correct one. What could be simpler than having nothing to explain? Based upon that principle alone, nothing at all seems more probable than any universe. But simplicity is actually a feature of the real nothing that exists, a quality of uniformity, the white world, the white canvas, an empty background. Expecting a nothing prior to existence to be simple is one of the best examples of how we confuse the qualities of real nothings with a nonexistence, which by definition doesn't have definable qualities at all, not even simple ones.

Where we get into trouble is in imagining nothingness is similar to nonexistence. We imagine nothing as uniformity, formlessness, emptiness, and then in creeps nonexistence, which actually belongs in a category of its own. We do so in part because we focus on "somethingness" as if it is the grand stage of reality, which is generally correct. But when we only identify with the magnificence of one's own little neighborhood we are failing to appreciate how we are inseparably part of a balanced whole. The eternal physically infinite Universe of universes isn't out there somewhere in the distance. It's here, it's there, it's everywhere. The infinite is right here within the one enormous moment of now we exist within.

In the use of meaning, the meaning that otherwise defines all things in all languages, a nonexistence cannot be described, simply because nonexistence cannot be. Nonexistence cannot be imagined or conceptualized. It cannot be signified, resembled, or symbolized. By its own definition, a nonexistence cannot even be inferred with any logical coherency. In this seemingly strange realization notice how we are confronting an anomaly, a single part of our thinking that doesn't fit in with the rest. Nonexistence cannot be described with the ordinary meaning of all language. What else in all of reality is completely without meaning? For this reason the term nonexistence does not actually belong as a member of any language. It cannot be represented by any symbol. Its use is a contradiction in meaningfulness. By definition, nonexistence cannot be.

All attempts to define nonexistence, even as the absence of existence or as the negation of being commit a fundamental semantic crime. It is true that the words absence and negation, or existence and being, each have real syntactic meanings individually, but when placed together they express a radical contradiction, since they attempt to define with meaning a non-something which by definition cannot have meaning at all. It is the very existence of the universe and its attributes that creates meaning, or at least meaning and existence are intermingled and inseparable. Only existence allows for there to be meaning, just as only meaning allows for there to be existence. The word nonexistence tries to refer to a single exception to that rule, and so far in history we have made a mistake in thinking how that one exception is acceptable, when it's nothing but destructive to our vision.

At the heart of the matter is that the total nothingness we imagine as an option to existence is merely uniformity, a real nothing that exists, and this real nothing doesn't relate to nonexistence at all. Almost every time we think of the word nothing we confuse meaning and non-meaning. If and when we fully distinguish between nothing and nonexistence, we can see that nonexistence is not really an option. It isn't possible. It isn't real. It never was. It never can be or could be. It could never have been. The meaning of the word doesn't even make sense. Nonexistence by its own definition cannot be. That is a very ponderous statement I know, but it pans out to be an extremely valid point when we ask, why does a universe exist? Nonexistence is simply not an alternative to the universe being here.

The simplest deduction in logic and the deepest intuition both lead to this same realization. There simply is no alternative to existence. The only reason we could ever find for why there is existence is that there is no alternative. Existence is the default. And so following a time line into the past to search for the beginning of existence can never reach any point of origin. There is no such precipice as the one we imagined earlier. Of course I am not suggesting that the big bang didn't happen, or that some creation of form didn't happen in the past. I am saying what Stephen Hawking is saying with the no boundary proposal. There is no boundary between existence and nonexistence. The only boundaries are the cosmic absolutes, which are boundaries more like that of extremes of character or personality. Certainly from our perspective the evolution of time has a beginning. If we should look far enough into the past we can find a point where the evolutionary changes of our space-time begin, but that does not exclude the past from continuing to exist even after our time begins and travels away from it.

Borrowing Meaning

Why is there a word for nonexistence? The reason is that other words successfully borrow their meaning from another meaning. If something is non-white, we know the color is some other, at minimum, off-white color. If we say a temperature is not cold, the reference is to being greater than cold, so something is warm or just right, or perhaps it's not cold, but extremely cold. Anything that is not, is something else, except non-existence. The term nonexistence is entirely unique from all other words in that it attempts to borrow meaning in a way that no other concept or idea attempts to borrow meaning, since the final product or attempt at meaning by definition has no meaning. Any other case of borrowed meaning refers to something not denied. If we refer to not-above, not-old, not-clear, non-Euclidean, or non-standard, all that is being referred to has meaning independently. All such terms have a place in reality. But the word nonexistence does not refer to anything independent. By definition it refers to what can have no meaning at all. No other concept needs to, or tries to, attempt such a reference. And so no other concept fails in this same way.

Nonexistence tries to specify the absence of existence, and without scrutiny this seems to work because all other such denials accomplish their task. It seems to work because in physical reality when some thing doesn't exist, there is indeed a void in its place, but the void left behind when we say there is nothing in the refrigerator is the real nothing, just singular form, and not a nonexistence. In fact the term nonexistence does not refer to anything that has any meaning whatsoever. There cannot be a non-existent type of nothing, because the meaning of non and existence when used together form a contradiction. Only the negating term non- and the word existence have meaning individually. Imagining that non-existence is a sensible concept is equal to imaging the phrase "being doesn't exist" can somehow make sense.

A Look at Things That Don't Exist

We accept the word nonexistence partly because it makes sense to say unicorns or square circles don't exist. And if there is nothing in the refrigerator, we can say, the milk doesn't exist in the refrigerator. But what we are really saying is that the milk isn't in this location at this point in time. We are saying that the milk is not enfolded out of the implicate order. We are saying "there is no imbalance here to take energy from". We know there is a basic principle considering time and space, that things exist in locations relative to one another. Take away the separate locations in time and everything exists in the same place, like ideas. The same ideas are located everywhere we try to think. The absence of all things that might exist in the space inside the refrigerator doesn't ever create a black hole of nonexistence. It merely turns that space into the seeming emptiness of a spatial singularity. Similarly, in the absence of magical unicorns the universe still exists, so all that we can really properly say is that imaginary things don't assume physical form here in the same way we do.

When we say dragons and unicorns don't exist we are only really saying first that we don't see them, and second that magical things that disobey the laws of physics don't exist. The first statement is obviously just an issue of time and place, since certain dinosaurs resemble dragons well enough to say they once existed, and in an infinite universe there are undoubtedly horse-like creatures with single horns living on some other planet in some other galaxy. Whether or not there are magical unicorns and dragons is quite another issue. Assuming for the sake of argument that magical creatures are purely myth and don't physically exist anywhere in any universe, what have we established? Actually we only are saying that matter and forces of nature are not ever fashioned in a way that make dragons and unicorns a real physical part of a space-time system. But when we claim something like unicorns are non-existent we are reversing our usual confusion, by adding qualities of the real nothing into a state of nonexistence. In the absence of magical unicorns the universe still exists, so all that we are really saying is that imaginary things don't exist in time. Square circles don't have form either, because there are rules to form that establish what is meaningful. And things that don't have meaning don't exist.

We know that form in some way establishes or relates to what is meaningful, and we know that there exists perfect squares and perfect circles and we know that a shape cannot be perfectly square and perfectly circular at the same time. The meaning of each separate form or idea contradicts the other when defined as one thing. So again there is a reliance of meaning on form. Nonexistence cannot by definition be form, and the term does not produce any more true meaning than a perfect square circle manages to have meaning. And sorry, a square with rounded edges isn't a perfectly square circle, but good try.

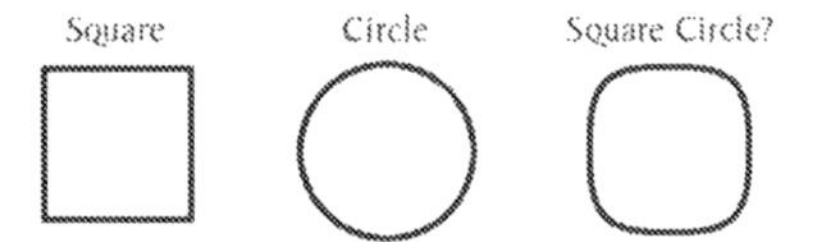

So what it all comes down to is that in the timeless world of meaning, we can invent a word to suggest non-meaning, but if the only way that we can infer non-meaning, is through the use of meaning, then we are just playing a trick on ourselves. The use of meaning, cannot give meaning to something meaningless. Given that there is meaning, then all there is, and ever will be, and ever could be, is meaning. There is no alternative. All there is, and ever could be, is existence. If the question is, *how did something come from nothing?* then the answer is simple. Nothing is balance and something is imbalance. If the question is, *how did something come from nonexistence?* then there is no real meaning to the question. It is just an erroneous thought pattern in our heads that eventually we will erase as we realize being is the default without any alternative.

A human being is part of the whole called by us universe, a part limited in time and space. We experience ourselves, our thoughts and feelings as something separate from the rest – a kind of optical delusion of consciousness. This delusion is a kind of prison for us, restricting us to our personal desires and to affection for a few persons nearest to us. Our task must be to free ourselves from the prison by widening our circle of compassion to embrace all living creatures and the whole of nature in its beauty... We shall require a substantially new manner of thinking if mankind is to survive.

Albert Einstein

Part Three

The Comprehensibility of All

In many years of conversations I have spoken to hundreds of people about infinity, including physicians, priests, professors, teachers, and various freelance philosophers, finding so many that carried with them a deep personal belief that the universe is infinite. Unfortunately, most people believe it is impossible to understand the universe if there are no limits, and so in suspecting the universe is ultimately infinite they consequently harbor a mistrust of the world, expecting that some ultimate chaos is more primary and more powerful than whatever supports our place in the world. It is unfortunate that in science very little focus has been placed on attempting to understand the infinite, and how we relate to it.

Of course the infinite is a bit intimidating. One could spend an entire lifetime visualizing strange science-fiction universes or mystical fairy tales, without ever making a dent in the enormity of what is possible. All the possible stories in all the possible places of dreams and tales seem completely inexhaustible and beyond summation. Our brain power seems so feeble when facing the infinite. And yet it shouldn't be overlooked that we can speak a single word and somehow capture all that exists throughout eternity.

The word Universe can include all things, all our thoughts, all that has ever been or will be. The word everything with equal power engulfs all being and all time in an instant. Either word can in meaning stretch to any length or size, becoming infinite to any measure that the universe itself is infinite. We sometimes take for granted these powers we wield, often assuming the map is not the territory, expecting that we merely attempt to describe an indescribable universe with words. Yet we wield ideas that are magical in strength and power. Then at the peak of our power, words assume their role in the meaningfulness of nature, and there our great powers end. Ideas themselves bound our imagination. Meanings bound our minds and our thoughts. No word can mean more than everything, or less than nothing. We think within a universe of bounded thought.

Words meaningfully manage to define all that exists and all that is imaginable. Einstein spoke of this eloquently as such, "The eternal mystery of the world is its comprehensibility" So why is it that the physical universe seems to exist in concert with the meanings that define the words and ideas we think with. Such profound questions are asked of science, even if they aren't ordinarily considered answerable with any great measure of certainty. We should at least consider that if the infinite universe is limited in some way, it is limited in concert with the very meanings that define language. So we should not be surprised to find that words, as fragmentary and ethereal as they seem, can define the ultimate boundaries of reality.

In previous chapters I have meant to create a solid foundation for anyone considering the timelessness of the greater Universe. I felt I should begin the book this way, by introducing the two orders which is probably the underlying theme of this book, even if the ideas branch out in all directions. The question we are about to ask is also critically important in laying the foundation for all the following chapters. We have been discussing timelessness and enfoldment. Now we can begin to focus on the meaning of infinity and descriptions of the infinite Universe.

When the doors of perception are cleansed,
man will see things as they truly are, infinite.

William Blake

Photo: Romantica Stroll © Mark Penny

The infinite! No other question has ever moved so profoundly the spirit of man; no other idea has so fruitfully stimulated his intellect; yet no other concept stands in greater need of clarification than that of the infinite . . .

David Hilbert

~~~

The essence of the power of thought is its capacity for the universal, and it cannot rest till it has apprehended the most universal idea of all the infinite.

James Orr

~~~

For after all what is man in nature? A *nothing* in relation to infinity, *all* in relation to nothing, a central point between nothing and all and infinitely far from understanding either. The ends of things and their beginnings are impregnably concealed from him in an impenetrable secret. He is equally incapable of seeing the nothingness out of which he was drawn and the infinite in which he is engulfed.

Blaise Pascal

Infinite Universe: Bigger than the biggest thing ever and then some. Much bigger than that in fact, really amazingly immense, a totally stunning size, real "wow, that's big" time. Infinity is just so big that by comparison, bigness itself looks really titchy. Gigantic multiplied by colossal multiplied by staggeringly huge is the sort of concept we're trying to get across here.

Douglas Adams

Chapter Ten

Infinity Means What?

In the winter of 1998, a team of scientists launched a probe to measure the oldest light in the universe in order to map the ever fading glow of energy leftover from the big bang. This wasn't a space-ship though, it was a balloon that carried a special type of telescope, a microwave anisotropy probe called Boomerang, into the ionosphere over Antarctica. The scientists knew this probe would likely answer one of the oldest and most profound questions the human mind has ever dared ask. What exists beyond the visible universe? Is space curved in a way that makes the overall universe enclosed and thus finite, or is the night sky our window to infinity? Scientists knew, if there were no surprises, the data from this probe would squarely resolve the ancient mystery once and for all. Is the grand arena of space positively curved, negatively curved, or flat? Sample templates had been prepared of all three cases, and the only surprise was how clear and succinct the results were. The measure of lumpiness within the early universe matched up perfectly with the flat template, indicating that the geometry of the large-scale universe extends infinitely in all directions.

Although these findings didn't make front page headlines in the daily news, it was quite a profound moment really in the history of humankind, even worthy of a note in the great annuls recording the history of the universe. The record found in the widely read *Cosmic Times* looked something like this.

> "At 13 billion, 712 million, and 390,284 years after the beginning of time, in the Andromeda sector, the human population on Earth, a small water rich blue planet spinning around an average sized star, became knowledgably aware of the infinity of space, uncovering what is perhaps the most intimate detail about the Universe and reality."

In science, the factual evidence indicating the infinity of other worlds has been piling up for some time. Once quantum theory was developed and its implications sorted out, it became difficult to be open-minded and believe anything except that we live in a Many-Worlds Universe. There are conservative hold outs to this day, but most of the high priests of science, physicists holding positions of tenure at the larger universities, adhere to the implications of quantum theory, just to name a few, David Deutsch the Cambridge physicist who pioneered quantum computing; Paul Davies, the popular author and Australian Professor; John Wheeler, the most famous American physicist; and of course the immortal Stephen Hawking who holds Newton's chair at Cambridge.

It is no longer difficult to find evidence that the Universe is infinite. The million dollar question is: "where do we put the brakes on?" Just how infinite is the Universe? What about other dimensions or parallel universes where the laws of physics are different than our own? What about other modes or realms of reality? What really exists out there? To consider the options, there are several different possible scenarios, arranged here as a series of levels. We shall now explore them one at a time.

...an understanding of the infinite tree of universes seems to be needed in order to make statistical predictions about the properties of our own universe, which is assumed to be a typical "branch" on the tree.

Alan Guth

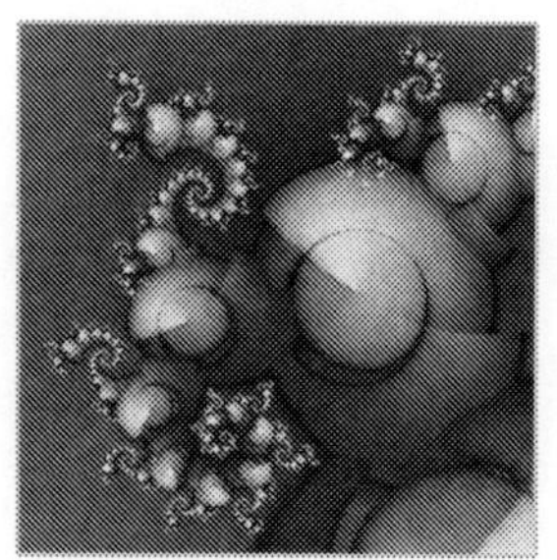

Fractal Art: Discus © Paul DeCelle

Scenario One:
A Branching Out of Many Worlds

Henry Brougham, a respected scientist of the Royal Society of London recognized the growing threat of Thomas Young's papers. A physician and a scientist, Young had recently published a paper in the proceedings of the Royal Society arguing that light travels as a wave through an invisible background medium called the ether. Young and Brougham probably should have been friends. Young was five years older, and both were seen as exceptionally gifted students when they each attended Edinburgh University. They were both accepted members of the Royal Society, but Young, a child prodigy who could read fluently by the age of two, appeared the more exceptional to those who knew of both.

And in his article to the society the young doctor was attempting to alter the scientific wisdom of Brougham's learning. Brougham himself had published a paper on light in agreement with the corpuscular theory. All the greatest scholars, Newton himself, had studied the various materials and behaviors of the physical world and determined carefully that light is corpuscular, constructed of tiny but solid particles. That theory effectively explained the refraction of light. So Brougham felt this idea that light is a wave traveling through an invisible ether could not stand. It confused the fact that the physical world is logically understood as solid material in motion through empty space. The idea that light might be a wave had been dispelled a decade ago. A wave would have to exist within some kind of mysterious medium. These arguments of Young's would only add confusion to the progress of real scientists. This was moving science backwards. Young had to be silenced. He had to be ridiculed so that others would not be swayed into similar senseless meanderings.

Young may have been the more gifted, but Brougham had at his power the *Edinburgh Review*, a highly popular journal he and several others had founded. Brougham contributed articles to his journal regularly. He knew in this case he should not publish his name as the author of the review. It would be best to print the review anonymously as if speaking authoritatively for the journal and other scientists. Some knew of the long smoldering envy Brougham felt towards Young. Seeming to be the larger voice of many, Brougham could criticize the Royal Society for printing Young's papers and ridicule this recent paper as speculative foolishness, all without fear of reprisal.

"It is difficult to deal with an author whose mind is filled with a medium of so fickle and vibratory a nature" wrote Brougham. "We have of late observed in the physical world, a most unaccountable predilection for vague hypothesis daily gaining ground; and we are mortified to see, that the Royal Society, forgetful of those improvements in science to which it owes its origin, and neglecting the precepts of its most illustrious members, is now, by the publication of such papers, giving the countenance of its high authority to dangerous relaxations in the principles of physical logic. We wish to raise our feeble voice against innovations, that can have no other effect than to check the progress of science, and renew all those wild phantoms of the imagination which Bacon and Newton put to flight from her temple....We demand, if the world of science, which Newton once illuminated, is to be as changeable in its modes, as the world of taste, which is directed by the nod of a silly woman, or a pampered fop? Has the Royal Society degraded its publications into bulletins of news and fashionable theories for the ladies who attend the Royal Institution?" The bitter attack carried on for several pages. "We now dismiss...the feeble lucubrations of this author, in which we have searched without success for some traces of learning, acuteness, and ingenuity that might compensate his evident deficiency in the powers of solid thinking."

Even without this flagrant assault from Brougham, the implications of Young's discovery that light waves interfere with one another may have gone unappreciated. It wasn't until over ten years had passed that others began to make the same conclusions. When Young first published his work on interference even he did not realize its importance, and consequently he was not very authoritative in his writing, which Brougham took full advantage of. Young tried to show that the iconoclastic Isaac Newton had not rejected the wave theory, an argument which Brougham harassed. After Newton the universe was imagined to be like a giant machine, like a clock with gears that produce a single precise outcome when moved forward or backward. Newton, the model pragmatist, even described light as tiny solid particles. So when physicists later focused on the atomic world they naturally expected to find the particles that Newton called corpuscles.

Today we understand that light behaves in some ways like a solid particle when it interacts with other particles, however, when light travels from place to place it behaves like an invisible wave of probability. Like a wave in an ocean it has no definite position. It is far easier to imagine a light particle has a distinct position in time and space; otherwise we have to explain how a particle could have the power to disappear and reappear elsewhere. Only ghosts possess such powers. Yet both energy and matter particles, even the particles that make up our bodies, regularly disappear between one position and the next. Particles only assume physical form when they interact with other particles. Otherwise they exist as indefinite ghosts in a ghost world of probability.

Part of what made Young question the corpuscle theory of light expounded by Newton was the way two light beams can shine through each other without the particles colliding and scattering as they cross paths. The two crossing beams of light remain invisible between source and destination. But Young also recognized that light behaves similarly to sound waves, which Young had also been studying. Young then devised ways of exposing how light waves interfere with one another just as sound and water waves interfere. Young produced a narrow beam of light that passed through a thin divider onto a screen. The divider splits the light beam in half. It would follow logically that if the beam of light was made from tiny particles then the result should just be two spots of light on the screen. But Young found the light multiplied into a pattern of light and dark lines, which modern scientists call an interference pattern.

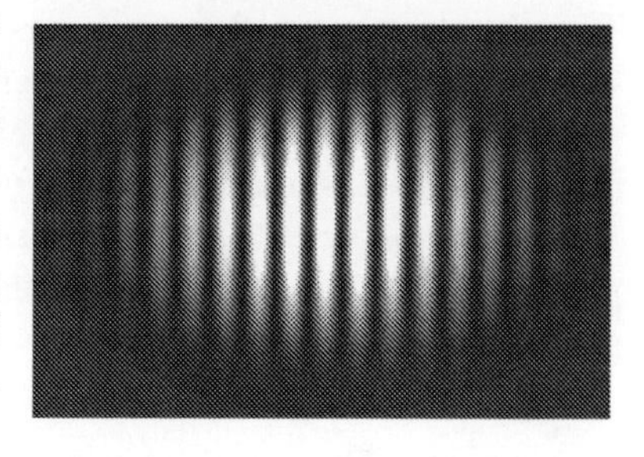

In Young's experiment the interference pattern results when the wave of each single particle is divided, then upon recombining the two waves interfere with one another. As indicated below, the resulting interference is virtually identical to the interference caused when two waves on the surface of water interact after they pass through two openings in a wall. The two waves colliding together add up in some places and cancel in other places.

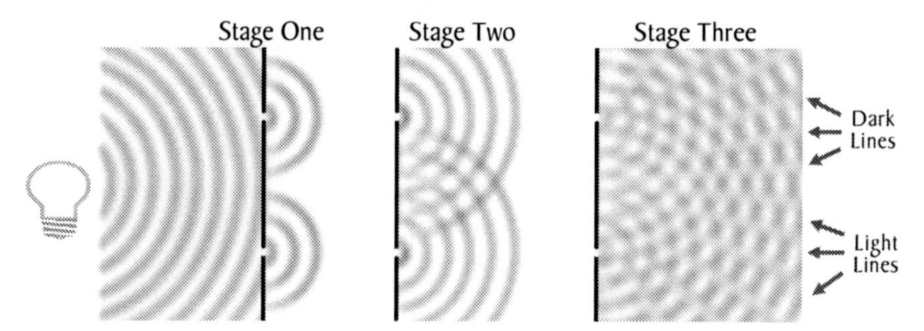

Figure 10.2: A light source sends out waves which pass through two slits causing two separate waves on the other side of the barrier. Those secondary waves are still part of the original probability which left the source.

The interference pattern is essentially the combination of all the many different possible paths that the light can travel between source and destination. When a particle travels, and this is true of all particles, both light and matter, the particle doesn't choose just one single path among all possible paths. Instead particles travel all the possible paths simultaneously. Then among all the paths a single outcome is selected. If instead particles of light and matter traveled from point A to B along a direct path we could of course predict precisely where they will end up, but in traveling as a wave there is no way to predict the final position of a particle. In fact, within the wave, which is merely the shape of probable positions for the particle to exist within, the final outcome seems to be chosen randomly.

Today interference is usually revealed using laser light which shines through two narrow slits onto a screen, but interference experiments have also been conducted with matter particles such as electrons being shot through two slits. The tiny electron particles of ordinary matter behave just like sound waves or water waves. Even if the electron particles are sent one at a time the interference pattern will still develop. This is because the wave of each single particle travels through both slits to interfere with itself. The interference pattern is not caused from many particles interfering with one another, it is caused by each particle passing through both slits simultaneously, so the particle interferes with the probability of its own destination.

Can we outsmart the particle and place some kind of detector at each slit, to determine which slit the particle is actually passing through? In fact as soon as an actual location of any particle is determined, as soon as it interacts with some method of measurement, the multiplicity of the waveform collapses, so the particle is no longer moving in the direction of the screen. Once we detect the particle in any way we have forced it to assume a single definite position in reality. Of course the mind naturally rejects the idea that something physically real can suddenly vanish into a ghostly world of pure probability. It seems more logical that we simply cannot observe the particle without disturbing it. This is indeed true, but there is more to acknowledge.

One of the founders of quantum theory, Werner Heisenberg discovered what is now a key principle concerning all quantum behavior. As a particle gives up information about its location, information about its momentum is lost in equal measure. This is called the Heisenberg uncertainty principle, which states that both the position and momentum of a particle cannot be known. The more we know about one, the less we know about the other. So as a rule, whenever a particle assumes a precise position in reality, in that instant it has no momentum. And whenever a particle is moving from one place to another, it has no specific location. Only when the particle interacts with something else does it then establish which physical reality we will experience, but in between interactions the particle exists in another type of reality, a sort of multiplicity where all possibilities are combined together.

How should we interpret the meaning of a single particle passing through both slits simultaneously? The most conservative explanation, called the Copenhagen interpretation claims the particle has no real existence in between the source and the screen. When a particle is in its wave-like state it is said there is only a probability of the particle's existence at a future destination. This led the physicist Neils Bohr to claim that the quantum world has no real existence, at least not one that can be described with ordinary terms. We might say, the mode of reality in which the motion of the particle occurs is ghostly, of a completely different kind of being unrelated to our own.

The ghostly scenario is really just a conservative bare bones description of what can be said for certain. It is the least amount of conclusion, and not really an interpretation. Any effort at interpreting the actual meaning or implications of quantum mechanics leads toward the extended conclusion that beneath the surface of our own world there is an enfolded level of reality, in which case the ghost world isn't ghostly at all.

Quantum theory was developed near the turn of the century and it wasn't until 1957 that all the possibilities within each quantum wave led a young graduate student of physics named Hugh Everett III to produce the now famous Many-Worlds Theory as his doctorate thesis. Everett was a student of John Archibald Wheeler, the renowned American physicist and longtime Professor at Princeton. The Many-Worlds Theory makes the simple conclusion that one probabilistic outcome is as real as any other, predicting an immense surplus of many-worlds branch away from each moment of now.

We can imagine an infinite number of copies identical to our present, but then in the next moment, in each copy there is one single particle that is in a slightly different position than all the others. The denser areas of probability in the interference pattern represent the more probable worlds, while the thin areas represent the least probable worlds. The areas outside the wave pattern that are completely dark can be thought of as worlds outside the realm of quantum possibility.

The Two Transformations

There are two basic transformations occurring in every moment of every day which are inevitably built into the unfolding of time. In the first transformation, the majority of bizarre futures one might find in the imaginary stories of science fiction, horror, and fantasy tales are erased from possibility. This boundary between what is possible and impossible is created by what we otherwise know as the laws and forces of nature. The forces of nature control and define the areas that matter particles can move about within. Forces are simply the shapes of probability waves, and those shapes bond particles together, in groups, in lattices, in symmetries. As if simultaneously aware of the positioning of all the matter in the universe, probability forces restrict particles to the possible rhythms of space-time. Once this first transformation is complete there are still many different tomorrows possible that abide by nature's rules. We can imagine all sorts of different events that could possibly happen in the unfolding of each new day. The quantum wave is essentially all the other worlds that are equal in possibility to our own.

This first transformation is probably best described by the Sum Over Histories method of mathematically determining the shape of the probability wave, which was developed by Richard Feynman. Feynman's original highly creative approach doesn't merely consider the expected classical paths of particles, it mathematically sums every conceivable path through time and space, even those that disobey the laws of physics, so a particle zig-zags from left to right, and loop d' loops irregularly in all directions, irrespective even of past and future. Yet when every conceivable path is considered and the amplitude of each is summed up, the irregular waves end up canceling one another, while only the expected waves that abide physical laws remain. For some unknown reason, when all conceivable histories are summed all that remains are the possibilities of everyday life.

Quantum mechanics led physicists such as David Bohm and John Bell to argue that the wave-particle duality signifies a deep interconnectedness in nature. Where Bohm developed concepts such as implicate and explicate order to explain the wave particle duality, Bell explained that no complete explanation of the quantum behavior of particles is possible within what we normally think of as the physical universe. Bell's theorem argues that quantum mechanics involves a non-local process. In other words, some non-local process shapes the world we live in.

The Second Transformation

The first transformation eliminates all the weird or abstract worlds we might find in a horror movie. The second transformation is when the multiplicity of possible worlds transforms into the single world we live in. We only experience one world at a time. But what causes this second transformation to happen? Exactly when does it happen? We naturally expect this second transformation occurs as a solid and definite past rolls into an indeterminate future. Only that isn't how things work at all. Instead the past is also a wave of probability.

Nothing reveals the inevitable existence of Many-Worlds more vividly than the story of Schrödinger's cat. To illustrate how the absurd multiplicity of the quantum wave effects the unfolding of physical reality, the physicist Erwin Schrödinger created a vivid thought experiment where he places an ordinary cat within a delicately rigged box. Inside this box there is a radioactive atom that is in a probabilistic state of decay. The atom has a fifty percent chance of decaying within a short duration of time, a half life, of about five minutes. If this atom should decay it will expel a single electron particle that will register on a Geiger counter, also inside the box, and upon detection a hammer will break a glass vial of cyanide gas, instantly killing the cat, that is, if the particle decays. So the door is shut on this contraption and we wait five minutes.

We know from quantum mechanics that until we open the box, in the same way the photon travels through both slits, the atom remains in a wave-like state of multiplicity where after five minutes, in half of all possible worlds it has decayed and in the other half it has remained stable. In fact, until there is a deterministic outcome which we create by opening the door and observing the contents of the box, the atom exists in both states simultaneously. Schrödinger pointed out that since the atom remains in a wave-like state of nothing but probability, as odd as it seems, the wave extends to the Geiger counter inside the box, which also exists in a wave-like multiplicity of having detected the decay and opened the cyanide in half of all possible worlds, and having not in the other half, which further means that the seemingly single cat we placed in the box is somehow simultaneously alive and dead.

Of course this cat-unfriendly experiment has never taken place, it serves only to scientifically dramatize how quantum mechanics, one of the most successful theories of science, extends to the larger macro-cosmic world of cats and people, and in so doing it forces us to dramatically change how we imagine the world around us.

The state of the cat can be interpreted in one of two ways, both of which define reality in dramatic fashion. In the Copenhagen interpretation we say the decaying atom is in a wave-like state and so is not real, it is just probability, meaning that it has neither decayed nor remained stable until we open the door of the box. But what about the experience of the cat inside the box? If the atom somehow isn't real when wave-like, then the cat isn't real either, it's neither dead nor alive. But is this idea acceptable? It is one thing to imagine a particle isn't real if a human being isn't observing it, and another to imagine a cat needs to be observed.

Believe it or not, there are some who egocentrically claim that only humans can create reality. The Copenhagen interpretation led one physicist to remark that the moon doesn't exist when no one is looking at it. Of course if someone is arrogant enough to claim that the cat's observations aren't independently real, the next hypothetical step would be to place that person in the box and start the experiment over.

In the Many-Worlds version of Schrödinger's cat we don't try to avoid the idea that beyond our vision two realities exist simultaneously, so we say the particle has decayed in one reality, and remained stable in the other reality, and two superimposed conflicting realities exist. But neither reality is yet connected in a definite way to an observer standing outside the box. Instead the experiment has created two cats, one is dead and one is alive. Then when we open the box to observe the dead or alive cat we collapse the wave and connect ourselves to one of the two realities. Note however something that is really strange here. The outside world as we open the door to observe the cat, and so we ourselves, split into two principle realities. In one we exist observing the cat alive and in another we exist observing the cat as dead. What this means is that we split into two different realities in the future because behind the door there are two different pasts.

Quantum mechanics doesn't merely reveal that time is branching away from the present. It reveals that the past is often just as indeterminate as the future. As physicist Thomas Hertog states, "Quantum mechanics forbids a single history". This brings the infinite so much closer than even an imaginative person is comfortable with. What is perhaps the most startling consequence of discovering quantum mechanics, is realizing that everything beyond our observations exists in a state of multiplicity.

When we imagine the future we usually see it as potential. Some people state that anything can happen although we don't expect much of anything will happen, we expect the ordinary, but we at least imagine that the varied possibilities of the future grow greater with a greater period of time. What might happen tomorrow is rather limited compared to what all might happen in a year or a decade. But we think the infinite possibilities are out there in the future. As it turns out, the past is also probabilistic. Quantum mechanics suggests that anything we haven't yet observed, everything happening in another state or country, everything happening a mile away, literally everything in the world beyond our five senses, is in a quantum state of multiplicity. The infinite isn't out there in the distant future. It is just on the other side of any door. It is just beyond our vision, our hearing, our touch, our smell. It is everything we don't know. If we haven't yet read the news of the day, all the possible newsworthy events are happening simultaneously in alternative universes waiting for us to turn each page of the newspaper, and only then do we connect with one of those worlds.

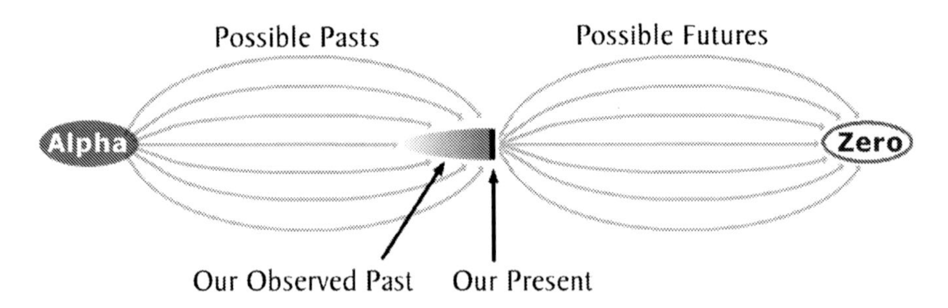

Figure 10.3: Only that which has directly defined our present experience is finite. The rest of the universe, even the past, is undetermined and wavelike. All possible pasts and futures shown here would agree with the observed past.

If this second transformation for some reason temporarily went awry, we might suddenly find ourselves experiencing more than one finite world. We would sense ourselves moving in every possible direction. Our arms would raise and lower, our legs would move one way and the other. In the next few moments we would inevitably sit up, stand up, lay down, and stretch from head to toe. The transparency of space would transform into a scenery of other worlds. In one branch of time we would turn on the television and simultaneously view every possible event that a news station could report. Our experiences would gradually blur into a synthesis of all possible events, with each path only slightly different than the next possibility. Ultimately, in this mode of hyper perception, if we looked into the deep past we could observe the entire infinite family of many-worlds all forming from the big bang event. We would see all paths of time originating from the Alpha state, from one single location in the whole realm of possibilities. And as our vision grew into the future we would see all the paths of time moving ever nearer toward the same destination, toward absolute zero, with every universe finally ending at Omega.

So if the Universe is infinite but restricted to the Many-Worlds scenario, what exactly is impossible? If the same laws of physics hold true everywhere in existence, we can rule out the possibility of humans evolving on a planet too near the sun, and rule out life as we know it having evolved on a planet the size of Jupiter. There is an interdependence between everything that exists in time. The form of any living being is dictated by its environment, however, if there is any real opportunity for life to exist in some particular environment, in the sands of Mars, in the ice fields of Europa, then life does in fact exist there, at least in a parallel world. There is certainly a version of the Earth where dinosaurs were not made extinct, and likely there are worlds where dinosaurs have evolved to become as intelligent as humans on a planet identical to our own. In fact these parallel worlds probably do exist just beyond the surface of this world.

Some scientists shrug at the Many-Worlds Theory and continue to believe there is something that makes quantum reality operate only at the subatomic level, and not at a macrocosmic level where we live. But the technological applications of quantum mechanics to chemistry and electronics have already had a tremendous impact upon society. In addition to television shows and movies where characters cross over into parallel universes, physicists are working toward a complete quantum description of reality. If a complete theory is ever accomplished, it will explain why certain things are possible while others are less so, and it will tell us what is impossible. Presently, the Many-Worlds Theory does not claim that other worlds with different laws and forces of nature cannot exist, but if the probabilities of quantum mechanics were found to be basic to nature then we would reasonably conclude the same laws govern all of existence.

Is the apparent irreversibility of all known natural processes consistent with the idea that all natural events are possible without restriction?

Ludwig Boltzmann

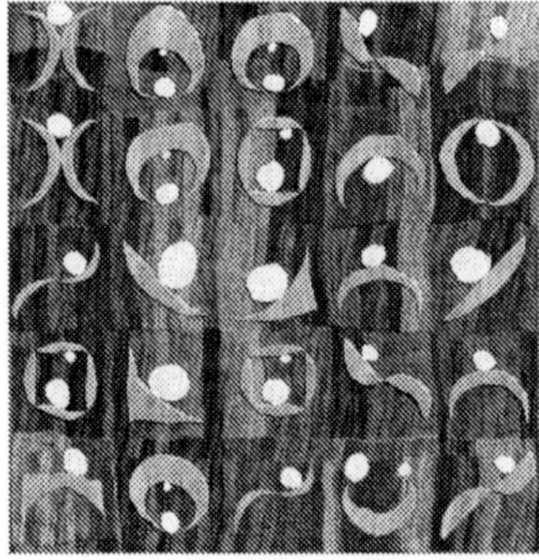

Forty Thieves: © Paul DeCelle

Scenario Two:

The Multiverse

What then if our laws of physics do not hold true in other regions of space and time? Would this mean that infinity is an incomprehensible chaos? One of the more radical views considered in science is that nature somehow explores all possible sets of laws and constants, creating universes that are entirely unlike our own, worlds for example which we could not visit, because our physical bodies could not exist within a different configuration of reality. The construction of these individual universes might span from infinitely heightened complexity, and orderliness to opposite extremes of randomness and chaos. In the multiverse all such worlds begin as our own did with a big bang, but shortly after time begins each temporal path spawns its own unique governing reality. We can refer to scenario two as the multiverse, where the big bang produces not merely other worlds, but many different cosmological systems.

Some scientists believe the dense heat of the bang may spawn universes like a fish laying eggs, each with a unique genetic code. And the uncertainty of quantum mechanics may be the key to what allows such universes to exist. Uncertainty during the big bang event might allow each bubble of space-time to write its own future, by arbitrarily developing unique constants, by setting unique magnitudes to the forces of nature, which for example cause space to expand at different rates.

The odd surprise is that this scenario might be an explanation for why our universe is so finely tuned for the existence of life. One of the greater mysteries concerning the known universe is that certain constants appear to be finely tuned for the creation of complex systems, which have led specifically to intelligent life. The stability of atoms in the myriad of forms we find in the periodic table of elements, as well as the formation of stars and galaxies, all obviously have led to the existence of the Earth. Some scientists acknowledge openly that this fine tuning may very well be sound evidence for intelligent design.

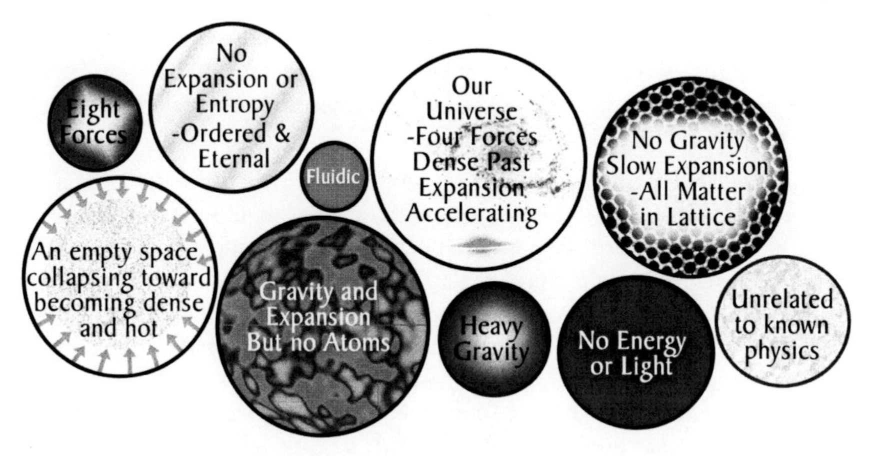

Fig 10.4: A tiny sampling of other imaginable universes.

I cannot help mentioning here that in later chapters we shall explore an idea that is not entirely dissimilar from intelligent design. The idea is that the ordered wholeness itself, the ultimate product of existence, shapes the evolving universe in a way to bring about itself. This would be described as backward causality, since the whole exists in our future. With that possibility noted, we must also consider the opposite possibility to explain fine tuning, especially due to how shockingly reasonable the explanation appears.

There is a theory of sorts called the anthropic principle which explains that we would only exist within a universe that contains properties compatible with the existence of life. An extension of the anthropic principle which we might call the totality principle argues that the existence of other dimensions with different laws of nature can explain why our universe is seemingly designed for the presence of life. The idea is that we find ourselves in this particular cosmos because among all the sets of laws and forces in other universes that actually exist, this is necessarily the one particular universe we would find ourselves in. Obviously, if all worlds exist, then we would find ourselves in one of those worlds and then ask questions about why the universe happens to be this particular way. In experiencing this one world it seems to us to be so extraordinary and finely tuned that it must have been intelligently designed, but the fact that we live in a world of blue skies and cherry sundaes might be because the extreme chaos of all other alternative universes also exist.

So the world we know might be sensible because there exists both sensible and insensible, and we are just temporarily fortunate to be living in the better part, since life would not come to be elsewhere. One reason scientists and philosophers sometimes resort to the idea that all possible universes exist is because in science today causality appears to break down at the quantum level. The governing of events is somehow non-local, meaning that the underlying physics influencing events isn't a product of ordinary objects. If the totality principle is the reason for non-locality then there may not be an answer to why our world is this way. This way may not be anything special. On the other hand, the more preferable hand, all the constants and laws of nature may have a deep purpose that we haven't yet been witness to.

It is inevitable that intelligent beings would experience a universe oddly harmonious to the existence of life because such qualities would naturally exist in any world in which there were such speculations by living organisms. The Anthropic Principle exploits this fact with a rather futile side note. The rather brutal consequence of an overall chaos is that no ultimate explanation of why a universe has a particular architecture is needed, or even possible, since each world is an arbitrary set of variable laws and forces. If this version of reality were true of nature, we might only hope to discover an accurate description or schematic of our particular governing system, what scientists call a unified theory, but that theory wouldn't provide any insight into "why" the cosmos is this way. The features of any particular universe would be randomly determined and without specific cause, since none would portray the universal governing dynamics of the greater Universe of all universes.

On a lighter note, we can tone down the totality principle and use it more specifically from within a many-worlds description of reality. The general reasoning of the totality principle is fun to play with as an explanation of why each of us exists in the moment that we do, as the one person we are. It even explains why we exist at this particular time in history. If there exists an infinity of other people living on Earth-like planets, in all other times, then the only reason we are each ourselves, is because if we were somewhere else in time, or somebody else, then that is who we would be, there wondering the same question, "why am I this person?" If we were somewhere else being anyone else, then that is who we would be, but the person we are now would also be. So, if not for the existence of you, existence would be awkwardly missing someone among the infinity of people that fill the whole of all possible people.

Nowhere exists an obstacle to the infinite
number of worlds.
Epicurus (341-270 B.C.)

~

Row, Row Row your boat
gently down the stream,
Merrily, merrily merrily merrily,
life is but a dream.

Fractal Art: © Doug Harrington

Scenario Three:

Many Realms

Contrary to popular belief, there is not any accepted theory in science as to what caused the big bang. The standard big bang model does not include a description of a creation event and merely recognizes the obvious implications of observations showing that the galaxies on the largest observable scale are flying away from one another due to the internal expansion of space between them. How a universe can begin is a paradoxical mystery rarely explored in science, and although there are sub-theories of the big bang, such as vacuum genesis, the only issue which there is any agreement on is that the early universe was once in a very dense state. In timelessness, it is possible for the origin of time to simply begin from a dense state, since time in this case is somewhat like a string passing through the pages of a book. The reason why our scenarios so far have only included worlds and universes that are born from the big bang event is only because we all see that event as the only known natural process by which time might begin and so exist.

In this third version of what the infinite universe might be like, we have to work hard to open up the gates of our imagination. Now we abandon the rule, and our assumptions, that a universe must originate from a big bang like event, and freely entertain the more radical possibilities of how other universes might be created and constructed. Perhaps there are multiple causes for why the same universe exists, or multiple causes that create many different universes. Mainstream science fiction has often explored the idea of universes with different rules. There might be universes where the inhabitants detect space is contracting rather than expanding, so that scientists in such a world conclude that galaxies originated from empty space, and thus they determine gravity, a force stronger than expansion, creates energy and matter out of curvatures in what in their past was a flat empty space (just try explaining to these alien beings that something cannot be created from nothing). Perhaps there are Euclidean worlds where time is absolute and space is truly empty and independent of objects, as Newton surmised incorrectly of our universe.

We can imagine a universe filled with fluids, or tiny strands of string tying objects together, or even a universe where celestial objects are square shaped and not rounded. None of these universes seem to be worlds we could physically exist within, even though we can somewhat relate to their properties imaginatively. But the greater test of our imagination is portraying universes which have utterly no resemblance to the properties of our own. There might be non-physical universes, or worlds without light or heat. There might be comparatively illogical worlds totally incomprehensible and unrelated to us. In such universes, not only would we be unable to exist physically, but we also would be unable to think or function conceptually. If there are other such realms, then what is mentally logical to one realm is completely chaotic and senseless in another.

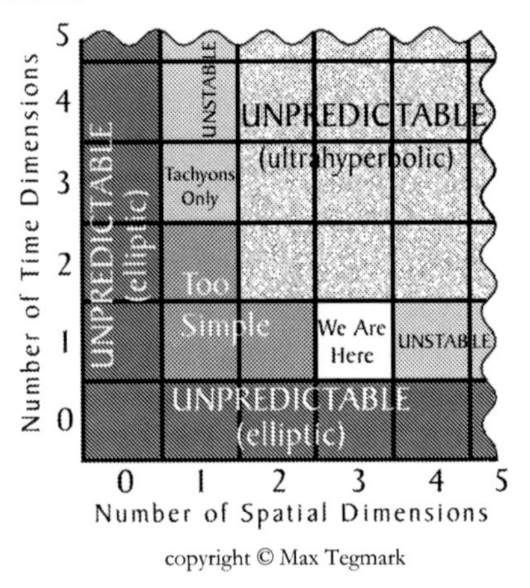

Fig 10.5: With this graphic Physicist Max Tegmark contemplates other worlds with different space and time dimensions.

One of the newer ideas on how this scenario might be possible was suggested by the physicist Max Tegmark, who made the proposal that each universe might be fundamentally reliant on a unique mathematical system. Tegmark admits, just how many other mathematical systems might exist is a difficult question to answer, but the suggestion also offers a possible solution to why nature might hold to a particular set of laws, that being if each physical reality is fundamentally reliant on a mathematical system. Further still is to hypothesize that concepts themselves, might supply the framework that produces universes. Each logical system of comprehensibility might be self contained. In other words, systems of ideas might create various realities.

If indeed the observer plays a role in creating reality, and if the past remains in a wave-like state until it is observed and becomes a factual part of the history of the observer, then it may be consciousness creates reality as a means of explaining itself. Presently, we really don't have any reason to assume the world we investigate, with its big bang past, is the only reality that can emerge from the quantum sea of possibilities. There may be a host of possible explanations for consciousness. Most would naturally be extraordinarily complex. As the *Merriam*

Webster Dictionary now defines the weak anthropic principle, "conditions that are observed in the universe must allow the observer to exist". Still it may be that only one set of many possible conditions is chosen and explored by some investigating consciousness. In principle, only memory would hold intact each explanation, for some period of time until the observer's memory degrades in some way, forgets or dies, and thus loses all record of the particular explanation it was exploring. The same primal consciousness would then move dreamlike into a new system of explanation, seeming to wake up to its own existence in a particular reality. Similar ideas are explored in Fred Alan Wolf's *The Dreaming Universe*, in which Wolf suggests that synchronicity, telepathy, out-of-body experiences, and near death experiences, are indications that the universe itself is dreaming. Why this of all dreams?

I believe in an open mind, but not so open that your brains fall out.

Arthur Hays Sulzberger

Demolition Man © Paul DeCelle

Scenario Four:
Absolute Chaos

Why the extended anthropic principle restricts itself to an even more radical extreme is probably related to its development and use by scientists, but there is an even greater version of an infinitude that would encompass not only every possible scenario of cosmological properties, but even realms that lay outside the bounds of what most scientists are willing to entertain as possible. The totality principle usually supposes the existence of ordered or chaotic worlds and stops short of considering imaginary realms, such as worlds where magical and inexplicable events regularly occur.

When confronted with the idea of other realities, kids are quick to imagine worlds where weird creatures resembling monsters can appear out of nowhere. In this imagination gone wild fourth scenario, fantasies, fairy tales, and nightmares are all part of reality. Reality is founded on conceivability alone so all our rational boundaries and our intellectual expectations are stampeded over by a parade of silliness. I imagine this model of the infinite as the artist's view of reality, reality as defined by the right brain not the left, where existence is as equally free to create, as our minds are to imagine.

For some reason, maybe it is because I loved to watch cartoons when I was young, I initially imagine this world primarily as an interaction of cartoon characters with the real world. Such visions have been explored wonderfully by movie makers in Hollywood, my favorite being the movie Roger Rabbit, where the barrier between the cartoon world and the real world, the world of solids, begins to break down. Could such fantasy worlds exist with some measure of being physically real as we define it? Certainly the ideas of such worlds do exist, that is, assuming that ideas exist before we think with them, and exist independent of our thoughts. Isn't it ideas that ultimately define reality? The question here appears to be, what is physical reality?

It may be that every story book tale we read as kids, Hansel and Gretel, The Three Bears, even Snow White and the seven dwarfs, is a true story. There is really a Santa Claus somewhere. Bugs Bunny is just as alive as we are, and up to his tricks, because physical reality may not be what we imagine it to be. Ideas, stories, dreams, may be what actually define reality. Our world may be a complicated dream among less rigid and fantastical dreams.

At first this artistic version of the infinite might feel freshly uninhibited and entertaining, compared to all those left brain scenarios we hear about, but in fact it should produce a mild sense of fear. The unfortunate consequence of the artist's version of the infinite is that crossing that border of allowing everything imaginary to be real doesn't stop with the fun cartoons, it also includes the horrifying and the chaotic as well. It would ultimately mean that an absolute and uncontrolled chaos reigns supreme. All the boundaries of normality and cohesiveness would eventually break down somewhere, creating an ultimate chaos of broken events. There would be no governing dynamics, no classical physics, no reliance on mathematical systems or natural laws. All our trusted consistencies would just be temporary.

Perhaps paradoxically, we can logically conclude that this scenario cannot represent the ultimate frame of reality. Once everything imaginable is real you have worlds which violate cause and effect and conservation of energy laws, and the natural laws which currently govern our own universe would be highly unstable, which we know to be otherwise. The reason for this instability is that the measure of irregular worlds where the laws of nature suddenly break down and allow imaginary events to occur, would far outweigh worlds that maintain themselves. So the chance of experiencing the one universe that remains sensible and systematic throughout its history would be miniscule. If the greater universe was indeed so unruly, the chance of us experiencing a universe where our surroundings at any instant decay into chaos would be virtually inevitable, far more likely to occur at every turn than the experience of the consistent universe we presently share. A simple analogy is a movie film. There is one series of frames where a movie film is shown correctly, where the pictures make sense and tell a story, versus a myriad of other cases where the frames could be shown completely out of order, or the pixels of each frame can be mixed up, in which case the movie has no discernable meaning.

If we place the partition between what is physically real and what isn't, too far away from the world that we know, then our chances of experiencing a universe that remains consistent and cohesive decreases dramatically. In two of Einstein's more famous comments he indicates his opinion that the same laws of nature apply everywhere. Indeed, simply due to the fact that this consistency is what we experience, it would seem possible that the whole of reality is universally governed in a similar fashion.

What actually exists outward and inward could perhaps be less than everything that can be loosely imagined. The imagination can easily create a lot of other worlds in fantasy. But this does not mean such worlds really exist, even if the universe is infinite. There may be limitations. There may be a distinction between what is possible and actual. Or, there may not be any distinction at all.

It seems that we have no way of knowing, when in fact there is one way we might know. If we discover clearly that our own universe is an elementary example of what exists, and we discover that what is ultimately possible is by nature restricted in some way to universes like our own. Then we would know with considerable certainty the nature of the infinite whole.

Again, it is easy to explain ways in which the universe could be infinite, the difficult challenge is to discover why part of what is conceivable exists while so much that can be imagined never made it to the big party. What limits existence and why do limits exist? An answer would set apart the improbable and the impossible. Discovering such boundaries is where the potential exists for explaining the why of our particular world's construction and its place within the greater scheme of nature.

Onward to the Timeless Domain

What all of the scenarios discussed so far share is the passage of time. In each scenario we have explored, all the while we have actually been asking: what does an infinite universe consist of should all of a certain group of conceivable time-worlds exist? However, could we be making a mistake in simply assuming that physical reality is ultimately described by time? What if the greater infinite universe isn't ruled by time at all? What if time is an illusion? What if time is a secondary feature of the universe, like color?

We are all so overwhelmed by our experience of change that we usually assume existence innately changes in time, and yet we always portray time as a sequence of still frames, such as a movie film or television show. We always break time up into still moments. No one has yet devised a method of portraying time as we assume time to be, as a linear unbroken progression of events. On one hand we assume things can only exist in time, as if time is necessary. On the other hand, when we use the word "timeless" we aren't using that word to suggest a non-existence, rather we are imagining things, perhaps basic forms or ideas, can exist without time.

As you have read each individual page until reaching this page, you have been surrounded by a space in which matter existed in one particular pattern, until reaching this place of space in time; or this single pattern, then another, then another. As you read the previous pages, in each scenario we have imagined time to be basic to existence, but now we are almost ready to begin to explore a very different answer to the question, *infinity means what?* We are about to begin seeing the world from outside of time. We have only one quick stop to make. There is still one scenario that we have to consider. Is infinity perfection?

Beauty, Goodness, and Truth, wherever they occur, are certainly clues; but they seem to be like cameras focused "to infinity" – we cannot tell how far and how great is the Reality to which they are pointing.

John Betram Phillips

Fractal arts.com © Doug Harrington

Scenario Five:

Perfection

At the base of most world religions we find the idea of a perfect God usually described as infinitely knowing, infinitely powerful, infinitely just, and of course eternal. The very strength of religion may result simply of the conceptual existence of the ideal itself. Certainly the majority of people regardless of religious influence at least vaguely recognize a direction toward improvement and goodness, even beyond what is instilled in them by their parents during childhood. The absolute extreme of intelligence, skill, power, imagination, awareness, and virtue can be imagined or discussed by anyone. Simply in recognition of the ideal, there exists within us a strangely deep-rooted sense of a direction or road to perfection.

Why do we have this strange appreciation for the perfect? I think it seems to be due either to a calculative or an intuitive sense of reality, not the everyday reality we experience, but rather our sense of the really big picture. Out there somewhere is either a path toward or the reality of perfection. In science we find a similar but more practical application of perfection in the notion of order. Scientists focus on physical order by studying and attempting to explain the universe, which is in a sense a degree of perfection rather than the imperfection of disorder and chaos. The evolution of time is widely considered to be an evolution of order, an order that is decreasing due to the breaking of past symmetry, which originally might have been a perfect symmetry. The quest to find that perfect state of order, where exactly it can be found, or whether any such state exists at all, is as much a controversy in science as the more general public controversy over the existence of God.

In those who are not so science minded, the notion of perfect order is still important, it simply isn't seen as limited to any physical state of the cosmos. Conservative scientists rarely appreciate this fact, probably because this deeper appreciation of order and perfection rivals the materialistic order of science. The mainstream ideal is a "perfect everything". God for example is thought perfect in every way. Or at least that is true of the common ideal. The religious translation of the ideal; how various religious groups specifically define God, is diverse, with each group believing their version of God is more perfect than others.

The only part of all this that we humans seem to find agreement on is the ideal itself. Perfection is out there somewhere. I honestly think this is why there is a common and universal sense of moral order. Even atheists have moral standards relative to their own sense of right and wrong. And if we are ever to gain any sense of what an infinite level of perfection is like, it is from that collective that we must build, but of course the assembly is a difficult process and the ideal of perfection changes as we grow older, just as it has changed considerably as humanity has evolved through the centuries.

George Fisher once said "When you aim for perfection, you discover it's a moving target" and taking this a step further the author George Orwell wrote, "The essence of being human is that one does not seek perfection." Many people find it better to let go of the perfect ideal and just try to be themselves. We have all heard warnings about trying to be perfect. And there is of course the extreme of simply rejecting the ideal of infinite perfection altogether. For example, perfect symmetry can be seen as an "empty" space, and empty space can be seen as nonexistence. Infinity can be seen as chaos, or the physical universe can be seen as purely finite, and finally, reality can be seen as godless or spiritually sterile. The reward of such beliefs is a considerable relief from one's own unrealistic expectations of being perfect. In a little book *Your God is too small*, the biblical scholar J. B. Phillips writes:

> Of all the false gods there is probably no greater nuisance in the spiritual world than the "god of one hundred per cent." For He is plausible. It can so easily be argued that since God is Perfection, and since He asks for the complete loyalty of His creatures, then the best way of serving, pleasing and worshipping Him is to set up absolute one-hundred-per-cent standards and see to it that we obey them. After all, did not Christ say, "Be ye perfect"? This one-hundred-per-cent standard is a real menace to Christians of various schools of thought, and has led quite a number of sensitive, conscientious people to what is popularly called a "nervous breakdown...If we believe in God, we must naturally believe that He is Perfection. But we must not think, to speak colloquially, that He cannot therefore be interested in anything less than perfection."

In considering perfection initially there seems to be an axis along which there are two directions, one where we strive to be perfect and consequently live uncomfortably with our own and others imperfections, and another where we in some way reject or ignore the ideal of perfection and so we arguably live with a less directional moral compass. It is somewhat like seeing the glass half full or half empty.

The great question is simply whether anything perfect or even near-so exists to guide the compass. If there were some hope of proof, then even the presently skeptical would be more likely to partake in exploring the philosophical implications of a perfect level or form of being.

We certainly know this one universe exists. It extends outward in every direction. And we know the other possible universes of quantum theory should exist as real as our observed world. If so and all of those many worlds are enfolded into our own space, then space itself, especially a perfectly symmetrical space such as Omega, would qualify as perfection.

If the universe is infinite then does the actualization of that infinity ultimately create absolute perfection? It is a profound consideration but the ideal of what we are talking about is real. And there is nothing unscientific about it. Existence alone may create a godly perfection. If literally all possibilities exist, then perfection would at least be part of existence, for it is one of the possibilities if not the whole of possibilities.

In regards to trying to be perfect, I think Joseph Addison said it best in pointing out that "It is only imperfection that complains of what is imperfect. The more perfect we are the more gentle and quiet we become toward the defects of others." And one step further might be to say that being perfect involves all at once appreciating and being our imperfect sides.

Our minds are often permeated by memories of the past or worries about the future. What gets missed is the present ~ and right there in the moment is the doorway into timelessness.

Ram Das

~ ~ ~

What we observe as material bodies and forces are nothing but shapes and variations in the structure of space. Particles are just schaumkommen (appearances). ... The world is given to me only once, not one existing and one perceived. Subject and object are only one.

Erwin Schrödinger

Fractal Art: Ice World © Kerry Mitchell

By convention, the arrow of time points toward the future. This does not imply, however, that the arrow is moving toward the future, any more than a compass needle pointing north indicates that the compass is traveling north. Both arrows symbolize an asymmetry, not a movement. The arrow of time denotes an asymmetry of the world in time, not an asymmetry or flux of time. The labels "past" and "future" may legitimately be applied to temporal directions, just as "up" and "down" may be applied to spatial directions, but talk of the past or the future is as meaningless as referring to the up or the down.

Paul Davies

~ ~ ~

It is utterly beyond our power to measure the changes of things by time. Quite the contrary, time is an abstraction, at which we arrive by means of the changes of things.

Ernest Mach

~ ~ ~

There is always another way to say the same thing that doesn't look at all like the way you said it before. I don't know what the reason for this is. I think it is somehow a representation of the simplicity of nature.

Richard Feynman

Just as we envision all of space as really being out there, as really existing, we should also envision all of time as really being out there, as really existing too.

Brian Greene
The Elegant Universe

Chapter Eleven

Time is a Direction in Space

The Many Spaces Scenario

It is not hard to imagine that all the moments of one's life between birth and death could conceivably be imprinted on the eternal fabric of reality, like a story within a book. And so regardless of the freedom the path of our lives may seem to enjoy, the whole of opportunities, and every possible experience, has already happened. Every breath, every observation, every feeling, are all occurring simultaneously. Does this mean that time as we experience it is an illusion? In one response to that question, Paul Davies in the Scientific American article *The Mysterious Flow of Time* wrote, "Most physicists would put it less dramatically: that the flow of time is unreal but that time itself is as real as space."

Indeed, time and space and form are real enough. Rather than say that time is an illusion, it is better to accurately define the meaning of time, and so realize that our ordinary sense of time is not caused by existence itself changing or evolving. Time is instead a collection of patterns, a connected series of many individual places or spaces which don't miraculously come to be and then just as quickly cease to exist. Time is very much like a series of photographs or the frames of a movie film. Time is many three dimensional spaces which when fused create a more complex kind of space, a four dimensional space, which we call time. This slightly more complicated form of space is as real as all the more simple directions in three dimensions such as height, width, and depth.

Everywhere you go, there you are. It is rather easy to imagine how each moment is just a different place in timelessness. And it is no coincidence that the best method of portraying ordinary clock time is to use a series of photographs. In a series of pictures a person walking across a courtyard is easily recognized as moving through both space and time. In each frame we can see how the configuration has changed, which indicates the passage of time.

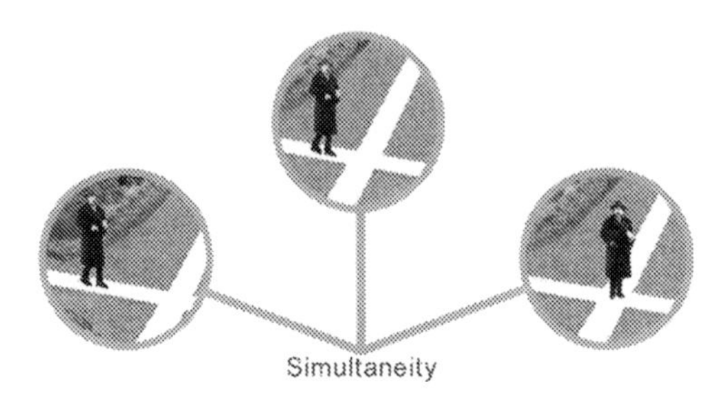

Yet as the person pauses at the center of the crossing to wait for a friend his position in three dimensional space is no longer changing. Standing still at the center of the courtyard a person is only moving through time. We can portray this odd type of motion, the movement through time only, with many successive identical photographs.

Although the position of the person is now stationary and so each photograph is identical to the next, we assume the lateral surroundings of this person are continuing to change. Other objects in the universe must change position in relation to the center of the square. The sun changes its position in the sky. The gears of a clock on a building are moving. The person who is being waited for is moving through traffic. These changes are what allow us to draw a distinction between each time frame and the next, because if the person were truly motionless, only lateral changes in the surrounding environment could define each photo as a different position in time. Only change gives meaning to the passage of time.

Changing time has two fundamental ingredients. First there must exist a linear string-like path through the landscape of possible moments, as opposed to let's say a single point or place. For a movie to accurately portray time it must be more than a broken sequence of unrelated patterns. Time, like any story in book, must include a binding which cohesively connects together the simultaneously existing moments. We might call this requirement *Linear Time*. We might envision linear time much like we envision a single direction in space passing from point A to B, except this path does not just travel through one block of space, it passes from one block of space to another block of space.

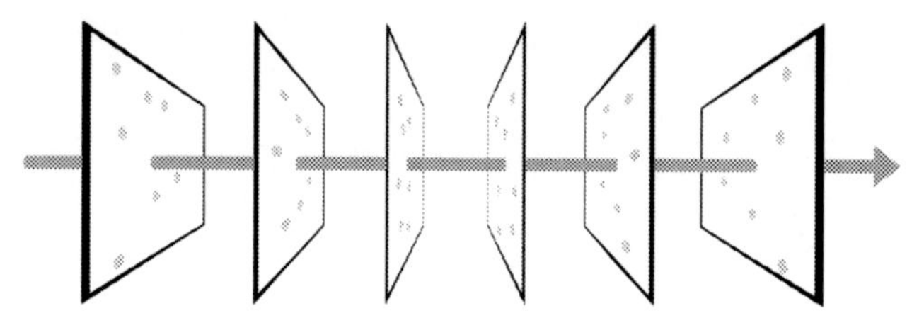

Figure II.3: Time is a special direction in space that travels through many three dimensional block-like spaces, portrayed here as two dimensional sheets.

Second and simultaneously, our story must include a series of distinct and unique patterns or conditions along that linear pathway. There must be a link between patterns but there must also be distinct differences in each pattern from point A to B, lateral to the linear direction of the story. The positions of objects must change in order for there to be measurable time. This means that each moment of time must possess a separate identity apart from other patterns along the linear time path. This is a very basic necessity for our experience of time, as it makes each moment of time feel physically real and distinct. The distinction of each moment is required in order to make us feel as if we exist only in an evolving present. This difference between each pattern and the next may be immeasurably small, perhaps infinitesimal, but without the distinction of each moment there would be no sense of change. Otherwise we would only exist across time, as if in one single moment, without past or future. There would not then be the illusions of time as a duration of existence. We might call this necessity *Lateral Time*, and imagine each static pattern to be like a solid block-like field of space.

In defining these two components of time we haven't derived some profound scientific or philosophical statement. We are just noticing something very simple. At any given tick of the clock the world is in a particular condition or state. Such is one ingredient of time. The second ingredient is the linking together of many states physically, a linear binding of some kind that cohesively fuses a particular group of possible states together. But these two components exposed, now we can reevaluate how we envision time.

In the usual way of looking at time, we imagine that the individual states that construct the linear progression of time suddenly manifest, existing only for an infinitesimally small instant, and then as quickly as they came to be, they disappear from existence. In another way of looking at time, we imagine that each moment of time exists independent of our sense of time and change. For example, each block-like moment of the past continues to exist even though we sense our experience has left those particular patterns.

Obviously the later seems the more simple explanation, as it does not require any process of becoming or unbecoming, which is perhaps the most inexplicable aspect of time as we typically imagine it. If instead each moment of time preexists then the world does not become real in time. The past and future already are real. The world of change that we experience overlays the preexisting moments of a timeless world. If indeed time is space then it isn't merely that we experience the moment, and we experience change, it is that we are the moment and we are change. The inseparability of our experience from change is clearly visible when we notice how little time we spend in each moment.

Many Spaces add up to Zero Time

Meaning to be humorous, George Carlin once said "There's no present. There's only the immediate future and the recent past." The statement is funny because it is insightfully true, or at least partly true. The single present that we imagine to be all that exists is very fleeting. Logically it seems to be an infinitely small region between past and future. Our experience is really of the recent past and immediate future.

The reason we think we only exist in an evolving present is first due to the way we are defined by the independent reality of each moment, but also because we are defined by many of those moments without spending time in any one of them. Most people's first inclination is that time is built up from tiny durations spent in each moment, but we actually don't spend any measurable time in one particular moment, even though we exist in each moment endlessly. The person walking in the courtyard cannot measure the time spent within each moment. Each moment is a static pattern. In order for a person to measure and experience time, their internal chemistry, the physical universe surrounding their consciousness, must change in some way. If everything in motion stops, so also does our sense of time.

Simply imagine how much time passes inside a movie when it is placed on pause. If we advance from one frame to the next, the hand of a clock can be in a new position, but how much time did the first frame contribute to the time within the movie? How much time does the second frame contribute? Obviously the time spent in each frame does not create clock time, only physical change, or the differences in each frame, create clock time. Clock time is change.

Objectively we can imagine the universe getting stuck momentarily like a barge in a river, so that the present exists in a frozen state for five minutes or forever, so one aspect of what we define as time would be occurring without the other, yet we would not be able to experience any duration of time in that frozen pattern. All clocks would be stopped. All the physical and mental processes that produce our experiences would cease. Trapped, a person would measure the length of time to be zero, as measured by a clock. We could exist there in between moments for an eternity but the duration of clock time would still add up to zero. It is only when we leap from that moment to the next that the clock moves, and the mind awakens to change, and that is fundamentally what clock time is, a secondary process, a change in perspective. We exist in a series of nows which always add up to the zero time of timelessness.

The Speed Limit of Time

All motion in our real cosmos causes time dilation. Einstein's relativity theory explains that time slows for whomever or whatever is in motion relative to the rest of the cosmos, even though the person in motion experiences time normally. The time dilation of ordinary motion is so insignificant that we never notice it at the speeds we travel in cars to work and back, but suppose in the future your new job is located on Mars, which at the time is about 150 million miles away. No worries, because you've calculated that your new fancy red spaceship in first gear, at 90 percent the speed of light, will travel the distance to Mars in about fifteen minutes. Of course you don't want to be late on your first day. Work starts at 9 a.m. so you hop in your spaceship at 8:30 a.m. and zoom off to the interplanetary highway, but when you arrive on Mars you're surprised to find that you're late on your first day of work. It feels like it has only been fifteen minutes, and the clock on the space-ship verifies that it is 8:45, but all the clocks on Mars say it's 9:05 a.m., and your angry boss says "hey you're late for work". You can't make sense of it (due to time dilation the clocks on the Earth and Mars were moving 2.3 times faster than your clocks in the spaceship) but you think maybe the clock on your spaceship is malfunctioning.

Of course the next day you don't want to be late again. Somehow the trip took thirty five minutes, not fifteen minutes, so this time you decide to leave at 7:30 a.m. and also decide to shift your spaceship into second gear and travel 99% the speed of light to Mars, even though your mechanic told you this uses ten times more fuel than first gear. But you can't be late again. The trip goes fast, and you carefully watch the clock this time, and in just under fifteen minutes you land and confidently board the Mars landing station, but you're astonished to find that once again instead of being early you're fifteen minutes late for work. The boss really chews you out this time, and you're really confused (time on the

Earth and Mars passed 7 times faster than your time in the spaceship) but you swear to the boss that you won't be late again.

When you get home your spouse just cannot believe you were late twice, but the mechanic said there was nothing wrong with the clock or the engines of the spaceship, so the next day just to be safe you decide to leave at 4 a.m. and decide this time to shift the spaceship into third gear, and travel 99.9% the speed of light, even though the mechanic has warned that you will again need 10 times more fuel than the day before just to accelerate the ship to a speed so near that of light. You just can't be late again. So you make the trip glancing at the new wristwatch your spouse bought you and after counting every minute you land in a little over fourteen minutes time. You plan to get coffee and a nice breakfast before work, and confidently you exit your spaceship with your body telling you it's still early morning, but you also glance a little nervously at the clock on Mars. In shock you stop standing there dumbfounded, staring at the time. The clock reads 10:05 a.m. At first you think it's a trick but eventually you realize that somehow, in the fourteen minutes it took you to fly to Mars, over five hours has passed. You're at a complete loss, how could this be happening, it feels like a bad dream (time on the Earth and Mars passed 22 times faster than your time in the spaceship), oh and by the way your boss says you're fired.

You may have lost your job, but it could be worse. If your job was located at the epsilon outpost in deep space and you had traveled 99.999% of light speed for one year to get there, you would find that 223 years had passed on Earth and Mars and the outpost, while only one year passed for you in your spaceship. Everyone you knew in your past life would have passed away, with the average lifespan being 190 as it will be, or as it was or, as it is then. Faster still, at 99.999999% the speed the light, nineteen years will pass for each day you spend so near lightspeed. One year at such speeds will make you 7000 years old. Of course the time difference isn't merely in reference to the Earth, the motion created by the spaceship is causing the time of the entire surrounding universe to accelerate, even though time within the spaceship seems normal.

Notice inversely that your clock time is slowing down more and more compared to the rest of the universe, and is actually approaching an extreme where the time of your clock would stop altogether at light speed. We might imagine this as if motion is using up the time we experience, as if time only allows so much motion, and so time is sort of a fuel that is drained by motion. That measure of possible motion is usually invested in the motions of your bodies atoms, but when we travel near to the speeds at which light and atoms move about, then we are stealing motion and so clock time away from the atoms, so they cannot move, which is why anything traveling near light speed is partially frozen in time. Particles traveling at light speed, such as photons, and even the galaxies beyond the event horizon, are literally in the stasis of timelessness.

The Holographic Space Movie

It is known today in science that the time that we measure with clocks has certain characteristics, many of which were unexpected, such as the speed of light, and time dilation. Time as a series of many spaces can be shown to create these odd characteristics. To explain we are going to imagine a very detailed holographic movie controlled by a very creative computer. This movie is like any other, except this one isn't made from two dimensional frames shown on a screen, rather it is made up of three dimensional spaces. Each space is a holographic pattern, part of a large library of holographic patterns stored in the computer. The movie will be a specially selected sequence of spaces, selected by the computer, which will allow us to interact with the hologram. So we now turn on the projector and suddenly a holographic person in the movie says "hello, how are you". Surprised, we say "hello" in return, and we begin to talk to this animated person, who seems so real and conscious of his surroundings that we decide to ask him about his experience of time.

He tells us that time flows smoothly, that the past disappears, and he says the future isn't real. He says, "The future is only potential and doesn't yet exist". He is convinced that only what exists as we speak is real. We jokingly decide to play some mind games with him and we increase the viewing speed of the movie by one notch on the dial, which speeds up his rate of time at what appears to be double speed. Suddenly our new friend walks faster, and the clock standing in his courtyard is visibly moving faster. And so we question him again. "So what is your time like now?" we ask smilingly. But to our surprise he replies, "Everything is normal, my time is working just fine", but looking out at us rather strangely he quickly says "but what in the world has happened to you, you're moving very slow. Has your time slowed down?"

Apparently he is observing our world as if it is moving slower. Why? The reason is more obvious if we turn up the speed of the movie even faster. So we turn the movie dial up about five settings faster and now our little friend is really moving fast. He experiences his whole day, he goes to work, to the store, has dinner, and then comes outside to talk to us again, and we have watched his whole day pass in only ten seconds. Quickly we shift back to normal speed in order to talk to him, and he says, "I kept an eye on you all day, and your clock ticked off ten seconds nearly as slow as the sun moved through my sky." He tells us an hour went by in the time it took our clocks to move forward one single second. Why can't we seem to affect his sense of time?

Inside the movie, time is normal because twenty frames are still required for his clock to show that a second has passed, and no matter what speed we view the movie, it still requires fifteen frames for our friend inside the hologram to make one step forward. We could reduce the total number of frames in the movie, decreasing the pattern difference between each frame, so there would be fewer individual moments in the movie, but the effect would be the same. That decrease would only change how many frames are required for one second to pass. In the other direction, we could increase the number of frames, and so reduce the difference between each frame and the next down even to being infinitely small, but the person's sense of time in the movie is still unchanged, because we are merely altering how many frames it takes to experience the same amount of difference.

We could try to limit the amount of difference between each two frames to zero, but that would only make each frame the same, in which case time for our little friend would stop. But suppose we turn the projector on the random display setting. If the series of frames don't follow an order, if each frame is dramatically different than the next, then there isn't a linear time direction, and the previously sensible images in the movie decay instantaneously into chaos. What all this reveals is that for the movie to mirror what we know as time there must be a reasonably consistent measure of difference between each successive rearrangement of space. There must be a limitation to how much change can occur in each successive moment, which limits the distance that objects can move in a set number of frames. This limit ends up being a speed limit for all moving objects, even though the limit actually is only controlling the tiny jumps made by tiny pixels in each frame.

In our real world there is an identical limit upon how much change can occur. The speed of light regulates the maximum amount of difference between one moment and the next. It is a measurable constant to which nature limits all change, managing that task by limiting the distance light particles can quantum mechanically leap from place to place. This naturally translates into a limit upon the speed which larger objects can travel in a measure of time. In every sense, the speed of light is the chief regulator of change.

Just as we cannot travel faster than the speed of light in a set period of time in the real world, objects in the hologram cannot move from one position to a different position beyond a particular speed. This change constant insures that the movie looks real and undistorted, both to us and even the person inside the movie. As we increase the speed at which we observe the moments of the movie, we do not change the overall number of moments which make up the movie. Nor do we affect the amount of change to each frame. We merely increase our remote viewing speed. And so even though we observe time moving faster from outside looking in, on the inside, everything appears normal and undistorted to its occupants.

If we slow down the movie film, slower than normal, to see what our friend says about that, he reports that our clocks and movements have become fast and crazy, and after getting dizzy watching us he requests that we return the speed of our world back to normal, so that both worlds can evolve at the same rate. So in conclusion we find that even though time becomes faster or slower in one or the other world, in both worlds the observer experiences the rate of time to be normal. Does this sound familiar? This is exactly how scientists describe the real time dilation of our own space-time universe.

Motion Eating Up Time

But why does motion in the real universe cause time dilation? Suppose our friend in the movie travels very fast away from his family who stay at home inside his house. Previously he was moving through a series of frames in order to experience time. So his time is really a sort of movement through many individual three dimensional spaces which creates a fourth dimension of space where changes take place. If our friend hops into his spaceship and travels to the moon then he is moving through space in two ways. In each successive frame he is farther away from his home and his position in each successive three dimensional frame is changing virtually at the maximum which the frames allow him to change position in the movie. In a way he is moving at or very near to the speed of time.

How then does his motion affect his sense of time? The problem for him is that his lateral movement through the patterns adds up to almost all of the change that is allotted to the movie. If he uses up all his time in order to change his position, moving away from his home and family, then he will not have any time left over to age as fast as everyone else. So if he could view his family, he would find that they are moving and aging very fast in comparison to his own sense of time. Time seems normal for everything that is traveling along with him in the spaceship. But his personal clock slows down in relation to places that are at rest in respect to the time rate of the movie. His motion is indeed using up time.

One way of describing this is to say that the spaceship's motion uses up most of the available time, which steals away time from the ship's internal metabolism. The occupants of the spaceship don't notice anything strange in the time leftover . Time seems to be passing normally until they look in the rear view mirror and see everyone on the Earth experiencing time twice, ten times, or a thousand times faster than themselves.

If more and more time were translated into the motion through the successive frames, then there would be less and less time left over for aging. Naturally then, as the traveler approaches the maximum speed of change (our speed of light), then his time slows and stops relative to everything else in the movie. He is then experiencing timelessness, the static existence of the movie frames, but as a result, the time of his outer world has escalated and all of history has passed by unbeknownst to him, in the blink of an eye.

Time as Space

The goal of this whole display was to show that a world created from a series of spaces has the same unique qualities as our own world. As moments pass by the measure of change allowed in each rearrangement of space determines the speed limit of all traveling objects. That speed limit also determines the speed of clock time when objects are at rest. Why is there a speed limit? We might conclude that the speed of light indicates that our real time is very much like a movie, made up of a sequence of static unchanging moments. Time dilation is real. The flexing of time has been proven in countless experiments of time dilation since Einstein theorized them. Atomic clocks were placed on the Apollo moon missions and the slowing of time was measured exactly as predicted.

The very fact that there is a speed limit, a metabolism rate to the universe, indicates that our experience of clock time is somehow similar to frames rolling through a projector. Suppose we ask, how many timeless moments it takes to get from here to there? The question sort of makes sense because we all know that the most obvious component of time is change. Generally, as a simple way of understanding time dilation, we can say that traveling at ninety nine percent the speed of light only leaves one percent of the overall metabolism of the universe to be invested in your body chemistry and the spaceship's clock. The amount of change in the ship's motion, and the amount of change invested in your body's chemistry, add up to the speed of light, which always remains constant, because it is the true clock of the universe.

It might seem at this advanced stage of history that the speed of light is a reasonable feature of the cosmos, but not long ago it was a totally unexpected and a highly odd discovery. No one expected a speed limit. Why can't objects move at any speed? We can see in the movie analogy how an ultimate speed limit is a requirement of linear and lateral time. If the speed of light did not limit the measure of available difference from one moment to the next, the cohesiveness of events would decay. The speed of light is just the maximum amount of change allowed in a given number of time frames. Without that limit time won't flow cohesively. Without that restriction time would turn chaotic.

Onward to a Map of Timelessness

Okay so we are ready to lay out the blueprints of time. We are ready to map all conceivable patterns, and find where our own universe fits into the big picture. I usually call this map *Pattern Space*. All conceivable patterns are assumed in this scenario to be physically real, as real as any moment. Look around you. The "now" of you reading this book is in the catalog.

In this timeless scenario of what infinity is like, patterns are more fundamental than time. Like numbers or ideas, pattern space just is, it is existence. It is all possible pieces of the whole pie. It is the seat of reality, the default state. Each pattern within the great bulk of all patterns has always existed and will always exist. Nothing causes them, nothing affects them. Relative to the flowing time we experience, we would say that pattern space exists at right angles to clock time and change.

In fact these elementary patterns have been hidden in our explorations of infinity all along, since in exploring the existence of any scenario of what infinity means, any of the evolving time worlds we imagined require that many of these individual moments exist at very least temporarily. Each imaginable moment of time in any time world is a template that could be utilized simultaneously by multiple time worlds. For example, the moment of now you are presently experiencing is utilized by the many different paths of time in the Many-Worlds scenario, since it branches out to many unique futures, and could have resulted from many unique pasts.

So in summary, when we try to explore what is possible within an infinite universe, we can try to imagine time worlds, or we can merely explore the structure of patterns, as if an infinity of patterns would satisfy the need for existence to be infinite. If the answer to "what does infinity mean?" is answered by, "all possibilities must exist", then infinitely many patterns may satisfy that principle, even without any time worlds existing at all. Any time worlds that do exist would then be secondary to the more primary existence of pattern space. Hence any time worlds that exist might be very select and limited. Given that we can understand pattern space, and explore its shape and structure, we might be able to learn to understand why only special time worlds exist. We would then have in our possession the blueprints of existence itself.

What follows transforms into what I have come to believe is the most simple and sensible version of an infinite universe, a version that can explain why time worlds such as our own are privileged to exist while others are not. The question we have now arrived at is, if time is made up of timeless patterns, what is the infinity of all patterns like? In the coming chapters we will create a precise model of all conceivable patterns. I believe what follows is the first complete model of timelessness.

It is probably true quite generally that in the history of human thinking, the most fruitful developments frequently take place at those points where two different lines of thought meet.

Werner Heisenberg

~~~

Appealing to everything in general to explain something in particular is really no explanation at all. To a scientist, it is just as unsatisfying as simply declaring, "God made it that way!"

Paul Davies

~~~

The vacuum is the most complex substance in the universe. Within it are all particles and forces, even those unknown to science. Physicists now believe that the vacuum – the emptiness in deep space, or even in a vacuum chamber – holds the secret to the newest question in cosmology: what is this mysterious [repulsive force], this anti-gravity force that flattens out the universe and pushes galaxies apart?

Charles Seife

~~~

The idea of nothing has bugged people for centuries, especially in the Western world. We have a saying in Latin, Ex nihilo nuhil fit, which means "out of nothing comes nothing." It has occurred to me that this is a fallacy of tremendous proportions. It lies at the root of all our common sense, not only in the West, but in many parts of the East as well. It manifests in a kind of terror of nothing, a put-down on nothing, and a put-down on everything associated with nothing, such as sleep, passivity, rest, and even the feminine principles. But to me nothing -- the negative, the empty -- is exceedingly powerful. I would say, on the contrary, you can't have something without nothing. Image nothing but space, going on and on, with nothing in it forever. But there you are imagining it, and you are something in it. The whole idea of there being only space, and nothing else at all is not only inconceivable but perfectly meaningless, because we always know what we mean by contrast.

Alan Watts

~~~

There is a theory which states that if ever anyone discovers exactly what the Universe is for and why it is here, it will instantly disappear and be replaced by something even more bizarre and inexplicable. There is another which states that this has already happened.

Douglas Adams
The Hitchhiker's Guide to the Galaxy

Fractal Art: Big Bang © Kerry Mitchell

The supreme task of the physicist is to arrive at those universal elementary laws from which the cosmos can be built up by pure deduction.

Albert Einstein

Part Four

The Great Cosmic Attractor

Over the past two centuries science turned inward upon the universe, with most of its focus placed toward the particles that collectively create the larger objects of ordinary experience. The underlying philosophy has been that particles explain the larger world, and it is hard to imagine that any approach could have been more completely successful at understanding and manipulating all the materials of the universe, the most effective level of comprehension being the atomic structures defined by the periodic table of elements. We understand the building blocks of the material universe, and we have extensively explored the chemistry of those elements, explored the physics of those elements, explored the cosmology of those elements, and successfully so. Never the less, and this is so important to acknowledge, the bottom-up approach consistently leaves the more profound questions unsolved. The bottom-up approach just isn't the right tool for asking why the universe exists, or why there is order, or even why time has a direction.

Most of us appreciate the world from the bottom-up. It is only natural that we see the world as many things all floating in empty space, fortunately kept from drifting away due to gravity. It is only natural that modern science originated as a study of the universe from the bottom-up. And there is certainly no reason to deny the fact that the bottom-up approach is the most important critical tool of science. As long as we remember there is more out there. There are other ways of seeing the universe, or at least for certain there is one other overall paradigm. There is the top-down perspective.

People who have looked beyond the known universe and tapped into the top-down perspective as they worked toward describing the greater whole include Richard Feynman, with the Sum over histories theory; Hugh Everett with the Many-Worlds Theory; David Bohm with implicate order; Stephen Hawking with his focus on Imaginary time and the No boundary proposal, and recently Julian Barbour with his triangle land model of Platonia. Such efforts can be detrimental to one's career, since so many scientists assume there is no top-down perspective. After all, there isn't a largest number. There isn't a top-down in mathematics. Infinity in mathematics is an incomplete never ending series. And according to the second law, there isn't a limit to the measure of disordered possibilities. There is just endless disorder. So how could there be a scientifically viable top-down view? Some skeptics think of top-down methods as pseudo-science, and consider the big picture beyond the scope of scientific study.

However, the top-down approach is becoming increasingly scientific in the twenty first century, as bottom-up information approaches stages of actually being complete and universal. It is then that top-down theories can be critically evaluated against all the bottom-up data. Recently the term top-down has become increasingly commonplace, where at times in the past those who thought in such terms were scrutinized, such as Piet Hut, a top-down thinker whose research position at the Institute for Advanced Study was at one time threatened until other scientists defended the necessity of academic freedom. Stephen Hawking himself has lectured on the need of a top-down approach in cosmology. Hawking recently openly defended the approach in an extensive scientific paper entitled *Cosmology from the Top-down,* which is available on the internet.

There are certainly both bottom-up and top-down elements within all the major theories of the universe, such as Einstein's general theory of relativity, and Barbour's triangles. Stephen Hawking has applied sum-over-histories to cosmology and acknowledged its implications for time. He also has focused on how a set of all possible states might influence the evolving universe. In his Top-down paper Hawking mentions:

> The bottom up approach to cosmology, of supposing some initial state, and evolving it forward in time, is basically classical, because it assumes that the universe began in a way that was well defined and unique. But one of the first acts of my research career, was to show with Roger Penrose, that any reasonable classical cosmological solution, has a singularity in the past. This implies that the origin of the universe, was a quantum event. This means that it should be described by the Feynman sum over histories. The universe doesn't have just a single history, but every possible history, whether or not they satisfy the field equations.

Usually a top-down theory or insight is the result of bottom-up scientific work, an extension or insight of hard science, but some scientists developed important science based upon top-down principles. Boltzmann's version of the second law is the most obvious example. Another, Barbour's belief that time is an illusion led him to conclude there must be a underlying structure creating the illusion of time, so he developed triangle land. Indeed, science is increasingly able to solidly answer ever bigger questions, such as whether the space of our own universe is finite or infinitely extended. We are slowly progressing toward a far different perspective of the universe where we view the big picture, the infinite universe, and then recognize our universe within that whole.

How does time begin? What is matter? Why is the universe systematic? Why are there four forces of nature? We need some other approach beyond the bottom-up in order to answer such questions. Instead of only exploring the universe from the bottom-up approach, we must also scientifically explore the universe from the top-down.

Exploring timelessness is the genesis of top-down science. And there is no more simple way of exploring timelessness than to consider the structure of all conceivable patterns. What could be more top-down than studying the entire infinity of what is conceivable. I have been thinking in terms of timeless patterns for a long time, roughly thirty years. And I can say with great certainty that what is missing in science today is a better model of the big picture. How Boltzmann modeled all possible states was a wonderful first step toward focusing outward on the structure of the whole but that step was made nearly a hundred and forty years ago, before most of modern science was even in place. Unfortunately, since then science has only narrowed its focus to include only what it considers possible within the restraints of a bottom-up (particle) driven space-time. We haven't built upon and advanced the basic concepts that Boltzmann practiced.

This brings us to the most evident missing ingredient in Boltzmann's original thinking; a state of absolute zero; the bottom end of physics. Even though today we are fully aware that the evolution of time is accelerating toward absolute zero, our modern vision of all possible states still does not include any representation or recognition of absolute zero. The time has come to admit an absolute zero state into the group of states considered available to the deep time of cosmological evolution.

As we continue we will be making several additions to how we presently model all possible states. I will present a number of dramatic modifications in how we see all possibilities, all of which result from integrating absolute zero into state space, and all of which impact our modern comprehension of time's arrow and the second law of thermodynamics. Integrating zero first leads us toward discerning the shape of reality itself, and then it leads to discovering the governing dynamics of the universe.

...in eternity there is no distinction between being and potential being.

Giordano Bruno

~~~

If we want to solve a problem that we have never solved before, we must leave the door to the unknown ajar.

Richard Feynman

~~~

Live neither in the entanglement of outer things,
nor in inner feelings of emptiness.
Be serene in the oneness of things
and such erroneous views will disappear by themselves.

Seng T'san The Hisn Hisn Ming

~~~

...how would things seem if time didn't flow? If we suppose for the moment that there is an objective flow of time, we seem to be able to imagine a world which would be just like ours, except that it would be a four-dimensional block universe rather than a three dimensional dynamic one...Things would seem this way, even if we ourselves were elements of a block universe.

Huw Price

~~~

Things which are put together are both whole and not whole, brought together and taken apart, in harmony and out of harmony. One thing arises from all things, and all things arise from one thing.

Heraclitus

~~~

Every man believes that he has greater possibilities.

Ralph Waldo Emerson

~~~

Lots of people limit their possibilities by giving up easily.

Norman Vincent Peale.

~~~

You and I are essentially infinite choice makers. In every moment of our existence, we are in that field of all possibilities where we have access to an infinity of choices.

Deepak Chopra

~~~

Those who do not stop asking silly questions become scientists.

Leon Lederman

The only way of discovering the limits of the possible
is to venture a little way past them into the impossible.

Arthur C. Clarke

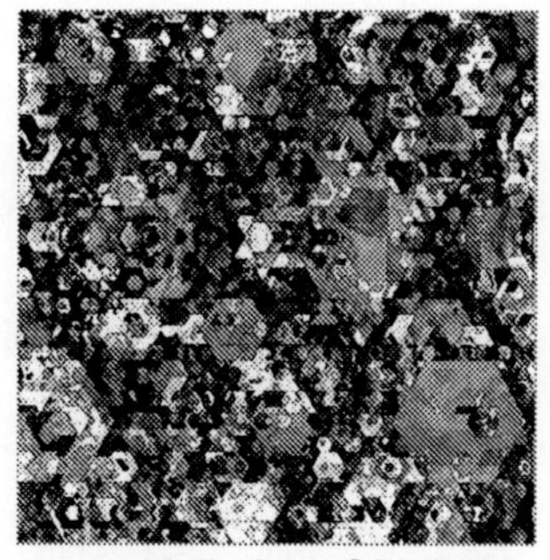

The Draftsman © Paul DeCelle

Chapter Twelve

The Shape of All Conceivables

How Grouping and Symmetry shape the Possible Realm

Whatever exists out there beyond the boundaries of our experience, that reality out there somewhere, out there everywhere, has a tremendous unseen impact upon our lives. The big picture is the key to answering all of the profound questions. Is the second law correct in implying that the universe is winding down? Does our existence fade away in time or does life continue beyond death? Is there a God? Let's face it, the big picture holds the secrets of the small picture, and not the other way around. The big picture decides our fate, the fate of the universe, and the ultimate purpose and meaning of life in general. The big picture really does matter intimately to each and every person.

Ludwig Boltzmann began what should have become a slow careful study of the shape of the big picture but a few years later when quantum theory came along scientists discovered that particles can only maintain certain energy states, and consequently they started thinking in terms of only what is strictly possible considering the structure of the atomic world, thinking only from the bottom-up. If one considers a number of gas particles in a closed container there are a specific number of energy levels available. Consequently the term 'possible states' has come to refer only to the select group of patterns that are available to a closed system of particles.

In science today there is a difference between all possible states and all conceivable states. An analogy of the difference is noticeable in how we might consider what is possible for sixteen balls on a billiards table. We quickly imagine the balls grouped together or spread out in a wide variety of patterns within the frame of the table. What we don't think to consider is all the different shapes the mass and particles of each ball could form within the frame of the tabletop. We don't consider that the balls might melt together and transform into other shapes since that never happens, it isn't chemically possible, and in science everything beyond what is chemically possible is overlooked. Why consider possibilities that would disobey the laws of nature as we know them? However, when we think big we realize there are two extreme possibilities. The matter of the balls could transform into one large ball, or it could spread out evenly across the table and therein become the surface of the table. In not looking at the big picture, in not considering every conceivable pattern beyond the ordinary, we fail to recognize the two directions of increasing order, and we also fail to see how extremes give shape to the overall realm of possibilities.

If we consider the possible states of a coin flip we recognize heads or tails as available possibilities. We certainly don't consider states where the coin when flipped into the air suddenly changes shape, disintegrates, or vanishes into a wisp of air. Nor do we imagine possibilities where the coin tossed in the air never falls to the ground but just floats upward into the clouds. Such events in time, although imaginatively conceivable, are not even science fiction, they're fantasy, simply because they would defy laws we know exist within time. The problem is that in limiting our imagination strictly to what is possible according to such laws, we are unwittingly cutting ourselves off from ever understanding those laws. We are thinking only from inside the box when we need to also see what's outside of the box (non-local) in order to understand the limits inside the box.

The full complexity and magnitude of all conceivable patterns we shall be assessing in this chapter is easily under appreciated. Just imagine the possible configurations of grains of sand on a beach, or the growth of trees in a forest. There is so much to imagine when thinking of the entire bulk of all conceivabilities that it can feel intimidating, appear foolish or senseless, and even seem a little scary. Just imagining all the various configurations possible of a deck of cards, the coin flip above, or a walk in the park, is mind boggling. However, what we shall discover is that the realm of all conceivable patterns is also often imaginatively over appreciated. It is wholly unrealized today even in science that the whole of imagined possibilities is in no way limitless or boundless as is commonly assumed. The big picture is in no way beyond description.

In the previous chapters, hopefully what has been conveyed is a deeper understanding that we are connected to and part of a greater whole, mainly from recognizing that with two kinds of order in nature there is a pattern in which everything forms a single undivided wholeness. But of course that 'system of ideas' isn't quite enough. In the face of our experience of a world of separate things, and the powerful illusion of existing only in the present, imagining the universe as a whole seems more to be just an ideal, something we can try to remember through the trials of life. And it will remain only an ideal until we lucidly recognize how our particular universe and life physically relates to that wholeness in a scientific way. There has to be a physical link. Then our connectedness with the whole would be more convincing. It would be factual. It would be science. It would be real and not just an idealistic intuition, or what might be wishful imaginative thinking. Just as the big bang led scientists to take seriously the idea that time has a beginning, principles such as wholeness don't impact us fully unless of course they become a part of the hard physical sciences.

We are now about to explore the broadest foundation of reality itself free from time so we naturally must explore the patterns of the possible realm. This new model can be referred to as the set of all possible states (soaps). Or it can be referred to as the possible realm, or pattern space, or configuration space, or superspace, or it could be thought of as a version of Barbour's Platonia. All that is important is that this new model of possibilities sheds light on why our universe is specifically this way.

Admitting to Zero

The first step in improving the wedge model is to integrate zero which means we must establish the location of zero. Where do we place zero within the wedge model? Obviously zero belongs opposite of Alpha. Zero essentially creates a boundary to what exists beyond the great bulk of diversity existing between Alpha and Zero. And this means the opening outward of the wedge does not expand indefinitely. We already know there is a boundary in the direction of the past, but there is also a boundary in the direction of the future.

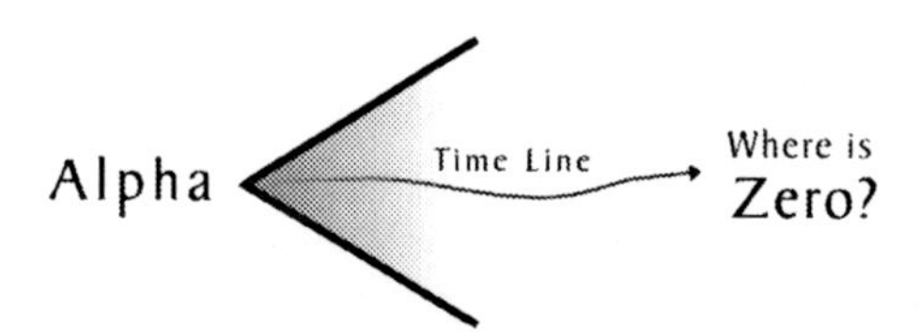

Figure 12.1: Science has long recognized the extreme of absolute zero, the point in physics where time would cease. But where is zero located in state space? Where is it located in the wedge model. Why doesn't the second law show any consideration of zero.

That first step is easy enough, but it immediately leads to a few new questions. What is the shape of pattern space near zero? Suddenly we have to completely reconsider the wedge shape as a description of all possibilities. In science we recognize there is a wedge shape to pattern space near Alpha. We know there exists an ever decreasing measure of states leading to the single Alpha state in our past. Those boundary conditions near Alpha actually are what define the wedge. They give shape to what is ultimately possible. Integrating zero doesn't challenge those boundary conditions. It just leads us to consider the boundary conditions that come along with zero.

It is only natural that the same principles that apply to the Alpha extreme also apply to a zero extreme. Zero is also a single extreme state, a single condition, a singularity. It is a uniform state. So as we might expect, in moving toward that single extreme there is another wedge, a reversed wedge, on the other side of the great bulk of diverse states.

Figure 12.2: Zero is out there somewhere beyond the great bulk of diverse states, and zero is an extreme single condition, beyond which no other possibilities exist, which means that the number of possibilities in the direction toward zero must decrease until one reaches the single state of zero.

Zero is perfect symmetry. It is perfectly flat. It is the same everywhere. It is the template of sameness and symmetry. This necessitates that there are increasingly fewer states that are zero-like, that are almost flat, that are almost uniform. Consequently as we move nearer to zero the shape of state space closes and narrows inward toward the single state of zero. The overall measure of possibilities decreases. This forms a closing of possibilities in the direction of our future.

Let's recall here Stephen Hawking's analogy of the puzzle in the box. Every time we shake the box there is a new pattern. And there is only one pattern where all the pieces of the puzzle fit together. But there is also only one pattern where the box is empty. Furthermore, if we consider patterns where there are just two or three pieces in the box, there aren't nearly as many unique patterns as there are with ten or fifty pieces in the box. As we take away puzzle pieces, there are fewer and fewer unique configurations, until finally there is just the one state of an empty box. It follows that the measure of possible states is naturally greatest when the box is not too full and not too empty.

The same is true for the universe outside of the box. We can imagine that a large measure of unique possible states exist for the amount of galaxies that presently inhabit the universe, keeping in mind there are wide expanses of empty space between those galaxies, and that empty space would remain a constant for each possible state. If we then imagine adding galaxies into the measure of empty space, so that there were more galaxies in the same volume, as is increasingly true of the distant past, then we can easily recognize there would be a larger measure of possible states, since there are more galaxies in the same volume to alter into unique configurations. It follows that the measure of possible states is greatest somewhere in the middle in between Alpha and Omega where the greatest measure of diversity exists.

In considering the distant future the same principles which define the shape of the wedge model in our past apply also to the shape of state space near zero. As matter is stretched flat and the density of space approaches zero there is an ever decreasing measure of unique patterns in that direction. The nearer we are to zero the fewer possibilities exist that are zero-like. A wedge shape that contracts toward Omega is the most important consequence of integrating zero into the set of all possible states. The shape of all conceivable states is defined by two extremes, not one. And now we can see that there is a wide spectrum of patterns ranging from infinite density to the zero density. This creates a density gradient.

Figure 12.3: The first step of envisioning state space clearly involves recognizing the density gradient between Alpha and Zero. We see the space of Alpha as curved to the extreme of a point, while the flat space of Omega zero opens up and extends infinitely in all directions.

The Future is Shaping its Own Past

Recognizing the reversed wedge shape of possibilities out there in our future is a big step forward. It carries with it some profound implications that are even immediately visible. Considering we are moving directly toward zero, even presently accelerating in that direction, a narrowing structure of possible states means that our universe is increasingly influenced and focused by the final condition at the end of time. It means that time will eventually be forced into a very limited and defining number of future possibilities. This funneling of time not only creates a considerable restriction to what is possible in the distant future, this funneling of time's direction would even be influencing our present at this stage of cosmic evolution.

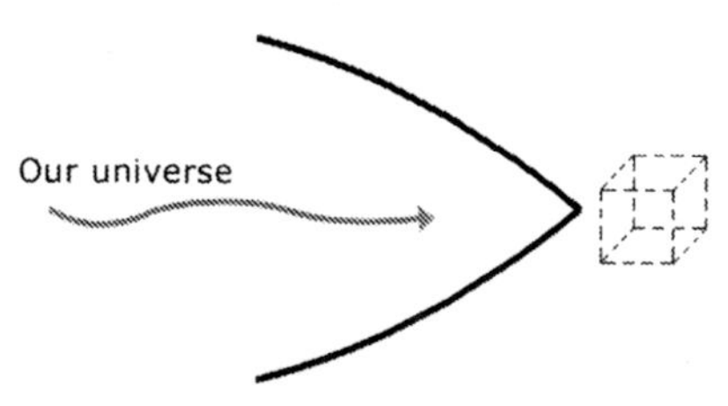

Figure 12.4: The shape of State Space near Zero. Acknowledging zero leads us to recognize that there is a decreasing measure of possibilities in the future.

Regardless of how many possibilities there are, if time is faced with ever fewer choices in the direction toward zero, then the future will naturally become increasingly determined or shaped by the availability of a fewer measure of patterns that exist in that direction. Considering how flat and empty the large-scale universe is at present, in the same way the past was focused outward by Alpha, as Barbour put it, "like a trumpeter's horn", our future is being increasingly channeled inward toward ZAT (absolute zero), so it is increasingly focused by, and made to be like zero, simply due to the narrowing of possibilities. Of course we are very near to zero presently, and with so much empty space in the universe, we have to expect that the final destination of time is already actively shaping our present. We are already in the stage of converging inward on zero.

Of course we are accustomed to the way zero is effecting our universe, the influences seem normal to us, or at least we see them as just the way the universe is. We take the influences of zero for granted. For example, the accelerated expansion of the universe isn't just the way the universe is, it is caused by the narrowing wedge of possibilities in our future. Further, in understanding that zero is the symmetry order extreme, and remembering how symmetry order supplies the component of balance to all patterns, we can realize that all the symmetries and increasing balances of our world are the result of moving increasingly nearer to zero. A great deal of the order of our world is the result of the tunnel of possibilities in the direction of zero.

This future wedge ends up being just as significant and influential as the wedge in our past. Up to now, the old wedge model has been our only means of understanding the difference between the past and the future. If imagining a single wedge-like structure of all states helped us in the past to understand the arrow of time, and thermodynamics, what can we expect from a far more accurate model? We have over thirteen and a half billion years of observable data concerning the past. Maybe an ultimate model could provide the mathematics as well as the reasons why time moves so directly toward zero, and why stars and galaxies and even atoms form. Perhaps a master blueprint of all conceivable states can explain why the momentum of time slows and then increases as the universe approaches zero. Such a model might also provide insight into gravitation or why the expansion of the Big Bang happened in the first place.

Omega is full and not empty. So what the narrowing wedge in our future means is that the universe and we ourselves in various ways are presently feeling the influence of the balanced whole. Zero is influencing the present universe in ways that increase balance. One force of nature in particular, the great balancing force of electromagnetism, is entirely the result of our moving ever nearer to the balance of zero. Later we shall give this focusing to the direction of time a great deal of further study since this influence from the future, combined together with the influence of the grouping order of our past, creates a wonderfully simple way of comprehending the forces of nature. However, before we are ready for that insight, we first need to more fully understand the shape of the possible realm. Now that we have considered the past and future, let's take a look at those possibilities that exist adjacent the present. Just how diverse is the great bulk of diverse states?

One of the deep puzzles is why the universe has become complex. Why has the biosphere become complex? Why has the number of ways of earning a living increased so dramatically? We have no theory about this overwhelming feature of our universe. I propose in *Investigations* that biospheres, on average, increase the diversity of "what can happen next", their "adjacent possible", as fast as they can without destroying the order already achieved. At least it is a possible start in this direction.

Stuart Kauffman

The Adjacent Possible

Ultimately we should expect that the structure of the universe we experience is directly the product of what is possible. In any scenario, a big bang, intelligent design, or creation, we should expect the overall big picture so to speak reasonably leads to the universe we experience, be that God, Mother Nature, or whatever that big picture is. If we take into consideration how the success of quantum mechanics and the second law both indicate that our universe is influenced by probabilities, then it starts to make sense that the really big picture is shaping every small part of what we experience. The really big picture shapes our past and our future, while both past and the future possibilities shape the present.

What we are going to do next is ask a question concerning the boundaries of a wedge. The classical wedge shape is conveyed by a top and bottom boundary expanding outward from a point. Julian Barbour diagrams his wedge with squiggly lines. Others have drawn straight lines. Here we are discovering a reversed wedge in our future. But what do the lines themselves represent? What actually defines the top and bottom edges of the wedge?

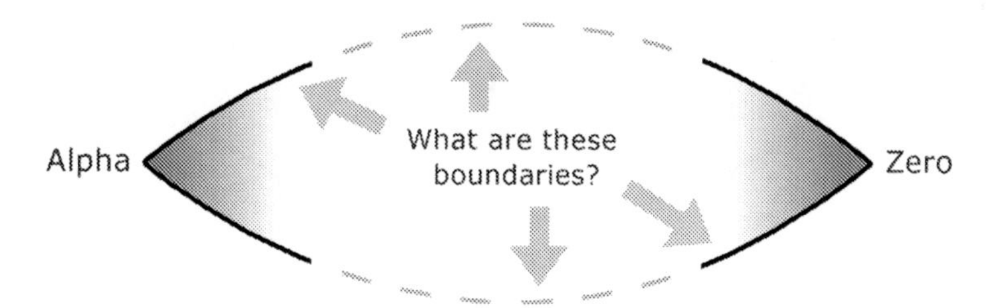

Figure 12.5: What defines the adjacent boundaries in state space?

The Inner Boundary of Smoothness

If we could watch the big bang event from a distance, a view from nowhere in particular, a question that would arise would be, why didn't the early universe remain perfectly smooth as it expanded? It is an old question in astronomy and cosmology. Instead of remaining smooth, very small variations in the rate of expansion produced minute differences, which gravity then amplified. The more dense areas underwent a phase transition and became matter, and eventually transformed into galaxies and galaxy clusters. Without these mysterious fluctuations in the density of space, the universe would have remained perfectly uniform. Gravity would never have been a distinct force of nature since gravity's influence would have been evenly distributed over the entire volume of spacetime. Of course life would not exist in such a universe to note the difference.

But the question of why the universe isn't smooth has at least led to a vague recognition in cosmology that a smooth configuration is an extreme of possibility. Physicists don't talk about double or hyper smooth universes. There isn't anything beyond perfectly smooth. It is just a natural boundary to what is possible at any given average cosmological density. We can easily imagine a smooth cosmos from beginning to end, as in, a universe where time starts from Alpha and ends at Omega, with conditions always remaining smooth. Such a universe would always maintain a uniform density and mass, without the lumpiness of particles, stars, or galaxies.

In this smooth universe scenario, time would begin smooth, always remain smooth, and end smooth, so it is a very simple, plain, and boring universe to envision, but it is a significant scenario in that it represents an extreme of possibility for the entire evolution of the universe. Every other possible pattern is less smooth. So in the same way Alpha and zero create boundaries for our past and future, these smooth states form a sort of outer membrane to the whole of possibilities, beyond which no other possibilities exist. This line in the sand even represents an outer boundary in the realm of all imaginable universes in the chaos description. It simply defines the outer edge of reality.

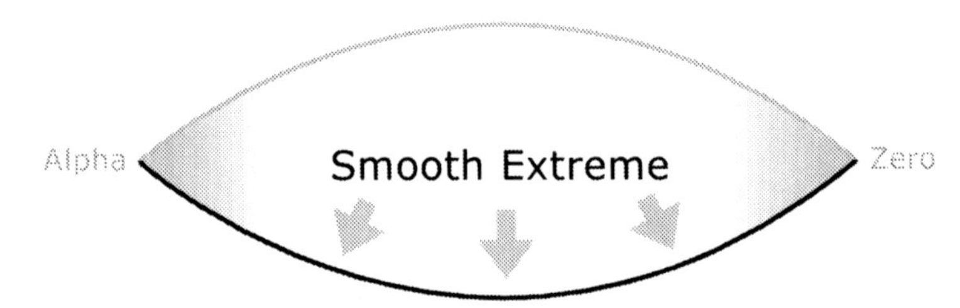

Figure 12.6: The Smooth Extreme.

The density variations in the early universe that became galaxies and us are usually attributed to quantum instability but there may be a simpler way of looking at the mystery, a way of understanding why there is quantum instability. The image above displays the smooth extreme as an outer boundary running all the way from Alpha to Zero. If we imagine the early shaping of the universe as if nature is choosing amongst all the patterns that are possible, then obviously the smooth path is only one possibility among the many alternatives where the universe doesn't remain smooth and instead becomes lumpy. In order for the early universe to have remained perfectly smooth for any extended period of time, the selection process would have had to choose the one perfectly smooth pattern for each and every new moment of time. The path of time would have had to pull that particular pattern out of the hat again and again, somewhat like the same person winning the lottery every day, every hour, every minute, every second.

If a universe is perfectly smooth then its directions of freedom are obviously limited to either staying smooth or becoming lumpy. And since there is only one perfectly smooth pattern and a whole variety of lumpy patterns, with so many lumpy patterns compared to the one smooth pattern, it is enormously improbable that the path of a universe would continuously choose the smooth extreme, that is, if time has a probabilistic nature. Note how this is the same basic reasoning behind the second law, except we are using it here to explain why the universe didn't remain in a perfectly equalized state. So we are turning the logic of the second law around to show that a perfect thermal equilibrium state for the entire universe is improbable. Smoothness is actually an outer edge of improbability.

The Lumpy Universe

The next step along this same way of thinking is to recognize that another extreme exists opposite of the smooth extreme. Adjacent the wide axis from Alpha to ZAT, opposite to the smooth extreme, there is an extreme case of lumpiness, even if such a state is initially difficult to envision. We know the early universe did not remain smooth. Small fluctuations caused the universe to become moderately lumpy and yet isotropic, meaning that matter was for some reason distributed very equally throughout the entire universe. The original variations in density were moderate and similar enough to create the distribution of over a hundred billion galaxies presently observed by telescopes. But how lumpy could the universe have been had conditions been allowed to be more extreme? What if the earliest fluctuations had been more violent? How might the universe have turned out?

Figure 12.7: The Lumpy Extreme. The smooth and lumpy extremes necessarily exist at any given measure of average density throughout the range of density from infinite to zero.

When our telescopes first grew powerful enough to see the galaxies, we might have discovered one side of our universe is totally empty of galaxies, while the other side is more massive. Some regions of the universe might have expanded during the big bang only for the first billion or so years, and then collapsed, forming ominous black holes unlike anything we observe today. We would know exactly where such a giant black hole was located because its gravitational force would intensely warp the large-scale topology of space, pulling millions or billions of galaxies toward it.

Can we imagine a universe that is still even lumpier? First we might imagine all the protons of the universe combined into a single giant proton particle, and all of the electrons combined into a super electron, and imagine this giant atom as a cosmos. It's a little hard to imagine and of course such an extreme division of charged particles doesn't seem normal or physically possible to us, but just consider if a basic law of nature was that like particles attract while opposite particles repel. Charged matter would then divide apart into two groups.

Imagine a single massive black hole existing in the center of a giant bubble of space-time, a black hole which for all intensive purposes would be the core of the big bang being maintained in that location. Seen from a distance, this obscure center of mass would not emit light like a star, but the event horizon might emit virtual particles that escape out into the surrounding space. I realize the lumpy extreme doesn't seem physically possible from our region of normalcy living in such a moderately lumpy universe, but keep in mind that what seems possible to us is a product of what we know to be probable. The whole point of this rendering is finding the outer edge of the possible, even if such a state is highly improbable or even impossible. All that matters is that the lumpy extreme is conceivable, which establishes a second outer membrane in the realm of all possibilities directly opposite the smooth extreme. And we of course naturally find ourselves in the middle between extremes.

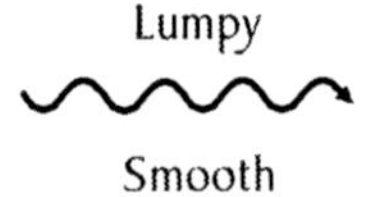

The flow of time for our universe is naturally trapped between the outer extremes of possibility. If time strays off course, if time moves toward the lumpy side, the group of states which are smoother than the present inevitably grows in size, so the comparison might become 60/40 percent, which means that there are more possibilities that are smooth, as compared to the present. Consequently the probability for time to turn toward the smooth side grows stronger. In fact the farther that time moves away from the balance point in between extremes, in any direction, the greater the probability for time to turn back toward middle ground. *The most probable location in pattern space is always the point of balance between extremes.*

The Contrast Gradient

The best way to properly appreciate the inevitable existence of the smooth and lumpy extremes and the gradient between those extremes is to compare them to the ordinary concept of contrast. In the most simple of terms these same extremes can be compared to all the possible degrees of contrast that exist for any image we might imagine. Contrast adjustments can be made to any television or computer monitor, causing color tones to either blend until they become a single color (low contrast) which is simply the average of the whole and is typically some shade of gray. Or the color tones of an image can divide apart into two opposing shades of light and dark (high contrast), which is typically black and white.

Figure 12.8: The two directions of contrast.

In the dichotomy of two orders, grouping order reflects the case of greatest possible contrast while symmetry order reflects the lowest possible contrast. A simple contrast gradient is an extremely useful concept, a sort of key, that allows anyone in the simplest of terms to generally envision the overall spectrum of possibilities. We have already recognized a density gradient that spans from the infinite density of Alpha to the zero density of Omega. Now we have found a contrast gradient that exists at right angles to the density gradient.

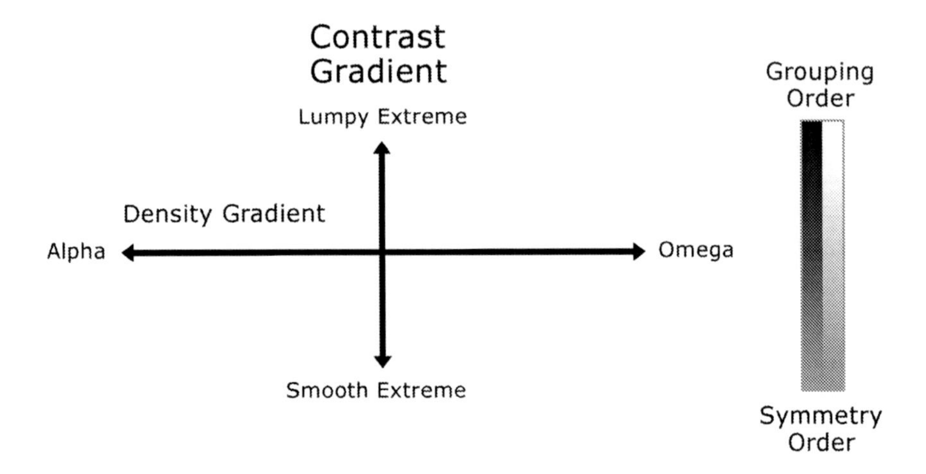

Note how the contrast gradient synchronizes well with the concepts of grouping and symmetry order. At any given average cosmological density, the lumpy extreme is also the extreme of grouping order, where matter and space, matter and antimatter, or positive and negative, are divided apart. The smooth extreme is also the extreme of symmetry order, where there are no such divisions and only sameness and uniformity.

These very fundamental boundary conditions define the possible realm and they are an important step in improving how we visually represent all possibilities. They also bound all possible states. The model shown below is infinite and yet bounded in all directions of possibility and conceivability. In between the four extremes shown we find all the varied patterns, a measure beyond imagination in quantity and variety, but not beyond imagination in its overall shape and structure.

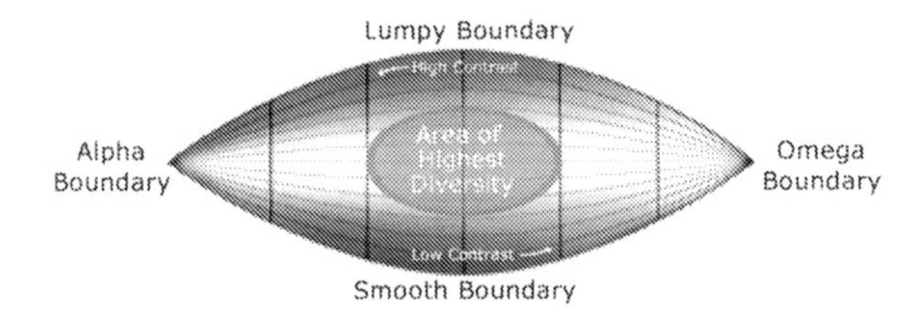

Figure 12.9: The four most basic boundaries in the set of all possibilities. With the smooth and lumpy extremes drawn in between Alpha and Omega, we have an oval shape similar to a football. Could there be a connection?

These rather simple boundary conditions can be derived from a basic knowledge of physics. Virtually everyone who has ever heard of the big bang theory at least intuitively knows there is an Alpha. And it hardly requires deep thought to appreciate that a perfectly empty space is also an extreme. Empty space can't get any emptier. Scientists have long pondered why the universe became lumpy. The condition of smoothness at a particular cosmic density is just like the condition of empty space. Both are extremes of uniformity. The only extreme that is somewhat difficult to imagine is the lumpy extreme, but we can certainly reason that there must be such an extreme, just as there is the Alpha extreme in our past. The lumpy extreme is a sort of carry over in the possible realm of the Alpha state, at lower densities. So now that we have an enclosed world of possibilities, what does it mean? What can we do with it?

Excess generally causes reaction, and produces a change in the opposite direction, whether it be in the seasons, or in individuals, or in governments.

Plato

Photos: Lower: Havasu Falls © Christophe Testi
Top: True Paradise © Philip Rocheleau

We are like a little child entering a huge library. The walls are covered to the ceiling with books in many different tongues. The child knows that someone must have written these books. It does not know who or how. It does not understand the languages in which they are written. But the child notes a definite plan in the arrangement of the books; a mysterious order it does not comprehend, but only dimly suspects...

Albert Einstein

To my mind there must be at the bottom of it all not an utterly simple equation but an utterly simple idea, and to me that idea when we finally discover it will be so compelling, so inevitable, so beautiful, that we will all say to each other, oh how could it have been otherwise.

John Archibald Wheeler

Chapter Thirteen

Everything Moves Towards Balance

Why is Time Approaching Zero?

In the science of studying the space of possibilities attractors are thought to influence or pull time in some direction. Sometimes attractors pull time in a specific direction while others pull time in a general direction. An attractor might be a single pattern or it might be a large group of patterns. For example, in Boltzmann's wedge model the arrow of time is thought to be caused by the greater quantity of disordered possibilities than ordered ones. This designates the larger body of disordered states as an attractor. As we understand possibilities today, the greater measure of disordered states is seen as the most dominant attractor in the whole of possibilities. The preponderance of disorder has long been imagined to be the only reason for a difference between past and future.

We normally assume the past is responsible for creating the present, but the second law suggests the future is pulling the present. The idea that the future has influence over the past and the present is clearly visible in the wedge model. If we estimate the present to be in some location within the wedge, this naturally splits the whole of possibilities into two groups, which we can refer to as past-like states and future-like states. Using Boltzmann's wedge as an example, below we are splitting possibilities into two groups, a group of states which are more ordered than the present, and so past-like, as well as a group of states which are more disordered than the present, and so future-like.

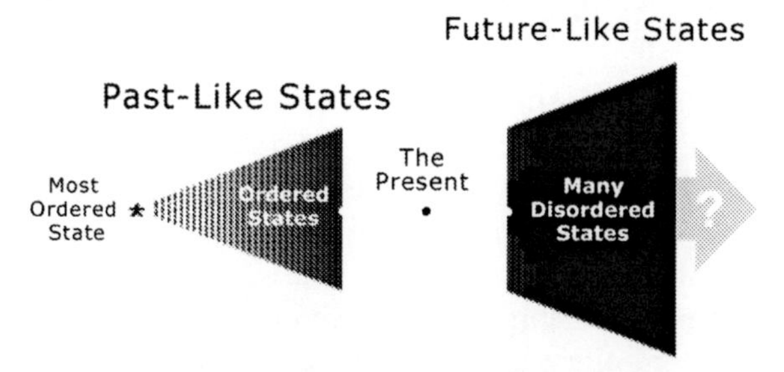

Figure 13.1: Even the outdated wedge model includes a past-like group and a future-like group of possibilities. Time is thought to move toward disorder because there are so many more disordered possibilities.

Now that we have integrated absolute zero and created the soaps model we can also portray that new model split apart by the present. Logically we have to place the point of our own present very near to zero, because we know the universe is only a few degrees away from ZAT (zero) and because there is already so much empty space in the universe.

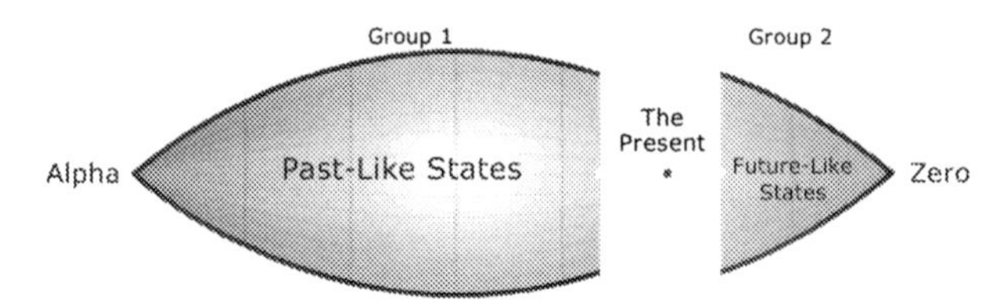

Figure 13.2: If the past-like set of states is larger than the future-like set of states, then why is time traveling toward a smaller group of possibilities.

Only now we have exposed a problem with the soaps description developed so far. We can no longer detect a reason for why time is traveling in the direction it is traveling. If we imaginatively transport ahead in time and look at what is probable from the standpoint of zero, then all other possibilities lie in the opposite direction toward Alpha. If time moved forward all the way to zero, all other possibilities would lie in the past. And with all states existing in one direction, absolute zero would be every bit as improbable as the original Alpha state.

The real magic behind Boltzmann's way of thinking about possibilities is the underlying approach of splitting up the whole into two large groups and then comparing those groups. In whatever measure that time is free to flow, the larger group of possibilities will naturally become the dominant attractor. Inherently this makes any extreme condition highly unstable. In the wedge model the Alpha extreme is highly unstable because there are so many states of lesser order pulling time away from Alpha. In the soaps model, the smooth extreme we just identified is highly improbable because of so many other states which are lumpier. But then the lumpy extreme is also improbable because there are so many states which aren't so lumpy. So naturally the probable area is somewhere in the middle between extremes.

In fact, the deeply simplistic principle that Boltzmann originally conjured up to develop the second law was the principle that the free flow of time will move toward whatever ultimate balance exists in the whole of possibilities. In the wedge model time is at least trying to find a state of balance, even if the flow of time cannot find a state of balance because there is an ever increasing quantity of disordered states. In Boltzmann's wedge the body of disordered states is indefinite, since it grows infinitely larger. So Boltzmann's idea that disorder is an attractor always holds true regardless of where the present is located in the model. It has been a mistake however to think that the potential for disorder is genuinely without end, one that scientists have become increasingly uncomfortable with in recent years, evidenced by a shying away from the order to disorder argument of the second law. Such ideas conflict with our observations that time is moving toward the extremities of zero. There is certainly no way to designate zero as the most disordered state, nor would a cosmic dominance of disorder probabilistically force time toward any outer edge in the realm of possibilities.

When astrophysicists sorted out the red-shifting of galaxies and determined the expansion of the universe was slowing, an expanding universe seemed generally consistent with the second law and the idea that the universe is winding down. In the more recent past, it was assumed the course of time must be settling toward some thermal equilibrium in state space short of absolute zero. There was then at least a possible correlation between the course of the expanding universe and how we vaguely imagined possibilities according to Boltzmann. Then in 1998 we discovered accelerating expansion and the big picture changed.

Today in being more aware that there is an absolute zero extreme in nature even in Boltzmann's way of modeling possibilities we need to realize there cannot always be a greater and greater measure of disordered states out there. If instead we imagine two distinct sets, we have one set that contains all states which are more disordered than the present, while the other set contains all states which are more ordered than the present. If the disordered set is greater then time will be pulled in the direction of increasing disorder, but as that happens the set of states which are less ordered than the present (future-like) will decrease in size, and the set of states behind us which are more ordered (past-like) will increase in size. And eventually the two sets will become equal. As the probability that order will increase grows, the probability for disorder weakens, and finally a balance between probabilities will ensue. The probabilistic flow of time will then be caught between two equal attractors. Whatever state exists in the perfect middle of those two attractors becomes a sort of cosmic equilibrium state. It becomes the great attractor.

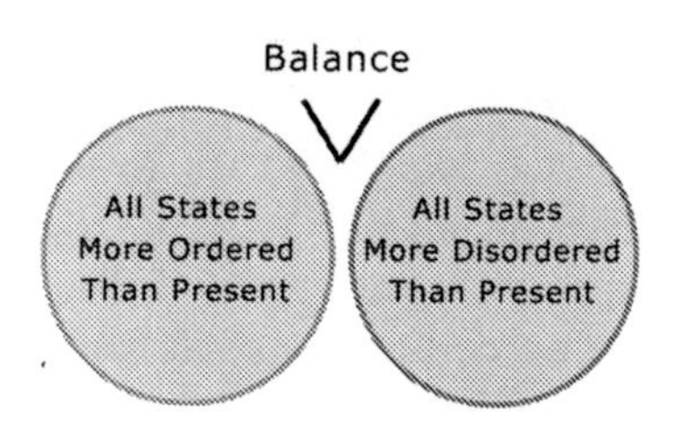

If such a balance exists somewhere in nature then that condition should be the primary attractor for all of time. In the new model we have developed so far, this cosmic equilibrium state would have to be located somewhere in the middle between Alpha and Omega, somewhere balanced between infinite density and zero density, but of course this doesn't agree with our observations either. Time is clearly not moving toward some middle point between Alpha and zero. Time is long past that center point and still is moving toward zero. Time is even accelerating toward zero. Why then is time moving toward zero? There has to be a very good reason. And it has to be something that makes perfect sense once we realize it, because we are considering something very basic about reality.

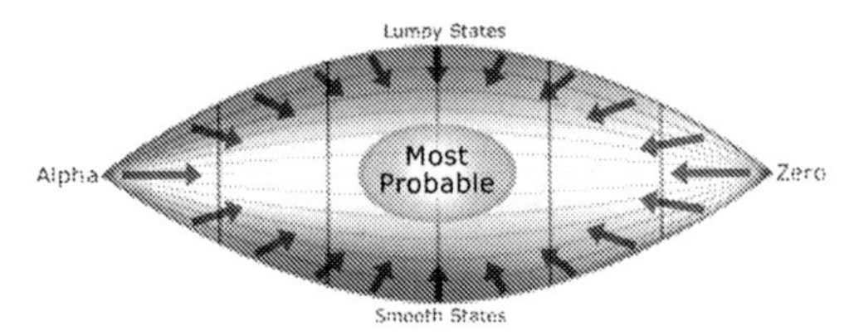

Figure 13.4. Logically time should be moving toward the ultimate balance point between Alpha and Zero, and Smooth and Lumpy extremes.

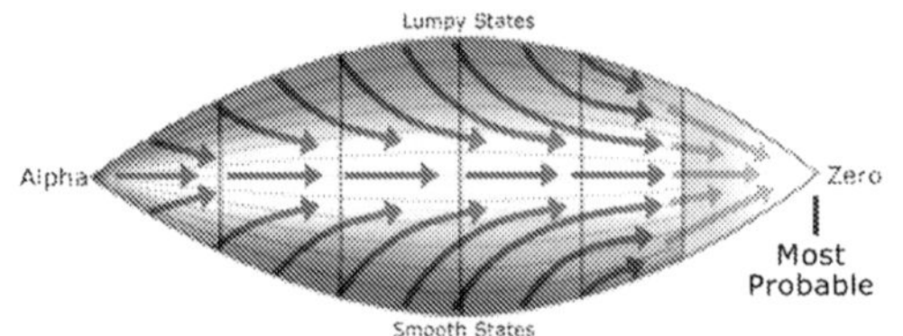

Instead we clearly observe time moving toward zero as if it is the ultimate balance point in the space of all possibilities.

Although it is not usually described as such, we can imagine that the universe has a momentum through pattern space. When time began the universe violently stormed away from Alpha through the vast sea of possibilities, in a general direction toward zero, but that momentum gradually slowed as the energy driving time was depleted. The momentum of time slowed at an ever decreasing rate nearly to a standstill before it began to pick up speed again. As if sensing ZAT was near, about six billion years ago the expansion began to accelerate, increasing approximately eighteen percent since then. What is the momentum of the universe telling us? Why is time behaving as if zero is the cosmic equilibrium state?

So here we are forced to finally ask a question that has been lying in wait since the discovery of accelerating expansion surfaced in the mainstream of science in 1998. Why is the arrow of time aimed directly at absolute zero, and presently even accelerating toward zero? What causes absolute zero to have such a powerful influence over time? What larger role does absolute zero play in the top-down picture of reality?

With the arrow of time so obviously directed at absolute zero, in essence our own universe is telling us that zero is the ultimate balance point in all of nature. So, we should probably listen and consider that option. How might ZAT be the center, the middle ground, the balance point of all possibilities? If zero is the center then obviously we must somehow expand the model we have so far created. Once we have reasoned this far the solution is rather obvious. The whole of pattern space includes both a positive and a negative side, in the same way real numbers have two sides. To correspond with what we observe the soaps model so far created needs to be expanded to include an inverse set of patterns, similar to the extension of negative numbers beyond zero in the mathematical plane of real numbers. Of course this mirrored set would be identical and yet opposite to the other. One side of pattern space would be positive, the other negative.

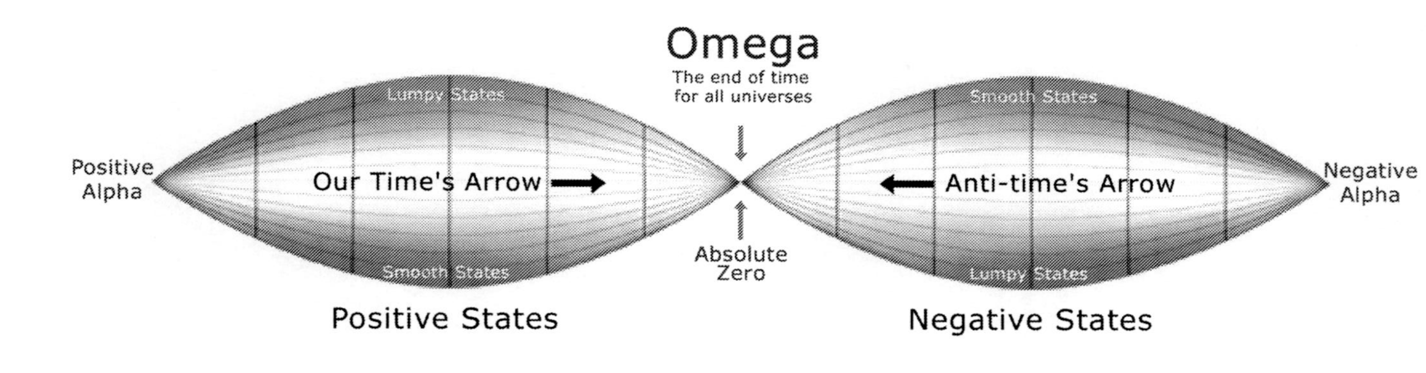

Figure 13.5: *The really big picture.* A symmetric model of possibilities balances out all that we know of physics and existence. Such an extension of known possibilities explains the true reason why time has a direction. Time does not move from order to disorder, time naturally travels toward the ultimate balance point in the whole of all possibilities.

Now that we have a complete model that portrays zero as the cosmic equilibrium state, suddenly the path of time toward zero makes probabilistic sense. And now we can say, "of course time moves toward zero". Zero is the mathematical balance between positive and negative numbers. When we stop to think about it, it makes perfect sense that zero is the balance of and center of all possibilities. Zero should be the center of superspace and of Platonia. We can also see right away, according to this model, that the Alpha in our past is a positively charged singularity. It is not a perfect symmetry because it is an imbalance. It is the most positive state in existence. On the far opposite side there is the other imbalance, an Alpha that is negative. And of course opposites attract, at least in time they do. In fact this model explains why opposites attract. It explains why our universe instantly shot away from Alpha like an arrow shot from a bow. The flow of time does not travel from order to disorder, it travels from extreme imbalance to perfect balance.

This soaps model is perfectly compatible with, and even a necessary extension of, the theory that there are two kinds of order in nature. Where the wedge model provided a reason for the arrow of time, this new model provides reasons why there are forces of nature. In the past we could not recognize any reason for why there are forces that create and maintain orderliness in contradiction to the trend toward disorder. But when all other possible states are balanced around zero the whole dynamics of how we understand possibilities changes. The most interesting difference is perhaps how this model portrays existence as having a natural structure which allows only the time worlds we recognize in the Many-Worlds Theory to exist.

This model now exposes clearly and succinctly that the second law is too simple. It is not a full explanation for the arrow of time. The logic of the second law as written completely breaks down if time is aimed directly at zero, evident particularly so at the point where the arrow hits the target. It is incorrect to say that the universe is winding down and dying, wrong to envision the universe is only unfolding after the big bang. It is incorrect to say the universe is moving from order to disorder, because there is another kind of order increasing toward the future, and because in the deep time of cosmic evolution the arrow of time is moving away from extreme imbalance toward the cosmic extreme of balance. Alpha is not where existence begins. It is just an inevitable imbalance existing within the ultimate balance of all. The fact that Alpha is so imbalanced is what caused the excitement of the big bang in the first place.

What was the initial state of the universe, and why was it like that...there doesn't seem to be a natural choice for the initial state. It can't be flat space. That would remain flat space.

Stephen Hawking

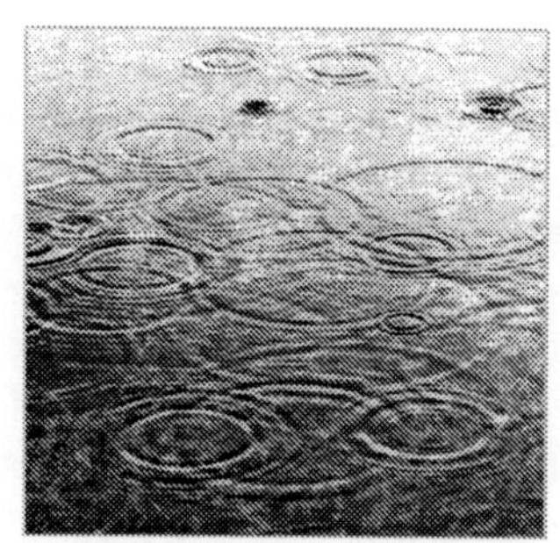

Photo: Rainy Days © Dave White

Chapter Fourteen

Equilibrium

Duality In the Pursuit of Balance

Reasonably speaking, there must be an ultimate balance, a common middle ground, where all opposites meet. We can conclude this intuitively even from ordinary experience. Positive and negative; red, yellow, and blue; matter and anti-matter, all indicate a middle ground. Two halves make a whole. Three thirds make a whole. Four quarters make a whole. And so if there is an ultimate balance there must be some kind of universal cosmic equilibrium in existence. Where it is, who knows? It may be just a way of looking at things, an idea or principle. Or it may be a physical reality, a stage of the universe. But in principle it must exist, there must be one single place where the whole of all times and places and properties exist in balance with one another.

But knowing this principle is just the first step. The next step is to technically reason that the balanced whole can only exist in the direction of our future. Balance is the most probable and natural state for any and all universes, because in considering all the other options, all directions of probability away from balance toward imbalance are equalized. For every probability to move away from balance toward imbalance there would be an equal probability in the opposite direction. For this reason a state of perfect balance is probabilistically dominant over any other condition. There would be no reason for time to make a large swing away from balance toward an imbalance in one direction or the other. It would never happen. There would be no impetus for such change. Once all of time achieved becoming a perfectly balanced flat universe, like a calmed body of water, it would remain in that condition forever.

However, this does not mean imbalances don't exist. The larger balance is inevitably a product of the whole, yet there is no whole without all the parts that create the whole. There cannot be fullness without all potential elementary content. The greater infinite cannot be without the lesser finite also being. It is even true that the principle that definitive form can be is predicated by the fact that nonexistence cannot be. Imbalances must exist, they just aren't a result of an ultimate balance temporally transforming or fracturing into them. Definitive things aren't the result of a broken perfect symmetry. They simply exist as part of the whole. Alpha exists as part of the whole.

The simplest and most cohesive explanation of a direction of time is that time begins from imbalance and travels toward balance. Once that idea is entertained one realizes it is really the only explanation that could ever fully make sense. Of course considering all that we know of physics and mathematics, how could it have been otherwise? A reciprocal negative set of states should extend in some respects beyond zero. There has to be an identical negative side of state space to balance out the side we are on. There has to be imbalances. And if there is a free flowing probabilistic time, which we know exists because we experience it, then the dominant direction of such time flows away from each side of any balance toward the big equilibrium state in the middle.

Consequently for every matter dominated universe like our own, ultimately there has to be a direction of anti-time, where a contrasting version our own cosmos begins from the opposite side of physical reality. Time across this great divide begins from an identical imbalance opposite to our own, i.e., the negative Alpha state, and that cosmos flows in synchronicity with our cosmos toward the same destination, Omega. There absolutely has to be a truly parallel universe where stars and galaxies are created from anti-matter rather than matter. Such ideas are awfully profound, but they are just facts of the big picture.

Probability Mechanics

There are very basic probabilities evident in this model concerning imbalances flowing toward balance which we can take a first look at with very little effort. This now complete Soaps model is slightly more complicated than Boltzmann's wedge model, yet we are now working more comfortably with definite groups of states, so we can realistically compare the groups of states which are more and less probable.

If we start by considering the probabilities of this model from position A in the diagram below, from Alpha, which is the most positive state in whole spectrum, we recognize that all other states are less positive than Alpha, including all of the negatively dense states and the negative Alpha.

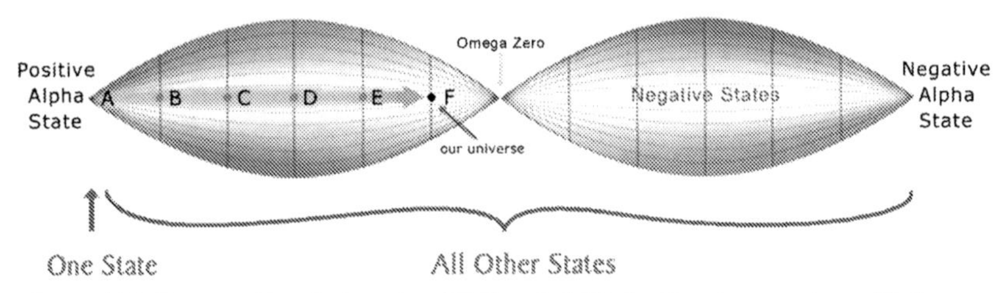

Figure 14.1: From position A there is a 100% probability for time to move toward balance and symmetry (expansion), then at each consecutive stage between A and F that percentage of probability decreases while a new probability grows for time to move backward toward imbalance and grouping (gravity). The direction directly toward zero is accented by the counter probabilities for conditions to move toward increased lumpiness versus increased smoothness adjacent the density gradient.

If we imagine ourselves to be a universe positioned at Alpha, then all other states create a large group shown below. If all other states besides the Alpha state form a collective group they represent a measure of probability which cannot be greater. All probability influencing time at this stage is for an evolving universe to move away from Alpha and pass through lesser dense states toward the negative side, toward a position of greater balance. Generally reading the model, this vast set of states acts as a powerful attractor, causing the direction of time to explode through pattern space as fast as the nature of time allows conditions to change. The decreasing positive density is invariably caused by an invisible influx of negative density given up by the parallel partner universe, making both sides more neutral.

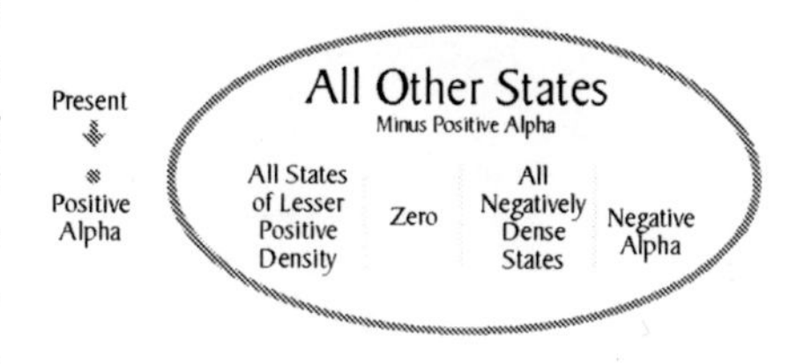

Eventually a universe evolves to position B and then C, which causes two considerably large groups of states to form, rather than just the one large attracting group. In addition to the future-like states there then exists a group of states behind the present which are more positively dense than the evolved present, forming a small group of past-like states, as shown below. The evolving present is moving through the Soaps spectrum toward zero, so the past-like states are growing in size and hence they begin to exhibit influence over the momentum of time, uniformly pulling backward against the strong flow of time toward zero. That same pull influences internal conditions within the cosmological system, such as creating the early fluctuations in the density of space that will later become stars and galaxies.

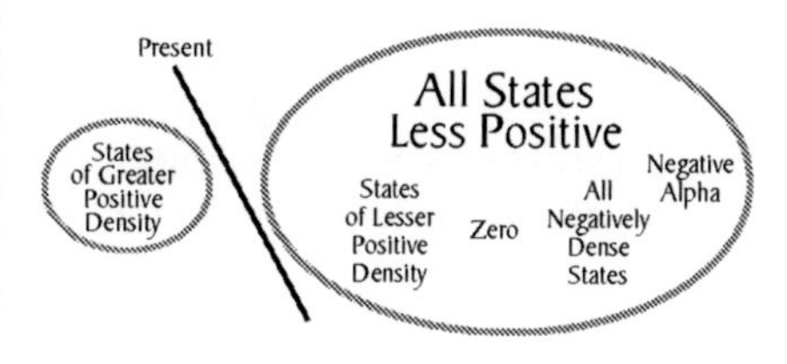

By general rule any evolving universe is influenced by the entire spectrum of possible states. No single condition is inherently more probable than another. The probability law which dictates that time moves toward a balance point is partly based upon this principle. So a system therefore responds to any influential group of probabilities, even its present state contributes to a predetermined dynamic of all the probabilities that influence time.

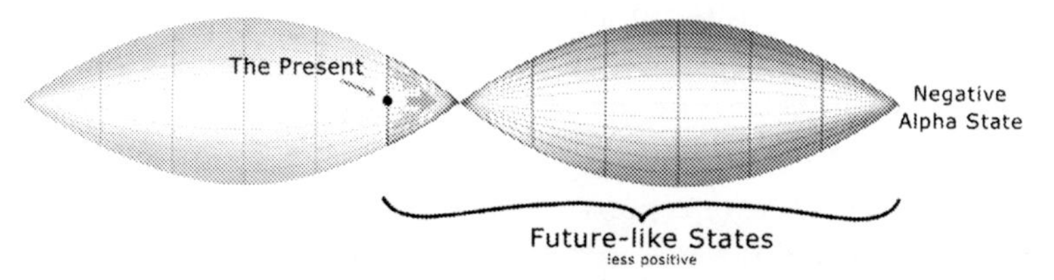

Figure 14.4: Future-like states are those states which are more negative and less positive than the present, a group which is always largest until time reaches zero.

The universe having evolved from point A through F, and finally reaching the present state shown in figure 14.4 is still dominantly attracted to the larger group of future-like states (all are more negatively dense than the present), but there is now a nearly equal group of past-like states attempting to pull time backward toward Alpha. As a universe evolves ever nearer to zero, that set of states grows ever larger and stronger, as shown below.

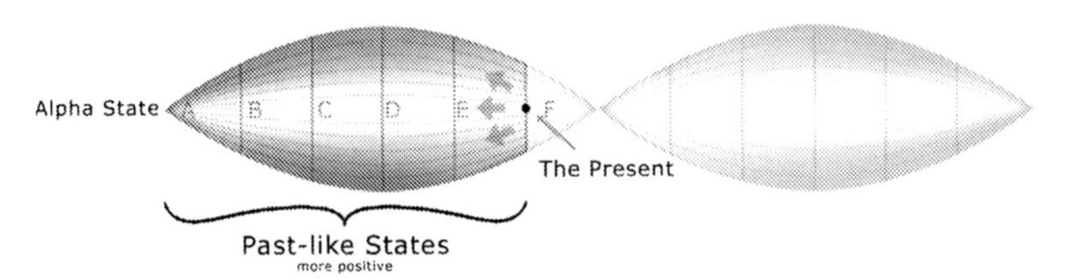

Figure 14.5: Past-like states are those states which are more positive than the present, a group which originally contains only the Alpha state as time begins at the big bang and then grows in measure and influence until it becomes equal to the Future-like group of states.

The British astronomer Fred Hoyle wrote, "The thermodynamic arrow of time does not come from the physical system itself... it comes from the connection of the system with the outside world." Indeed, time is influenced by the whole body of all possibilities. Therefore, we mustn't overlook the fact that probabilistically it is possible for time to travel backward, even here in our own cosmos. Time does not have an inherent direction which the entire universe obeys. Time freely flows backward and forward. It flows into more dense past-like conditions and less dense future-like conditions, which is why there are both dense areas in the cosmos where time has traveled backwards, and nearly empty areas where time has progressed forward. After the big bang, as the past-like set grows it also becomes an attractor and stubbornly opposes the more powerful future-like group of states, even though the larger group continues to pull time toward the negative side of the spectrum, and thus toward zero. We know this backward pulling group most commonly as gravity. Gravity is essentially the past pulling time backward.

Probabilistically the future-like directions cancel out a measure of the past-like directions, and what remains is a specific measure of probability for the flow of local time to travel forward and backward. The universe as a whole, only feels this backward pull as a global slowing of expansion, but locally or internally the universe feels this backward pull in the form of gravity.

As Richard Feynman recognized in his sum over histories approach to quantum mechanics, all the directions of time sum to a specific set of probabilities. In guiding the universe toward the specific destination of balance, those probabilities necessarily restrict and control what is possible in time, while the remaining freedom within this measure of control makes room for diversity.

If it were not for the narrowing of pattern space near zero, conditions in a universe approaching zero might become extremely varied and unstable. A free flowing time in the wedge model for example would simply decay into chaos. In the Soaps model however we recognize the future-like states are always the larger dominant group so they pull time into a decreasing measure of available states near zero. This forces conditions to move increasingly toward the perfect balance and uniformity of zero in various ways, the most obvious being the balancing force of electromagnetism. Electromagnetism causes opposite particles to attract and like particles to repel. Electromagnetism constantly works to increase balance throughout the atomic world. That force or control over what happens in forward time simply reflects the power and influence of a perfectly balanced and uniform zero state.

Accelerating Expansion

The shape of state space near zero does explain why expansion is accelerating, although the reasons aren't plainly visible. The expansion of the universe is accelerating for two reasons. Firstly, Omega is a hyper-full space which we approach in stages. This inevitably requires that our curved space expand or inflate, which is a flattening of space. We necessarily view this expansion increasing at an accelerating rate because the final state is perfectly flat and full, which from our perspective is in a state of inflation. But the final state is not inflating, it is simply inflated. Secondly, or at least a detail of the same process, is that the momentum of the cosmos toward zero increases because the measure of states adjacent our present state decreases considerably as we enter the closing tunnel of pattern space on our way to the single state of zero. Our sense of time is a reference to the speed of light which is really a measure of the individual states that we evolve forward through, so in converging to zero where there is a decreasing measure of adjacent states in all directions, our sense of time then perceives accelerated changes in the form of accelerating expansion.

It is important, particularly for scientists, to appreciate that some laws in physics are relative to our position in state space. For example, just because we cannot cool matter to absolute zero presently does not mean the cosmos cannot reach zero in the future. When we cool matter toward zero it maintains a non-zero temperature by pulling in energy from its surroundings, but in the future the expansion of space won't allow that neighboring energy to be transferred. Bose-Einstein condensate already indicates that matter and energy can become space-like. In the end the curvatures of matter simply flatten out.

Just as a pendulum comes to rest at balance, time comes to rest and ends at the balance of zero. Also like a pendulum, time accelerates away from imbalance, but unlike a pendulum the momentum of time slows half way between Alpha and zero, then accelerates as it approaches zero. Time doesn't swing past zero because the cosmos is essentially the two halves combining together into one whole, and all energy to swing past zero is spent in the journey, translated into space. In a way the physical cosmos is created by the slow collision of both time directions, sort of like two pendulums colliding at zero.

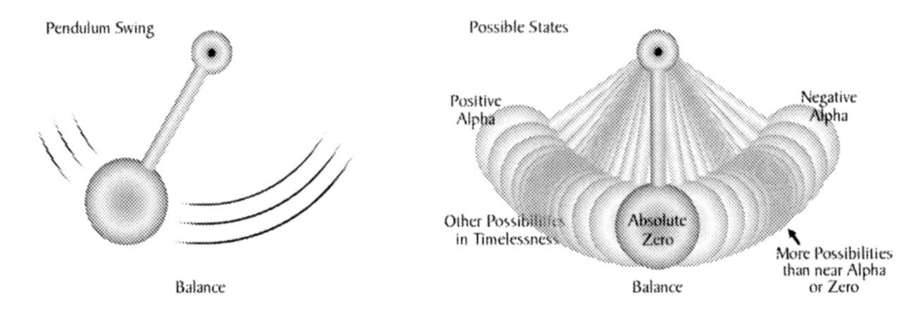

Figure 13.6: In some ways cosmological time is like a pendulum except the swing down to zero is neutralized by an inverse pendulum swing. Also there are more possible positions at the mid-point of the swing so the momentum of the swing initially slows from potential directions in time adjacent the path to zero, then once beyond the mid-point the momentum accelerates as the measure of patterns that surround the present decrease.

The groups of past-like and future-like states are in a constant state of flux, always in conflict, each pulling at time. The future causes the universe to expand and balance out while in contrast the past attempts to counter that influence by gravitationally pulling time backward, literally attempting to recreate itself. This results in a tension between the past and future, and so it results in a tension between the two kinds of order. In the clash of time and anti-time, in the battle of two orders, the cooperation and complexity we cherish blooms like a flower.

The future-like group invariably remains stronger until time reaches zero, so globally the ultimate attractor is the one single Omega state itself. In discovering accelerating expansion, scientists have found reliable evidence that gravity will never prove to be the greater force and reverse cosmological expansion. Once we discard our assumption that arbitrary forces can exist, we can then appreciate electromagnetism as the strongest evidence found in nature that time is moving directly toward the balance of zero. In fact there is no possibility that gravity will ever win the battle and collapse the universe in the future, which can be said true for our universe as well as all other universes.

The Privileged Worlds

A main tenet of this book is that Boltzmann invented the ideal method of understanding why the universe is this way; he just wasn't able to create an accurate model of all possibilities in the 1800's. This aggregate model of states should feel intuitively satisfying in part because it replaces Boltzmann's outdated asymmetric model with a symmetric superstructure. If we look at the present state of our own universe, the amount of grouping of matter and the balanced distribution of galaxies since the big bang is certainly congruent with the river of high probability flowing from Alpha to Omega balanced between smooth and lumpy extremes. This channel of high probability inevitably selects a special partition of states that are utilized by the natural flow of time. This special group is very narrow compared to the whole of possibilities. In probability science the path through this group is called a basin of attraction. I call this section of pattern space the many-worlds partition.

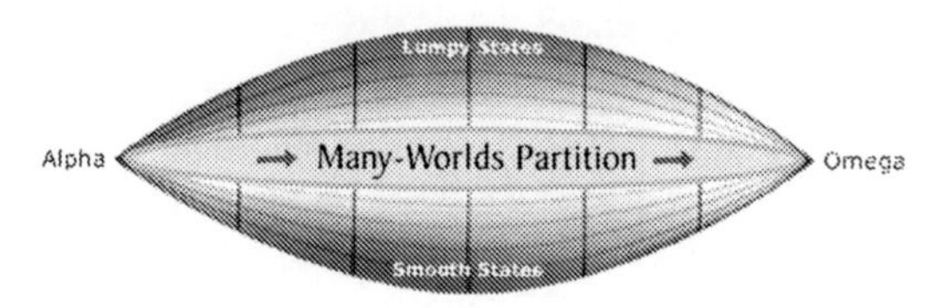

Figure 14.6: The more probable paths toward zero define a special partition of states, the Many-Worlds partition, identical to the many worlds now predicted by Quantum theory.

Probabilistically, all physically real temporal universes must begin from either the positive or the negative side of the overall spectrum of patterns, from a positive Alpha or a negative Alpha, and both directions of time travel inward toward the same balance point, with both directions of time ending at zero. And since the natural course of time, or probabilistic change, is necessarily always toward the balance of zero, it follows as a universal principle of nature that time in any universe inevitably traces backward and originates from an ever greater imbalance.

We could flash back to the chess game which begins with two players, one on each side of the board. The game pieces are divided apart from one another. Each side is purely one color, as if their color makes them oppositely natured. Likewise time as we know it begins from one side of the board or the other, originating purely positive or purely negative.

In hindsight we could have predicted time begins positive or negative from the two kinds of order theory. For any universe fundamentally constructed of protons and electrons, the extreme of grouping order would necessarily result in all the positives being divided apart from all the negatives. So the identity of the first moment of time would have to choose to be one charge or the other, just as a player of chess must choose one color or the other. We happen to be on the positive side, although the designation could be negative, as in, we wouldn't know the difference of one charge from the other. After all we are not talking good and evil here, or normal versus the strange mirror universe, just identical positive and negative opposites; the two players of a game called Cosmos.

The improbability for an entire universe to vary from the regular course from Alpha to Omega becomes so great that most of the universe scenarios, even though they are imaginable in fantasy, are in fact so improbable that they are genuine impossibilities. One can recognize that there are scenarios so highly improbable that they can be deemed impossible, even being newly introduced to this model. The smooth and lumpy paths, and world lines anywhere near those extremes, would clearly seem to be probabilistically impossible. But whole cosmological systems turning around and collapsing inward in a big crunch toward Alpha are without question probabilistically impossible if this model of a great divide does accurately represent the possible realm.

Everything learned in physics and cosmology, all we know of quantum physics and the forces of nature, as well as a trip to a casino, suggest probabilities somehow distinctly guide the path of time and shape the world we know. That knowledge in modern science, and what we intuitively know in our experience of chance events, combined together with this model of possible states, strongly indicates the extraordinary conclusion that worlds outside of those that travel from Alpha to Omega within the highly probable many-worlds partition simply don't exist, at least not in the same sense that our universe is physically real. We can safely conclude that the universe we live in isn't some fortunate arbitrary fluke of nature. Our universe does not require an absolute chaos of worlds, or many realms, or even a multiverse of worlds ruled by different laws of nature. It appears that we are physically real because we are quite privileged. Our own universe is an example of what is naturally probable in respect to the big picture, in respect to the whole of imaginatively conceivable universes.

I
can explain the basic
reason for why the universe expands
very rapid at the beginning of time and then slows
down, using this diagram, with the sentence length representing
the measure of increasing possibilities that exist at each stage. With more
words in a sentence it takes longer to read each sentence. Time slows because
there are so many more directions for time to travel, in the middle stages. Then
once convergence begins there are fewer directions distracting the flow of time
away from its general goal of reaching balance, and so slowing
expansion turns to an ever increasing rate of expansion
simply because the measure of possibilities
decrease rapidly near
zero.

Existence is now, all at once, one and continuous... It is not divisible, since it is all alike; nor is there any more or less of it in one place which might prevent it from holding together, but all is full of what is.

Parmenides

The Parmenides Principle

Ordinarily we imagine the universe exists only in the present. If we take a whole other approach, if instead we imagine that possibilities are actualities, and so all possibilities exist in a timeless unchanging way, not simply in an ethereal form but each exists as physically real as any other, as real as the moments we experience, then, and only then, do we suddenly have a simple and reasonable solution to why our universe originated in the past from a dense and ordered state. If all the possible patterns of the past and future exist simultaneously, totally independent of the passage of time, then the history of any temporal universe moving through those patterns will inevitably trace backward to the extreme imbalance, and hence extreme order, of Alpha.

We know from Boltzmann's general approach that the arrow of time is built into the realm of possibilities. Time naturally evolves from the improbable to the probable. That is the very basis of the second law. So then ask yourself why it seems strange to us that time begins in an improbable location? The reason is because a non-physical realm of possibilities wouldn't have any obvious influence upon where some universe pops into existence. So it seems strange that time began from Alpha, unless...unless all patterns are physically real. Then time would invariably originate as it does from the greatest imbalance in all of nature, because the direction forward for time is inevitably toward balance. If the direction forward for time is always toward balance, then traced backward, that same path invariably originates from imbalance. Thus we can conclude that any path of time embedded within a physically real pattern space would naturally exhibit a history that originates from the single most improbable location of all, which happens to be Alpha. If pattern space is physically real then the flow of time is built into physical reality, causing probable time worlds to exist, while extremely improbable time worlds do not.

A similar principle is evident in the way that we write stories in books. If we pick up a book and study the story, we will find a flow or plot that originates in the front of the book and ends in the back of the book. In order for there to be orderly sensible stories in books, there must be a forward progression or evolution to the story, which is ultimately an evolution of patterns. Likewise, if time is embedded in a greater space of all physical states, then at any given time that any observer examines history, time will be seen as originating from an improbable state. When that observer examines the future course of the universe, it will be found to be moving toward the most probable state. As it turns out, there are two states which are improbable, the positive and the negative Alpha, and one state which is the most probable, Omega. So a universe like our own that is more positive than negative (the proton is 1836 times the mass of the electron), will have an increasingly positive origin, finally originating from the purely positive Alpha.

I don't know of a more compelling argument in support of the idea that the greater universe exists timelessly. With all the paradoxical problems surrounding any kind of creation event, a beginning to time, and a beginning to existence, there is alternatively one truly simple solution. All possibilities exist. The infinite universe is actualized, not in the form of all time worlds, but rather as a timeless world of patterns from which we borrow each moment of now.

The philosophical premise here is that all possibilities must exist, because nonexistence cannot be. Existence is inevitable, but what does that mean? What is existence? What does exist? First we must at least conclude that whatever is possible to exist does exist. In this rather natural scenario, we could conclude further than any actualization of what is possible would be infinite. However, we don't need to assume that some chaotic set of time worlds are real, because the existence of all conceivable patterns satisfies the necessity of what exists to be infinite. In this case time worlds like our own are real as well, but they are secondary and subject to the more fundamental physical reality of timeless patterns.

We actually know that the passage of time must pass through patterns, whether they are ethereal or real, since we know the physical moments we experience are themselves patterns, so the option that all patterns actually exist shouldn't really seem any more spectacular than the seeming miraculous existence of each moment. We know our universe exists, so the notion that all the patterns that create time exist independent of time, should in no way seem existentially absurd. It should seem a necessity.

We might also recognize here that a fundamental existence of patterns is a very good thing. That fundamental existence appears to lead to the hierarchy of atomic elements, stars systems, biological life, awareness and consciousness, and perhaps finally intelligence and wisdom. If time and change were not restricted to the probability arrow of time built into pattern space then anything could happen and would happen. The secondary realm of change would be chaotic. Reality itself would be chaotic because reality, physics, reasoning, is so relative to cosmic structure. The big picture would be chaotic and life would not exist.

If time worlds were primary, if they somehow existed independently, insubordinate of the probability for balance to ensue in pattern space, then the existence of any one or some set of such universes implies the existence of all imaginable scenarios, since if one universe manages to force its way past the borders of improbability, why wouldn't all others find the same loophole.

We can become ever more certain that a physically real pattern space supplies a critical foundation for a hierarchy of many-worlds and many minds, as we continue to discover that our own universe is an elementary example of what should exist, in other words, if the features of our universe are predictable from pattern space. If we continue to discover that we can understand time and many of the various characteristics of our universe, such as gravity and electromagnetism, the proton and electron, the stars and galaxies, as consequences of an accurate soaps model, then we could reasonably expect that what is possible is by nature restricted to universes like our own.

Modified Photos, Thanks to: Paths © Andrew Kazmierski, Trail Sign © Noah Stryker - fotolia.com

Why is nature so ingeniously, one might even say suspiciously, friendly to life? What do the laws of physics care about life and consciousness that they should conspire to make a hospitable universe? It's almost as if a Grand Designer had it all figured out.

Paul Davies

~~~

When I despair, I remember that all through history the ways of truth and love have always won. There have been tyrants, and murderers, and for a time they can seem invincible, but in the end they always fall. Think of it - always.

Mahatma Gandhi

~~~

...it is the nature of synchronicity to have meaning and, in particular, to be associated with a profound activation of energy deep within the psyche. It is as if the formation of patterns in the unconscious mind is accompanied by physical patterns in the outer world.

F. David Peat

~~~

You are here to enable the divine purpose of the universe to unfold. That is how important you are.

Eckhart Tolle
~~~

"And the truth is that as a man's real power grows and his knowledge widens, ever the way he can follow grows narrower: until at last he chooses nothing, but does only and wholly what he must do...."

Ursula K. Le Guin *The Earthsea trilogy*

Chapter Fifteen

Convergence

How Could the Future Influence the Past?

Imagine you decide to create something complex and beautiful out of wood or stone or metal. Perhaps it is a musical instrument, or a statue, or some kind of vehicle. Of course no one can instantly transform raw materials into a violin, a fine piece of art, or a bicycle. We have to carefully plan and organize and put a great deal of work into our creation. We have to affect the world around us in specific ways to make our future creative goal a reality. Essentially we have to influence the present in order to make something happen in the future. In this same way the universe itself has intent, it has a specific goal in mind, a future destination, and that future in order to make itself happen has to reach into its past and influence events.

An inevitable future has a wonderful way of making things happen. Imagine that at exactly noon in one month's time you will be at the top of the Eiffel tower in Paris. That one point of time in the future is set in place. It cannot be avoided. Imagine it has already happened. But of course you are not aware of your predestined future. You don't know that some distant place is calling to you and pulling you in that direction. At least initially, everything seems normal because your immediate future is still open to the usual range of events. Almost anything can happen early in the month's time, as long as it's not something that will stop you from eventually making the future journey.

Everything seems normal, yet behind the scenes, suddenly your life is anything but ordinary. The usual probabilities that govern your life are being redirected. Suddenly an event in your life must be designed and planned and made to happen. Some particular series of events must organize itself in a coordinated effort to bring you to Paris. Of course there isn't just one way to travel to Paris, so behind the scenes, in a very non-local processing of potential events, all the various ways of traveling to Paris are probabilistically competing with one another. While simultaneously, all the futures in which you don't visit Paris are disappearing from possibility. The chance of visiting Paris was always there in your life as a possible future, but now that journey is the top priority of your life, and consequently the chance of unrelated events happening to you is narrowing dramatically as the date draws ever nearer to your singular destiny.

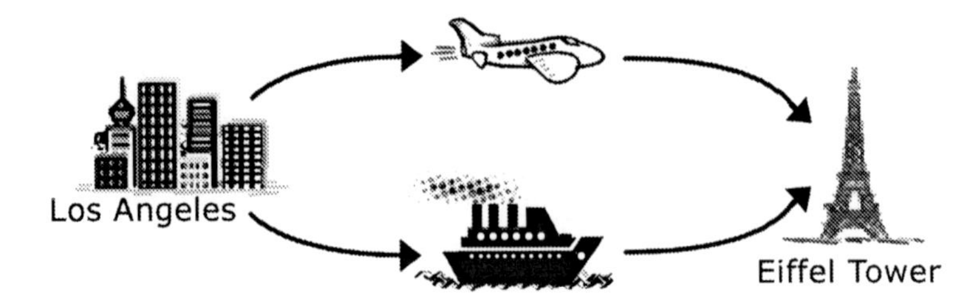

We can imagine many of the unique events that could arrange and bring about your future visit. You might just decide you need a vacation and your spouse conveniently wants to go to Paris. Or it might be that you win the lottery and want to celebrate. Or you might find a job in Paris. Or your visit might be a planned romantic interlude with someone you met online who lives in France. It isn't difficult to imagine unique scenarios that will bring you to Paris. They may even seem endless to an imaginative person. They are all possible here. Of course which specific scenario you experience isn't important for our discussion. All that is important is how one moment in your future can cause a series of specific events to unfold in a way that brings you to a specific place in physical reality. An inevitable future must shape its own past to bring itself about.

If winning the lottery is what brings you to Paris, you must first buy a lottery ticket. Then you must realize you won. And then you must collect the prize. If it's a vacation that brings you to the tower, you will have to accrue some vacation time at work. Your employer will have to approve your vacation. You will have to plan and schedule the journey. If a romantic interlude is what brings you to Paris, you will have to meet this person ahead of time, and build the relationship, and it will have to feel important enough to make the trip seem worthwhile. Every possible reason, every possible scenario of you going to Paris will have to set itself up. And so your inevitable future will need to make dramatic changes in the course of your life.

In that the natural course of time inevitably ends at zero, that one single moment of time dramatically influences our lives presently. In the same way that the noon event in Paris makes events related to the future more and less probable, the destined future for the whole universe, even though it will not occur for many billions of years, is currently shaping our present. It literally determines what is allowed to happen in our present. The condition of Omega must set itself up; it must arrange the past in order to happen in the future. In a strange way the future has to construct itself. The physical universe has to evolve in a way that will allow itself to become the Omega state. So in order for our small portion of universe to become perfectly balanced and unify with the greater whole, the future Omega reaches into its own past to organize events in a way that will make that one future happen.

Synchronicity is an unusually coordinated series of events. You may not notice the synchronization of events that carry you to Paris, depending on the makeup of your particular past and present life. You may have always wanted to go to Paris, and may travel regularly around the world or regularly visit Paris on business, so the trip may not seem at all out of the ordinary to you. On the other hand, Paris may be the last place on the planet that you would visit, or you may not have the means to go, in which case the arrangement and character of events that convince you, or help you, or force you, to arrive in Paris may seem outlandishly arranged.

At first the invisible attraction toward the Eiffel tower is mild. Your immediate day to day experience is still rather open to events that are unrelated to the Paris journey. In the first week it is not yet necessary for anything out of the ordinary to happen, although travel or preparations for the trip are more probable than they would be otherwise, while other events which lead you further away from Paris become increasingly improbable as each moment passes. Behind the scenes of your life, your predestined future necessarily eliminates all alternative futures. The possibility of remote travel, for example, into the wilderness of Alaska, becomes impossible, since it would interfere with the journey to France. Perhaps a trip to Alaska was a very unlikely event in your life, but now it is virtually an impossibility in your life. If you were scheduled to vacation in the Alaskan wilderness something will happen to negate such plans.

An evolving universe experiences a similar increase in organization and coordination due to Omega. When time first launches away from Alpha the immediate future isn't very specific compared to what becomes necessary in the later stages of cosmic evolution. Initially the potential pathways of time diverge away from Alpha into an expanding or widening parade of possible worlds. We must imagine here all the different galaxy formations that are possible in the vast expanse of space we see around us, realizing that what we observe is merely one of those. But when we begin to consider strange or even highly organized star and galaxy formations, lined up in rows for example, we are outside the boundaries of what is possible within the natural flow of time. Even in the first moment of time during the big bang, the attraction of Omega is a dominating force that eliminates the vast majority of possible configurations that aren't aligned probabilistically with the cosmic journey to zero.

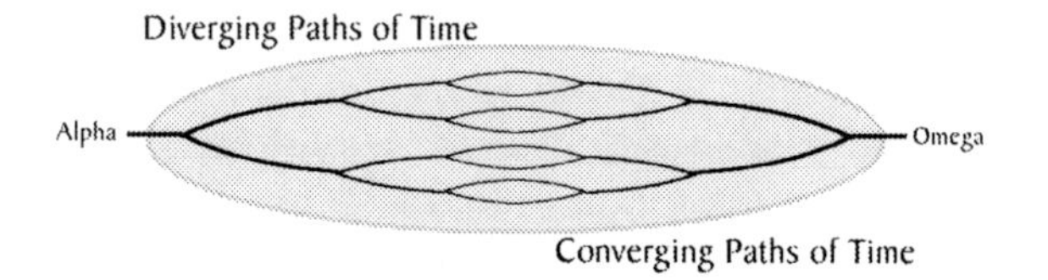

Figure 15.2. Time paths diverge away from Alpha and converge toward Omega.

In a sense the universe knows where it is going from the very first moment of time. So the matter distribution in stars and galaxies throughout the entire universe has to be within tolerance of that goal, within the degree of randomness and freedom of that particular juncture of time. Within tolerance obviously would not include being the lumpy extreme or the smooth extreme. Within tolerance means the structure of particles and atoms that govern the stars and galaxies and matter will eventually produce Omega. For example, there has to be an equal number of protons and electrons in the universe, so that near the end of time a balance between time and anti-time can ensue. All such planning and coordination is the long arm of the future reaching into the past.

Generally, we can see in the trip to Paris analogy that a specific future event being inevitable dramatically increases the probability of some method of travel leading you toward what we have defined as a predestined event. And as the date of the event approaches, making the journey to France becomes increasingly more likely. You might leave early and tour all over the European continent for work or for play, but the probability of each event is aligned with the precise future moment of noon at the month's end.

If the journey is delayed, and not instigated early on in the month, then faster modes of travel become ever more probable. A journey to Europe by boat for example is no longer an option. In fact as time passes, the width of the unique events is dramatically decreasing in your life. A large number of events are continuously becoming less likely and finally moving into the impossible realm. Yet the future moment of you standing on the tower is still not dictating a specific means of travel at a specific time. There are still many different pathways of getting to France, different airline flights for example, although the cloud of possibilities is ever more shrunken and focused.

Finally toward the end of the last week, events in your life become increasingly coordinated toward creating your one future. If events earlier in the final week had worked against your trip, later in the week events will seem increasingly synchronized and planned, as if everything is going your way, as if the whole universe wants you in Paris. If you don't have the money for the trip, it would be a good time to take up gambling. Your destiny is calling you. And the more the defiant or random events occur to resist or divert you away, the stronger forces grow toward aligning you back up again with your course to the tower.

If you still haven't yet left for Paris, at a specific moment, a certain number of hours before the tower event, all flights except those directly to France become impossible because other routes will no longer get you there in time. Absolutely powerful forces beyond your control now come into play. Some particular direct course across the Atlantic toward the tower now becomes absolutely necessary. You might be kidnapped or arrested and extradited to France, or your plane traveling elsewhere will be hijacked. And finally, at a specific point in time, the scenario of you not having left for Paris is no longer possible. Such worlds don't exist. In the final hours it is inevitable that you are on your way, because in the future you are already there, so failure is not an option. Resistance is futile.

Listen as the wind blows
From across the great divide
Voices trapped in yearning
Memories trapped in time
Sarah McLachlan

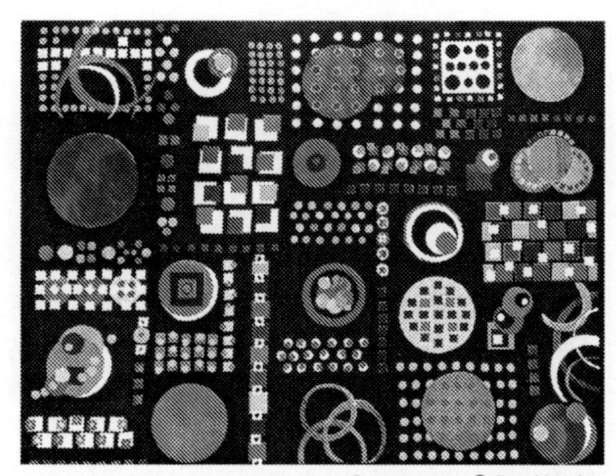
Convergence © Carol Taylor

Chapter Sixteen

The Big Bloom

The Reason Why Space-time is Systematic and Orderly

It has seemed evident for a long time that cosmological expansion was just the universe coasting outward as a result of the initial explosiveness of whatever originally caused the Big Bang. Expansion has been seen simply as a consequence of a past event, not a force. An educational page at NASA reads, "The shape of the universe is determined by a struggle between the momentum of expansion and the pull of gravity." It has also seemed apparent that gravitation was nature pulling back, resisting the initial explosion, as if to say "this wasn't supposed to happen!" Gravity seems to be nature wanting things returned back to the way they were before all this turbulence, before a bit of something managed to escape from an otherwise stable nothingness.

The brightest scientists intelligently expected the existence of a timeless perfect symmetry and they naturally looked for it in our past. The seeming vacuum before time was seen by them not as non-existence, but rather as an ultimate harmony or balance with no quantifiable properties, a physically real nothingness. In one form of this approach some theorize that gravitational energy is negative to the positive energy of matter, which from a God's eye perspective means the total energy and mass of the entire universe is equal to zero. It has seemed to some physicists that the existence of our universe apparently all comes down to a inexplicable flaw in the fabric of an un-quantifiable perfect symmetry which led to a broken symmetry. This tear in timelessness by chance produced the few arbitrary constants we needed and the known universe was born.

Now in this new century and new millennium as we peer more clearly into the future, scientists are being led toward a new vision, brought on by the discovery of accelerating expansion. New revolutionary ideas are surfacing as the unknown future is resolving into a distinctly visible future. Now it appears that cosmological expansion is not merely a consequence of the initial big bang explosion. All properties of the universe are not simply a consequence of some explosion or initial fracture to an ultimate perfect symmetry that once existed in the past. Which means the universe is not simply winding down from a past event. There are other sources of force and energy. There is a force emergent in the future. Space has an energy of its own.

The influence of the future was recognizable even in Boltzmann's view of possibilities, in the attraction of disorder. Under the second law, one is led to imagine there is this huge body of potential disorder and chaos out there pulling time forward. No one has ever proven a clear connection, but the second law clearly suggests expansion is a product of the alleged great probability for disorder and chaos. Expansion allows matter to break apart and then scatter and disperse. Without the expansion of space, there wouldn't be a second law. The dense singularity at the beginning of time would have just remained dense. Instead, with expansion now accelerating it appears there is a newly emerging force. Space has its own energy, its own force. So we wonder, how is the old expansion that was so rapid and powerful at the beginning of time which has now decreased to a standstill, related to this new expansion that is accelerating and growing increasingly more influential? How might they be the same force?

I believe we are slowly discovering that the future plays a major role in the big bang. The universe was never pushed outward toward zero, the universe has always been pulled by zero. As we discover the implications of an absolute zero future, the most dramatic shift in our understanding of time and the cosmos is the recognition that we are now being drawn into a decreasing measure of possibilities, rather than an ever enlarging body. In fact the direction of time has been facing an ever fewer number of choices for many billions of years.

The known universe is like a giant cosmic foundry. It is as if Omega has something in mind, a goal and begins with matter heated up in the kiln of Alpha. Then suddenly out comes the universe into the cold air, and as the matter cools it is guided into shape by the future. Slowly the hot plasma crystallizes into worlds, galaxies, all shaped by the great powers of grouping and symmetry, by the past and the future. This foundry is after all timeless. As the original potential of each time line is hammered out the future interestingly becomes itself, a sort of cosmic blooming that couldn't be any other way. What ultimately exists, the greater Universe, the 'infinite but bounded' whole, cannot be changed. But when the whole is understood, who cares. Compared to any other scenario, this innate guidance system creates a pretty amazing local universe.

Stage One: Divergence - The Increasing Possible Futures

In spacetime's evolution toward the balance of zero, there are two unique temporal periods of change. The first phase of spacetime; the initial burst of change, is the period of Divergence. Divergence marks the period when there is an increasing number of unique pathways for time to move in. When time accelerates away from Alpha, it faces a rapidly expanding number of unique futures. For example, using the trip to Paris analogy, a person in Los Angeles at the beginning of the month can take a plane and travel in any direction, north, south, east, or west, since every direction generally leads toward Paris. They might travel first to Asia or Africa. Initially the variety of possible futures is very wide. Divergence generally defines a period of the universe when what is uniquely possible overall is vastly on the increase. But cosmically speaking, eventually the expanding diversity in the direction of greater balance gets used up and begins to decrease.

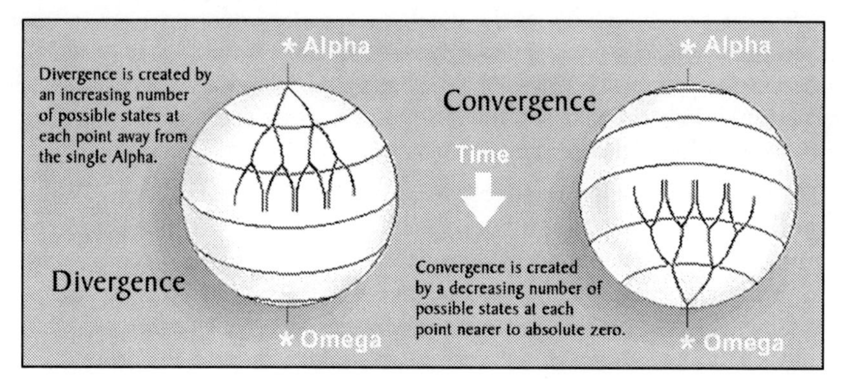

Figure 16.1: The upper half of the globe represents the way pathways of time diverge away from one single state; the north pole of time. The lower half of the globe reveals the rest of the story, explaining why time has a definite direction and a consistent nature. All space-time paths, after diverging away from one pole, are bent inward toward a common single inevitable future, the south pole of time.

Divergence ended long ago, possibly even during the early stages of the big bang, or it might have ended about six billion years ago. The outset of accelerating expansion, which was roughly seven or eight billion years after time began, may have marked the changeover between stages of divergence and convergence.

Stage Two: Convergence - Decreasing Possible Futures

If we imagine how many unique tomorrows there are in comparison to the one single present, then it seems like what is possible is an expansion of possible pathways for time to travel in, and not a decreasing set of pathways. But when we imagine the person from LA has one day left to get to Paris, we can see that most of the different ways of getting to Paris which were available early in the week are no longer a part of what is possible. The remaining unique routes to Paris still branch outward from a single specific present or location, but there is no time to drive across the states to New York and fly from that location. There is no time to take a boat through the Panama canal and cross the Atlantic. Only an airline flight will allow the person to meet their date with destiny. There may be many different airline flights leaving from the local airport, but as each hour passes there are fewer options. So the whole of what is possible is decreasing, even if there are still many unique pathways into the immediate future.

Pattern space, the infinite realm of possibilities, in that it is bounded by extremes in all directions, imposes severe limitations on the flow of time, limitations which establish a sensible reality. Just as spatial directions that travel away from the North Pole travel toward the South Pole, all time directions are guided by Alpha and Omega. This is visibly why physical reality itself is not chaotic. It is visibly why the universe is comprehensible. It is visibly why we are able to experience a sensible reality as we do. If possibilities were themselves unrestricted there would be no guidance system, so there would be no sensibility relative to physical reality. But since time has a goal, what was, what is, what will be, is meaningfully coordinated. Randomness and irregularity are kept in check, and the unfolding and enfolding universe remains systematic and organized.

The universe is not just a path of time, but an evolution of content. The evolution from one order to the other is likely the most important single piece of information we will ever know of the universe. Seeing both the divergence away from grouping order and the convergence toward symmetry order changes everything. A grand cosmic evolution makes the universe vibrant, purposeful, and alive. Time becomes a growth process, not a decaying or dying process. This knowledge awakens us to the universe and our own cosmic significance.

Figure 16.2: Divergence and Convergence forming the Many-Worlds Partition.

Of course a converging future returns us to the question, "infinity means what?" and all the different scenarios we discussed earlier. Objectively, scientists would imagine all the possible methods, and times, of how a person might travel to Paris as a cloud of potential, and beyond that probability cloud we can imagine seemingly normal events which are virtually impossible purely because the one future moment at the tower is destined to happen in a specific measure of time. A single destined future appears to draw a distinct line between events that are possible and events that are impossible even in the multiverse.

There is a distinct line being drawn here that not everyone will want to agree with, however, it is simply not enough to recognize certain events as highly improbable. If ZAT is acknowledged as the universal attractor for time, one has to conclude that there are for example no universes in which the big crunch scenario occurs. There are no parallel universes where gravity overcomes the expansion of the universe, where a whole universe ignores the attraction of the cosmic balance point and returns to Alpha by collapsing. Such a universe is so improbable that it doesn't physically exist in time.

Convergence means there are fewer possibilities available today than there were yesterday, but a single future moment being inevitable is both limiting and liberating. The collision between past and future does initially dramatically knock down the more radical set of possibilities, but it also can create complex and improbable events. For example, for a person who is a homebody and would otherwise not visit Europe, it causes exciting and unexpected events to happen in their life. This may seem to be increasing what is possible, but if we consider the entire range of events that are to some degree remotely possible in that person's life, no matter how slight the possibility, then we can see how a predestined future actually reduces the size of a much wider range of possibilities. Time will find a way to cure the person of their otherwise more probable tendency to stay home, leading the person out into the world of greater experience. But it will do so by eliminating their tendency to stay home, the option that would otherwise dominate the person. It will make things happen, it will cause a person to learn and grow if necessary, by eliminating all other options rather than by creating new options that weren't possible before. The options of a France trip were always there in the person's life, if remotely slight, they just become more probable due to a destiny, as other options such as those we would expect to be more ordinary are ruled out as being impossible.

In this way convergence is an enfolding process in-synch with the idea that the symmetry order of the universe is continually increasing. During convergence each time frame has an increasing multitude of possible histories. Many pasts are ending up in the same future. They are merging together and not diverging apart. This is most evident at the stage where all paths of time reach zero and space becomes flat. All universes and so all possible pasts have arrived at that one time frame. All universes, both the positive and the negative directions of time, have merged and enfolded into a single whole.

The Handshake between Past and Future

Both Alpha and Omega pull at the structure of the world. The location of the person whose future is inevitable, whether it be Los Angeles, or Beijing China, or Sydney Australia, is naturally a defining factor for the course of the future. Also the momentum of events, especially the more recent flow of time, a person's immediate past or history, influences the types of travel and times of travel. The personality, identity, health, wealth, and fortunes of each person we might imagine in the Paris scenario add all types of complex variables to what is uniquely possible and impossible, probable and improbable, for each person in each tapestry of time. A person may have a fear of flying that makes other modes of travel much more probable. They might be more likely to drive to the East coast and then boat across the Pacific, than fly in a plane.

In the same way that the Eiffel tower moment requires specific events to occur weeks before a person could ever arrive at the tower, the specifics of the past dictate what is possible and probable also. The deep and recent past, and the near and distant future, come together in a sort of collision to define the present. What we know as the forces of nature are part of the guidance system. Forces pull us backward toward the past, while greater forces pull us into the future, and the collision between past and future can be seen as creating all the refined motions of atoms, rhythmic motions of the planets, the falling rain, and waves crashing on an ocean shore.

There is one popular physicist who has for many years recognized the influence of the future within the physics of quantum mechanics. Not too surprising perhaps, there is already a well known interpretation of quantum mechanics, called the Transactional Interpretation developed by John Cramer, which recognizes the influence of the future. Cramer, a long time Professor of Physics at the University of Washington, columnist for Analog magazine, as well as science fiction author, is famous for developing one of the more interesting interpretations of quantum mechanics. Originally based upon what is called absorber theory originated by John Wheeler and Richard Feynman, and the time symmetric Lorentz-Dirac electrodynamics, John Cramer's Transactional model describes all quantum events as a "handshake" executed through an exchange between past (retarded waves) information and future (advanced waves) information, which causes both probabilities to collapse into the present. Cramer himself explains, "The absorber theory description, unconventional though it is, leads to exactly the same observations as conventional electrodynamics. But it differs in that there has been a two-way exchange, a "handshake" across space-time which led to the transfer of energy from emitter to absorber."

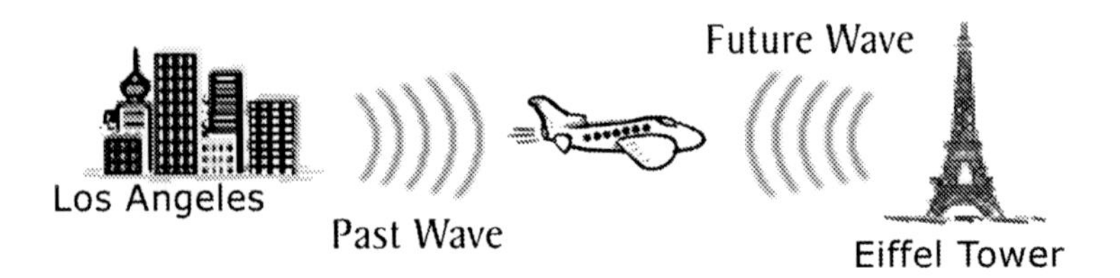

Figure 16.3: We normally assume the present is created by the past but actually both past and future create the present. Without the influence of the future most of nature's systemization and control of events would not exist.

Time does not simply roll into the future. From our perspective in the present, we recognize cause and effect in the larger motions of things but to everyone's amazement there is no such motion in the small world of particles. Instead there is a mesh of probabilities where possible pasts and futures meet which scientists correctly call non-local because those probabilities are produced literally from the whole of what is possible. As Cramer explains, "The transaction is explicitly non-local because the future is, in a limited way, affecting the past." Cramer writes: "When we stand in the dark and look at a star a hundred light years away, not only have the retarded light waves from the star been traveling for a hundred years to reach our eyes, but the advanced waves generated by absorption processes within our eyes have reached a hundred years into the past, completing the transaction that permitted the star to shine in our direction."

There is great power behind a model of understanding which recognizes the influence of the future. The river having a specific ending explains why, what scientists call the wave function of the universe, is so specific. If what is possible is thought of as coming from the past only, there is no reasonable explanation for the control of probabilities, such as the wave density of atomic particles. A river only from the past would be flowing outward into chaos. But if time is like a river that flows in between two lakes, from Lake Alpha to Lake Omega, then the universe has a natural guidance system. The universe has a goal.

Limitations

Naturally, life would be more probable in a coordinated temporal system, but we can even suspect that convergence encourages the existence of life. If we consider the implications of Omega as an enfolded totality of all being, we can conclude that Omega contains at very least everything that exists in time, all the many worlds of quantum theory, and so all the lives and minds throughout an infinite collection of universes.

Omega is a great oneness. In the past it may have seemed to the skeptic that people who imagined a God or a cosmic omniscience were being foolish if not just idealistic, but with all great seriousness, if we study and acknowledge the cosmological big picture honestly, or if we acknowledge the enfolding aspect of symmetry order, it appears there is really a place in nature definable as perfect or God-like. I think it would be unscientific at this point to deny that. This isn't to say any of our present human conceptions of this highest realm of existence are accurate or even do justice to a summing of all life and being. I can't stress that point enough. One might imagine from what we have discussed so far, that theology is like science, a developing part of our own growth and the growth of human conception. Hopefully understanding the two kinds of order will lead as much toward an appreciation of our finite form, as it does toward a new appreciation of a deeply profound future. With all wishful thinking set aside, it turns out there is really an ultimate purpose to the universe.

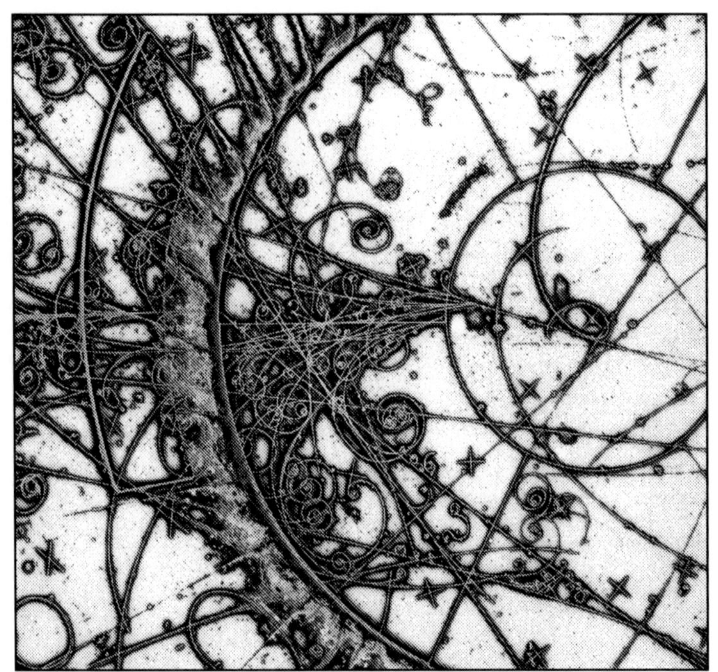

Particle Tracks in the Big European Bubble Chamber © CERN

We are all terminally ill. We suffer from a disease known to physicists as the second law of thermodynamics.

Christian Theodosis

~~~

More than any other time in history, mankind faces a crossroads. One path leads to despair and utter hopelessness. The other, to total extinction. Let us pray we have the wisdom to choose correctly.

Woody Allen

~~~

A cynic is not merely one who reads bitter lessons from the past, he is one who is prematurely disappointed in the future.

Sydney J. Harris

~~~

I arise in the morning torn between a desire to save the world and a desire to savor it. This makes it hard to plan the day.

E. B. White
~~~

A successful unification of quantum theory and relativity would necessarily be a theory of the universe as a whole. It would tell us, as Aristotle and Newton did before, what space and time are, what the cosmos is, what things are made of, and what kind of laws those things obey. Such a theory will bring about a radical shift - a revolution - in our understanding of what nature is. It must also have wide repercussions, and will likely bring about, or contribute to, a shift in our understanding of ourselves and our relationship to the rest of the universe.

Lee Smolin
Nobel Laureate

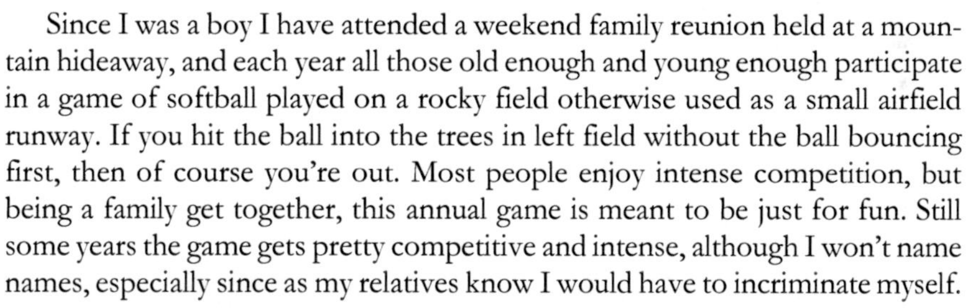

Part Five

The Second Law is Too Simple

Since I was a boy I have attended a weekend family reunion held at a mountain hideaway, and each year all those old enough and young enough participate in a game of softball played on a rocky field otherwise used as a small airfield runway. If you hit the ball into the trees in left field without the ball bouncing first, then of course you're out. Most people enjoy intense competition, but being a family get together, this annual game is meant to be just for fun. Still some years the game gets pretty competitive and intense, although I won't name names, especially since as my relatives know I would have to incriminate myself.

All human sporting events involve competition. Classically two sides, two groups, or two individuals, compete with one another. Most people assume it is the competition that we find attractive. But as I am sure others have noticed also, it is actually something else that makes sports enjoyable and beautiful to us. What we less consciously appreciate in sports is cooperation. I would even say it is actually the measure of cooperation within each team, or within an individual, that we enjoy observing.

In everything from professional baseball, to golf, to car racing, besides the intense competition, there is at times an almost immaculate form of cooperation, between team members, but also cooperation of one's own body and mind. There can be so much preparation and planning invested in each designed play. There can be so many defensive adjustments. Along with the intensity of the competition, the necessity of cooperation builds as winners play winners toward a championship. The rivalry and competition, the product of winner and loser, is really merely the means to increased levels of finely tuned cooperation. And we watch or take part in sports purely because of the psychological fruits of such cooperation.

The yearly softball game at my family reunion greatly accentuates the comradery of the whole weekend. It plays an important role in bringing an extended family together, many of which only see one another once a year at best, making all feel more like a single whole. Cooperation unifies a team, it unifies all the components of an individual, the physical and the mental, producing a combination of form and symmetry we find very attractive. The same combinations of form and symmetry attract us to art and music. In books, in writings, we appreciate the coming together of ideas or characters into a comprehension or plot that is complex in form and yet connected and logical.

In cosmology and science today we imagine the order of the past is competing with the disorder of the future, and most think disorder is winning. As with sports, in science we terribly under appreciate and under explain all the cooperation taking place in the cosmos simply because we don't understand why it is taking place. Today mainstream science is comfortable in assuming all the cooperation evident in galaxies, in the periodic table, in the emergence of life, in the gradation of intelligent life, they all exist as just odd fortunes of a nature that has no intent or purpose. Who would expect otherwise? Why would cooperation be built not simply into the nature we experience, but why would it be built into reality and existence?

Abstract Quilts by © Carol Taylor www.caroltaylorquilts.com

The very fact that the universe is creative and that the laws have permitted complex structures to emerge and develop to the point of consciousness – in other words, that the universe has organized its own self-awareness – is for me powerful evidence that there is "something going on" behind it all.

Paul Davies

Disorder may indeed be the negation of order, but this negation is then the implicit affirmation of the presence of the opposite order, which we shut our eyes to because it does not interest us, or which we evade by denying the second order in its turn-that is, at bottom, by re-establishing the first.

Henri Bergson

Chapter Seventeen

Away from Order toward Order

Seeing Balance and Imbalance rather than Order and Disorder

Deep within the efforts of human science the hope of finding reasons and making sense of the universe is still alive, but such hopes have taken a real beating in modern times. Our old philosophical faith in a master plan of nature, and the search for meaning through science, was severely crippled by the second law. It's not as if scientists viewed the glass as half empty. Somehow we came to accept the idea that the universe magically arose from nothing in an almost perfect state of order. By chance alone, with no purpose whatsoever, in complete defiance of the greater probability for disorder, a dense oscillation of matter and energy cleverly transformed itself into an array of complex atoms and galaxies and star systems. And this cosmos emerged and formed while simultaneously winding down toward only two prospective futures, both of which are dramatically short of any altruistic meaning.

In the most preferred version of the future the cosmos ends crushed and melted by its own gravitational collapse, drawn backward into the ultimate black hole and destroyed. At first this big crunch was expected to oscillate, expansion followed by collapse followed by expansion, except then we were told that each new event harbors less initial energy as it rebounds, so eventually all temporal activity winds down, and finally physical existence ends fruitless.

Absolutely no one preferred the other possible future, the Heat Death scenario, where the universe expands at a decreasing rate. Once known to be our more likely future, it was brutal to everyone who learned it. In this model the universe just continues to expand, becoming ever thinner, becoming boring and lifeless; never collapsing, never quite ending, just an ever weakening measure of expansion, complimented by ever diffused gravity and increasing cold. Eventually all the stars burn out and the universe goes dark. Yet this final stage of near-death was believed to continue indefinitely, eternally, never ending. All traces of life would be erased in this frozen wasteland, and yet there would never again be even a final nothingness, no peaceful finality, just an absolute death of purpose and meaning.

This age, this twentieth century paid a heavy price for knowing so much about the past and so little about the future. Science was meant to free us and enlighten us to the true wondrous universal truth and beauty of nature. Instead we seemed to discover that no matter what was achieved in the time that we have, there could be no ultimate meaning or purpose to our existence. In both possible futures the complexity and beauty of nature was washed away in the sands of time. We were just a fluctuation in the void. That was the message and we all felt it even if we didn't accept it without question in respect for the miracle that is science.

Scientists themselves weren't immune to the meaninglessness of such theories, nor was the populace that stood listening in the background. Such despairing prospects were engraved into our left brains and were continually reinforced by Boltzmann's version of the big picture, which ridiculed any attempt of a deeper comprehension of cosmic evolution. We might have all been told horror stories. Hell could hardly have been worse than either imagined future. It seemed there was hardly any meaning even to the present. And yet it was all wrong. It was all wrong. It's almost like waking up from a bad dream.

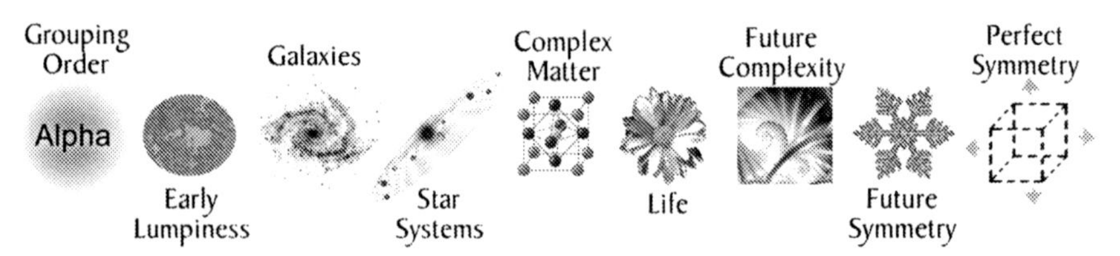

Figure 17.1 If we look at all that we know of the universe from a top-down perspective, if we consider all the order that we observe, and how a simple kind of order has evolved into a systematic and complex type of order, why do we imagine that the universe is running down, generally moving toward disorder or chaos, and in the process of dying?

Having understood two orders for a long while, I recognize that historically, not having recognized the two types of order and instead our seeing the world as ordered or disordered will eventually be seen as the most debilitating paradigm of human history. We have been living in the dark ages. Many people, perhaps the majority, have turned away from science because they feel it lacks a depth of vision and heart. It is no secret in science and it is worth noting here that Ludwig Boltzmann himself committed suicide, and although there were certainly many factors in Boltzmann's mental health and the despair that took his life, we know he must certainly have made the conclusion from his own theories that the universe was ultimately meaningless. Many scientists have mistakenly made that same conclusion. The Nobel Laureate Steven Weinberg in this now infamous passage from his book *The First Three Minutes* writes:

> It is almost irresistible for humans to believe that we have some special relation to the universe, that human life is not just a more-or-less farcical outcome of a chain

of accidents reaching back to the first three minutes, but that we were somehow built in from the beginning.... It is hard to realize that this all [life on Earth] is just a tiny part of an overwhelmingly hostile universe. It is even harder to realize that this present universe has evolved from an unspeakably unfamiliar early condition, and faces a future extinction of endless cold or intolerable heat. The more the universe seems comprehensible, the more it also seems pointless.

The source of this black cloud hanging over science can be traced directly to the second law of thermodynamics. I have always felt that critical thinking plays an important role in science but have never understood or appreciated skepticism, and in my mind the second law is the ultimate example of a skeptical bias present in the religiousness of science. In hindsight it seems absurd to me that the second law received so little scrutiny, and that the boundary of absolute zero was ignored for so long, although it is easy to be astonished at how people have viewed the universe in the past. In any case, so little thought given to Boltzmann's version of the second law has tainted our vision of everything else learned about the cosmos.

With zero added, the realm of possibilities completely transforms, it is made sensible, with clear and distinct boundaries which reveal that the universe is not decaying into endless disorder and chaos. In seeing the evolution from grouping to symmetry order the universe isn't thought to be meaningless at all. The future is beautiful and meaningful beyond imagination. In hindsight we will weigh the harmful impact the second law has had on humanity, with its sense of emptiness and futility, and shake our heads in dismay. Young minds looking to science for answers have been either profoundly disappointed in the news or wisely dissatisfied in general with science. Fortunately, there is reason for change in the coming years. One of two parts of the second law is absolutely wrong.

Two Theories in One Law

It is important to understand clearly that the second law includes two related but different theories. One of those theories is almost unquestionably correct. We will call it theory number one, which states that there is always an overall deterioration of usable energy (increasing entropy) in the cosmos. The total amount of stored up or concentrated energy is always decreasing. In the terms of two orders, the dense "grouping" of energy evident in the past is decreasing with time, so the overall cosmos is slowly running out of condensed fuel. This part of the second law will almost certainly never be overturned.

This loss of usable energy does not mean energy itself is ever destroyed. According to the first law, the amount of energy in the universe never increases or decreases. When most people learn that energy is never destroyed, which is contrary to what most expect, they immediately wonder, what then is changing with the passage of time? In what way is the universe transforming or evolving? What makes the future different from the past? Theory number one explains that energy is lost or spent whenever it is dispersed out into the larger environment, like the smoke and heat of a wood fire, or the light and heat given off by the sun. When burned in a fireplace, the energy stored in wood isn't destroyed but merely spreads out into a larger space. This diffusing of energy is what heats a cold room. But it's also why the room cools when the fire goes out. The heat invariably passes into and through the walls of the room and then escapes out into the larger environment. When energy spreads out in the environment it becomes difficult or impossible to ever regroup and re-use that same energy. It becomes difficult to collect the particles or the energy back together again into an ordered group. In fact attempting to regroup the lost energy would expend far more energy than what can be gained.

The reason for this loss of usable energy (entropy) is simple. Energy is a potentiality; it is a probability for something to happen in time. In the first moment of time the universe was extremely hot, it was full of energy potential. All that energy is created by the great probability for time to move toward balance. Alpha represents the greatest possible imbalance in nature, so it is also the most energetic state in nature. As time moves ever nearer toward balance that probability, that energy, decreases, as it is used up by the evolution of time. Once time reaches zero, all the energy causing change is depleted because at that time the cosmos and the Universe exist in a state of perfect balance and symmetry. There is no longer any energy because energy is a product of imbalance.

So there is a continual overall loss of energy in the cosmos. This much of what the second law claims, all of theory number one, is correct. We are all familiar with our everyday experience of theory number one. Things grow old. They rust or break down. We constantly have to invest energy into maintaining the temperature of our living environment. We constantly use up our stored energy. What is the reason for these physical facts? Is there a reason why nature works this way? Why can't available energies naturally increase?

The second theory forming the second law, theory number two, attempts to explain the why of the first theory, by stating the reason for entropy. The reason usable energy is always decreasing is because there are fewer patterns where things are grouped together, and so ordered, than patterns where things are spread out, and so disordered. This is Boltzmann's statistical addition to the second law and I have explained this thoroughly but it has to be restated here.

In essence, the reason why entropy increases is presently said to be because the realm of possibilities (configurations) includes fewer patterns where energy or particles are in a concentrated usable form, than there are patterns where energy or particles are not as grouped or in usable form. So the second law ties the thermal loss of usable energy together with the loss of physical order, stating the universe in general is evolving from order to disorder because ultimately there are so many more disordered patterns for nature to choose from, which make the potential for disorder greater.

Brig Klyce also divides the second law into two theories which he describes as thermodynamic entropy and logical entropy. Klyce, the author of *Cosmic Ancestry*, maintains an excellent article on the world wide web that identifies many others who have recognized two distinct theories in the second law, most notably he includes Richard Feynman, noting that Feynman discusses the loss of usable energy in one chapter of his venerated *Lectures on Physics* published in 1963, then discusses the configurations of logical entropy in a completely different chapter. Interestingly in the later chapter Feynman writes:

> Suppose we divide [a reference frame] into little volume elements [like a checkerboard]. If we have black and white molecules, how many ways could we distribute them among the volume elements so that white is on one side and black is on the other? On the other hand, how many ways could we distribute them with no restriction on which goes where? Clearly, there are many more ways to arrange them in the latter case. We measure "disorder" by the number of ways that the insides can be arranged, so that from the outside it looks the same. The logarithm of that number of ways is the entropy. The number of ways in the separated case is less, so the entropy is less, or the "disorder" is less.

I do find it quite interesting that Feynman chose to use molecules designated as black and white to represent a state of order, when he only needed to portray a single group to explain Boltzmann's version of the second law.

In a book, *The Mystery of Life's Origin* by Charles Thaxton, Walter Bradley, and Roger Olsen, the same two parts of the second law are referred to as thermal entropy and configurational entropy, which are the terms I myself would prefer, thermal in reference to the increasing entropy of theory number one, and configurational in reference to the reason why entropy increases, seemingly explained by theory number two.

If we were to define a separate law-like statement for theory number two, that part would state that the overall amount of order in the universe never increases. Klyce in his article states similarly that theory number two would simply state, "things never organize themselves". Would that law be true? In the following paragraph Klyce points out a few examples of things that never self-organize before explaining that there clearly are ways that things do organize themselves:

> The rule that things never organize themselves is upheld in our everyday experience. Without someone to fix it, a broken glass never mends. Without maintenance, a house deteriorates. Without management, a business fails. Without new software, a computer never acquires new capabilities. Never....But this ignores the plain facts about life and evolution. Life is organization. From prokaryotic cells, eukaryotic cells, tissues, and organs, to plants and animals, families, communities, ecosystems, and living planets, life is organization, at every scale. The evolution of life is the increase of biological organization, if it is anything. Clearly, if life originates and makes evolutionary progress without organizing input from outside,

> then something has organized itself. Logical entropy in a closed system has decreased. This is the violation that people are getting at, when they mistakenly say that life violates the second law of thermodynamics. This violation, the decrease of logical entropy in a closed system, must happen continually in the neo-Darwinian account of evolutionary progress. Most Darwinists just ignore this staggering problem. When confronted with it, they seek refuge in the confusion between the two kinds of entropy.

Unless we assume some form of creation, the cosmos is arguably one grand example of self-organization, in everything from the selection of constants and forces of nature to the choice of particle and chemical structure, to galactic spirals and solar systems, the rate of cosmological expansion, and particularly the origin and evolution of biological organisms. The universe is a big parade of self organization. Here again that skeptical bias mentioned has probably been influential. When someone doesn't acknowledge the miraculous conditions of the cosmos, and instead attributes all the systemization and orderliness that we observe all to thermal energy, as if we should expect energy has some innate potential to create orderliness, it stinks of a kind of repression of the profound for no other reason than to avoid the philosophical implications.

Theory number two can be made to seem reasonable enough, the seeming sensibility of the argument has baffled us for years, but it still doesn't deserve a free ride next to the certainty of theory number one, and a more careful study of pattern space shows it to be absolutely wrong. There is a trend in nature, an organized flow to the cosmos. Patterns are transforming in a very specific way. But the real trend is for increasing balance, not disorder. We can state as law that the overall amount of usable energy never increases. We can state as a law that the overall density of the cosmos never increases. And we can state as a law that the overall measure of grouping order in the cosmos never increases. But we cannot say the general measure of order, or the overall measure of orderliness, never increases. Quite the contrary, if we evaluate grouping and symmetry orders carefully we end up realizing that Omega is the true supreme state of order. And thus orderliness has increased since the beginning of time. It isn't that the order which the two Alphas represent is inferior or even purely derivative of Omega. Form is every bit as primary as the infinite whole. It is just that wholeness, how should I say this, beats two of a kind. Symmetry order is a higher form of order than grouping order. If I use hands in poker as an analogy, a royal flush is the ultimate hand in poker because it is a straight, a flush, and the highest cards in the deck, all combined in one hand. Cosmologically speaking, Omega is a royal flush.

The theory that time moves away from order toward disorder has seemed viable because it is a keyhole view of what is really taking place. It is a fraction of the truth. I would call it a half truth but it isn't really even half of the story. As mentioned before, the problem all along with the second law's representation of all possible states is that zero is missing from the equation. The actual shape of pattern space is decided by the simple fact that if there is an absolute zero in physics which we certainly know to be the case, then there can't always be more disorder than order. Since the underlying principle (the fraction of truth) of theory number two is that change moves toward whatever balance exists in the whole of possibilities (time would move toward disorder only if it were trying to find balance), if such a balance actually exists then time will find it. In Boltzmann's ideology, eventually the set of ordered states left behind will attract time in equal measure to the remaining states of disorder up ahead in the future.

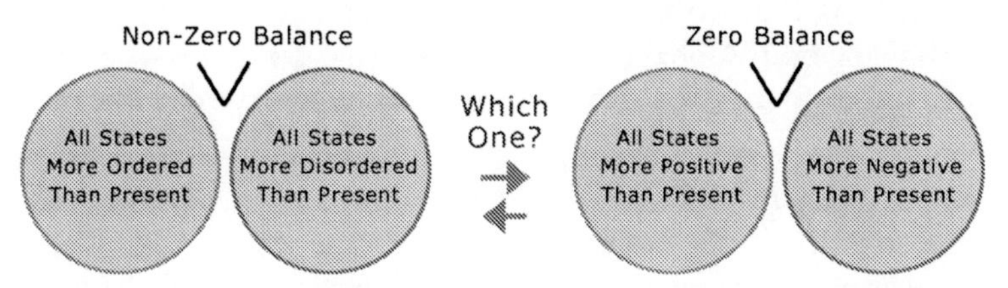

Figure 17.2 Which model of states correctly describes the great divide of reality, the asymmetric model of Boltzmann or a model where zero is the universal balance of all?

Is time moving toward the balance between order and disorder, or the balance between positive and negative? Once we acknowledge zero as a boundary state we recognize time must be moving toward balance. And once we acknowledge time is moving directly toward zero we know precisely where the universal balance exists.

What has been overlooked in the past is that decreasing usable energy (theory number one) can also be explained by a trend toward balance, actually much more effectively (and more meaningfully) than a trend toward disorder (theory number two). The second law today incorrectly correlates entropy with increasing disorder, when the actual correlation exists with increasing balance and symmetry. The direction of time travels away from the most extreme imbalance possible, away from grouping order, and moves toward the order of balance and symmetry. In fact once we correct theory number two by recognizing zero as the great attractor of time, then pattern space explains a great deal more about the cosmos than just the reason why of theory number one.

Integrating Two Orders and the SOAPS Model

In now accurately seeing the whole of possibilities we not only recognize that there is a previously unrecognized order in our future, we can also understand more clearly the order of our past, and thus the whole evolution of order. In the life span of our own cosmos we can recognize a neutral future state and a positive past state, so we know there must also be a negative state somewhere. We know the extreme of grouping order creates two opposing sides, a positive and a negative side. We can realize now how that division of everything into two groups is not the supreme order where all the pieces of the puzzle fit together. There is a perfect state, but it exists in our future, not our past. The order in our past is only half of the puzzle; merely half of a larger duality; one side of the great divide; one side of the coin.

Our Alpha is one player of checkers as the game begins, when all the pieces of one color are divided apart from the pieces of the other color. Our Alpha is half of the most extreme case of grouping order, while together both Alphas represent the highest possible measure of imbalance and contrast. The following diagram shows how the two kinds of order integrate with the soaps model, necessarily showing both Alphas split apart existing at opposite ends of the full spectrum, even though the two Alphas might better be represented side by side.

Time travels from positive to neutral, while anti-time travels from negative to neutral, however, keep in mind that time and anti-time are virtually identical. We are in a sense in both worlds simultaneously, but to appear sensible I have to speak as if we are only on one side, which we might designate as the positive side. Still we are only assuming our side to be the positive designation. The inhabitants of each universe consider their own system to be the positive side, and consider their native universe to be made of ordinary matter while the other side is made of inverse-matter. Amusingly not only is it impossible to say which universe we are actually in, we may even be alternating back and forth, with no way of telling when we are in one or the other. Obviously it is a challenge just to discuss the inverse universe, but I will now try to explain how I believe the two worlds relate to one another.

Inverse Matter

The physicist Paul Dirac predicted the existence of anti-matter in 1928 and it was discovered experimentally in 1932. Richard Feynman viewed antimatter as ordinary matter traveling backward in time. I believe there are two types of anti-matter. If matter is evolving forward in time toward Omega, then the anti-matter we observe or create here within our direction of time is matter evolving backward in time (rather than inversely in time) in the direction of our own positive Alpha. This would explain why anti-matter is unstable on our side of the cosmos. However, a stable kind of anti-matter exists on the other side of our cosmos in anti-time, which is traveling inversely in time rather than backwards, away from the negative Alpha toward the universal Omega. A direction of anti-time has been predicted in several other scientific works, and anti-time has also been speculated upon by science fiction authors, most notably, having watched every episode ever created, is Gene Roddenberry's portrayal of a technologically advanced science in Star Trek.

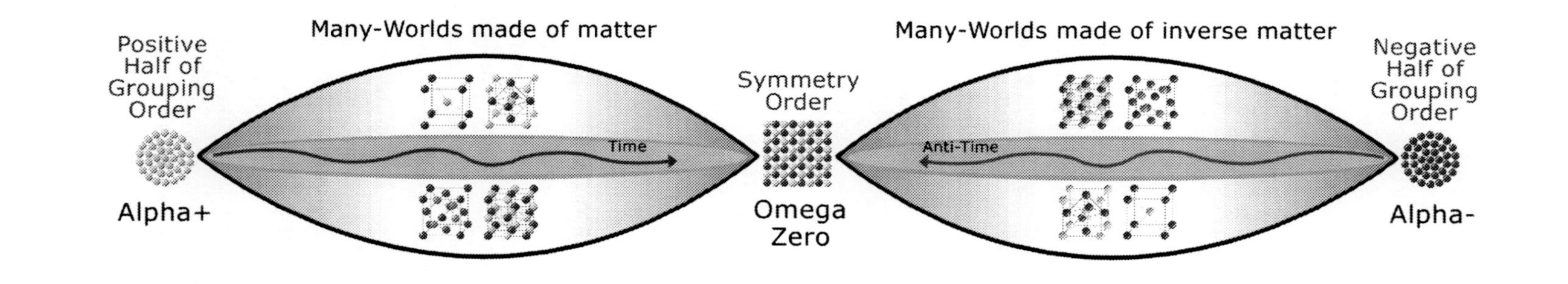

Figure 17.3 The integration of Grouping and Symmetry order with the Symmetric model of possible states reveals that the Alpha state in our past is actually one of two parts. Like the invisible dark side of the moon, the many-worlds of anti-matter are as ordinary and as real as the side of physical reality we know. Mathematicians in the future may one day be able to more precisely represent the large-scale shape of pattern space, as distinctly as we now represent the small-scale probability density of atomic orbitals.

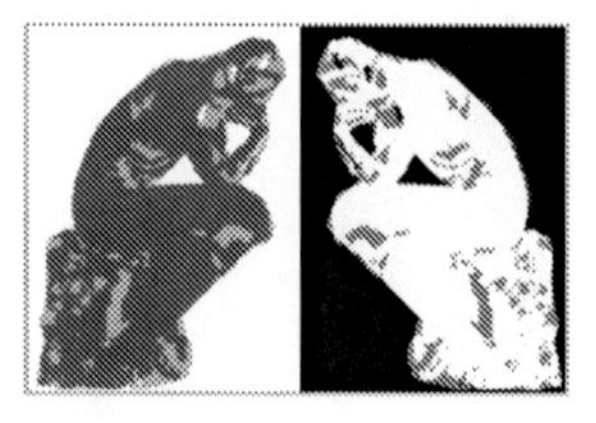

If time is a direction in space, then the space that we exist within is special. It isn't a three dimensional space in which things move about, it is instead what Einstein called space-time. My own belief is that space-time is a product of special "fourth-dimensional directions" in space that travel through a series of many three dimensional block-like spaces, jumping from one to the next. This creates a new volume of space which is purely dependent upon the fourth dimensional directions.

Exterior to the three dimensional spaces, the fourth dimensional volume that we exist within changes and transforms, it evolves because the majority of the spatial directions are moving away from Alpha toward Omega, due to the probability of balance. This special volume is what collapses in the post big bang phase of accelerating expansion (Big Rip) even though the actual universe is expanding. This dominant probability makes our volume (our space) positive in reference to the volume of anti-time which is negative. Anti-time is a kind of negative space, or negative space-time, which is properly referred to as negative volume.

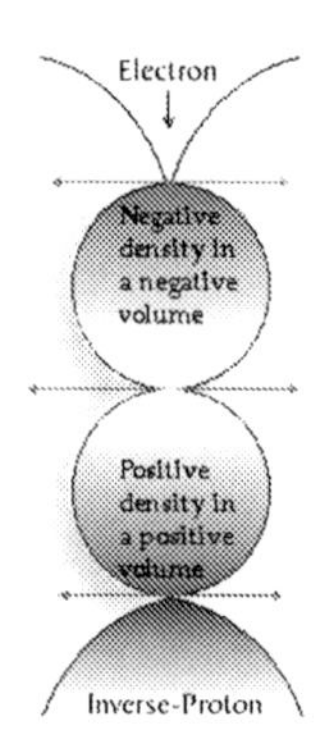

As mentioned, inverse matter can only exist in the negative volume of anti-time. As an example, we cannot observe the negative density of an inverse-proton in our positive space, since our positive volume would collapse before the inverse proton could exist spatially extended, that is, on our side of the dividing line between worlds. This is why the negative electron is just a point particle. Our positive volume collapses before the negative density that gives the electron a definite mass can exist spatially extended. If that isn't enough of a mind twister, I strongly suspect the electron is one side of a inverse-proton in anti-time, which means the back side of our protons are inverse electrons, and our electrons are the back side of someone else's protons.

We cannot observe negative density, it simply cannot exist in a positive volume, but we can certainly detect it, in fact what I know with great certainty is that behind the scenes the whole progression of time is driven by a slow cosmic influx of negative density leaking into our positive volume from the other side, a slow merging of the two sides is occurring, which is the only way for the two opposite sides to move toward balance. They must slowly and gradually unite, which from our perspective causes the cosmos to grow larger and expand.

In the first step away from Alpha, time is stepping away from the uniformity and sameness of a perfect positive state and is necessarily moving into the beginning of variety and diversity. Time has the option of moving in so many directions, toward so many different futures. However, initially there are only two basic ways that the negative side can intrude on the positive side. It can invisibly embrace Alpha on the macro-scale, intruding on the whole of Alpha, therein slowly neutralizing both sides, which causes a smooth overall expansion slowed only by a smooth overall gravity. Or the negative can visibly invade Alpha by tunneling into its positive space, either at large scales creating the fluctuations we observe today in the background radiation, or at the micro-scale creating negatively charged point particles.

The Cosmic Plans

In order to control the cosmos, in the necessity to bring itself about, Omega must rigidly regulate the early cosmos even in its first moments. As if planning to bring about complexity and life, the early fluctuations in expansion are kept moderate, the masses of electrons are all kept equal, and the overall ratio of electrons to protons is inevitably kept perfectly even. With the precision of a watchmaker, the inevitable future carefully designs the stable particles with properties that will eventually produce the table of elements, the molecules, even DNA.

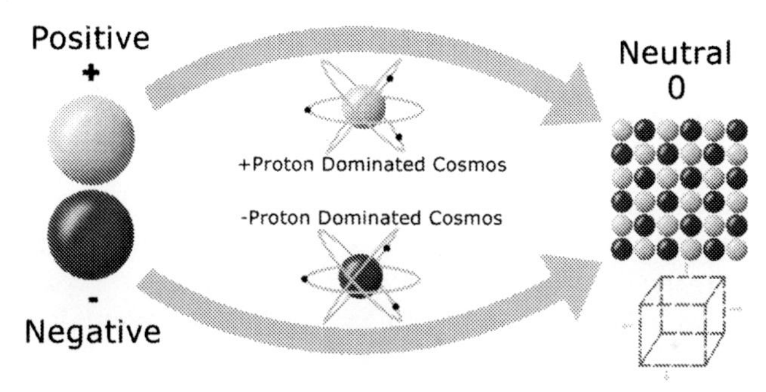

Figure 17.4. Our universe and all of time can be understood to be a grand evolution from one order to the other.

The two Alpha halves are inseparably linked together presently and during their entire evolution. They are interdependent and essentially part of the same cosmological system. In simply referring to a cosmos we are referring to the larger symmetry of both. We can imagine our own path of time launching from the positive side synchronized with an identical cosmos launching from the negative side. In that moment, the perfect symmetry of our fully positive Alpha is broken due to an influx of negative density, while inversely our positive side invades the negative Alpha. The two time paths are merging together toward becoming one, and consequently depleting their probabilistic energies, not canceling, but combining together into something measurably greater than the two individual halves.

The birth of particles in the first seconds of the big bang event is unquestionably the most crucial stage in the formation of physical reality as we know it. It is the stage at which intelligent design seems most evident and necessary.

Among all the myriad of conceivabilities, why the proton and electron? Why just two stable fundamental particles? Why the marriage of the two, considering how different they are? Why are they matched with equal charges? The selection of those two fundamental particles among all other conceivable scenarios marks the most critical point of influence by the future. These two particles are not arbitrary selections. They are forged with a specific goal in mind.

All the Pieces of the Puzzle

We know intuitively that what we think of as a material universe is really an intricate interplay of imbalances. All that we know are positives and negatives, protons and electrons. And following time backward only increases imbalance. In seeing that time originates fully positive, and slowly becomes less positive, until it becomes neutral, we realize there actually should be great imbalances evident in the world around us. There should be even greater imbalances evident in our past. The proton should be heavier than the electron. We shouldn't observe an equal number of anti-matter galaxies. We shouldn't expect there to have been equal amounts of matter and anti-matter near the beginning of time.

Anyone can express doubt about the actual existence of a parallel universe that would balance out what we observe. Anyone can express doubt toward the infinity of worlds possible under the same laws of nature, of infinite many copies of oneself, each only slightly different than the next, but once we work beyond our surprise that we ourselves exist, there is nothing left to be surprised about, since it is only the existence of anything at all that makes everything seem improbable. What we imagine to be ordinary and natural is intimately related to what is most probable or what we expect. The more an event seems improbable or strange the more it also seems unnatural. And of course what seems most *natural* is balance. Balance just seems to be the way things should be, at least ultimately. In fact an ultimate state of balance seems so logical, so natural, so inevitable to us that all the beautifully complex features of life on Earth, all seem supernatural in comparison to the uniformity of a zero balance, even though they are all inevitably parts of an ultimate balance.

The question that has haunted us for so long, "why is there something rather than nothing?" is deeply related to the question, why is there asymmetry? Why are there imbalances? We are learning here perhaps more lucidly than it has ever been known before that form and material things, substance itself, are products of the imbalances that exist within the cosmic balance. Everything we know is a great tapestry of imbalance. And when we ask the question *why are there imbalances?* instead of *why is there somethingness?* then it seems quite simple and unavoidable that imbalances exist. All possible imbalances naturally exist inside any overall balance. Our finite world could only be an imbalanced system.

We exist because a balanced whole exists which we incorrectly see as nothing, and worse still confuse with nonexistence, instead of knowing it as the timeless everything or one thing. The universe isn't moving toward disorder, it isn't merely moving away from an ordered past. Our future is actually the state which Stephen Hawking refers to as the one pattern where all the pieces of the puzzle fit together. As unbelievable as something that wonderful can seem, it is true, it is even good hard science, so eventually, as the implications of symmetry order are understood by everyone it is going to have a considerable impact on both science and society. I hope scientists take a long look at past and present skepticism now that we are breaking through to see such a deeply meaningful cosmos and Universe. Maybe some introspection by disciplined scientists will encourage the more emotionally intuitional and socially minded leaders in religion to take a look at themselves and be more concerned with dogma. We should all feel the importance of being accurate in what is portrayed of the really big picture. The universe is moving towards the most ordered state in all of nature, a quantum superposition of all universes and all life. There is no longer a need be fear what we might find. We need not any longer be concerned about a disappointing scientific truth. Maybe we can all be more aware of what we can accomplish together as a whole, in cooperation with one another. There is such great potential in the balance between science and spirituality. We can all appreciate a future convergence of all time and life into a single great wholeness. Understanding both our individual distinctiveness and our unity with one another and the world, even our connection to the timeless whole, is undoubtedly an essential ingredient of our future survival on this particular planet.

That space has so many dimensions and the dynamics is so unconstraining that after any deviation we should surely never expect to get back to where we would have been.

Matthew J. Donald

~~~

Most physicists feel that the ultimate theory should have no input parameters, no fundamental dimensionless constants, and that all the masses of quarks, and all coupling strengths, should be predicted by such a master theory.

Heinz R. Pagels

~~~

I would like to state a theorem which at present cannot be based upon anything more than upon a faith in the simplicity, i.e., intelligibility of nature: there are no arbitrary constants; that is to say, nature is so constituted that it is possible logically to lay down such strongly determined laws that within these laws only rationally completely determined constants occur.

Albert Einstein

~~~

Imagine you can play God and fiddle with the settings of the great cosmic machine. Turn this knob and make electrons a bit heavier; twiddle that one and make gravitation a trifle weaker. What would be the effect? The universe would look very different so different, in fact, that there wouldn't be anyone around to see the result, because the existence of life depends rather critically on the actual settings that Mother Nature selected.

Paul Davies *A Brief History of the Multiverse*

~~~

...for each mass there is a gravitational limit at which the opposing forces are equal. Within this limit the gravitational force exceeds that of the space-time progression and the net movement is inward. Beyond the limit these relations are reversed and the net movement is outward.

Dewey Larson

~~~

...the world was made, not in time, but simultaneously with time. For that which is made in time is made both after and before some time - after that which is past, before that which is future. But none could then be past, for there was no creature, by whose movements its duration could be measured. But simultaneously with time the world was made.

St. Augustine

~~~

All great advances in science have by definition the effect of reducing the prestige of the "experts" in the field in which the advance is made.

Frank Tipler

What is time? If nobody asks me, I know; but if I were desirous to explain it to one that should ask me, plainly I know not.

Saint Augustine

Photo: Koi © Linda Bucklin

Chapter Eighteen

Multiple Arrows of Time

The Many Directions of Time Travel

Many physicists abruptly shy away from the idea of intelligent design but they have no scientific reason to reject that option, because beyond the idea that we experience this universe because all imaginable universes exist, a reasonable solution to why the complexity of the universe has come to be, or why the universe is uniquely this way, has never been found. Specifically we have no idea why there are forces of nature.

The reason we do not yet understand the forces of nature is because we haven't yet realized there aren't actually any forces of nature. There really aren't any initial conditions. There isn't even a single direction for time which the whole universe follows. The world around us results of a free flow of time. The forward direction of time is dominant, but the general flow of time is constantly moving both backward and forward, and also time moves in various directions at right angles to the past and future. Time is flowing in all directions simultaneously. We actually feel the freedom of time pulling us around. We call the various directions of freedom that govern time *the forces of nature.*

What makes the soaps model so compelling is how it becomes easy to understand why there are forces of nature and what causes them. Where Boltzmann's approach was correct enough to provide a simple explanation for the general direction of time's arrow, an improved picture of what is ultimately possible provides clear and simple 'reasons why' for gravity, electromagnetism, the strong force and even the weak force. The forces of nature and the various directions that time can travel in are really the same thing.

All forces are probabilities. In the very same way that certain events in our lives are possible and impossible, probable and improbable, the four forces of nature that govern the physics of the universe are themselves probabilities. The forces of nature are simply the most predictable events we experience. For example, the chance that gravity will hold you firmly to the surface of the Earth is one of the more predictable events in your life. Gravity is so dependable that we tend to categorize it as a constant of nature, but gravity, like all three other forces, is just a large group of possibilities attracting the present. In fact, gravity is the probability for time to travel backwards.

Gravity is trying to recreate the past. You've probably have never heard such a statement before but once you think about it, the idea that gravity is trying to recreate the past is nothing but common sense. When noticed it seems self-evident. In an expanding universe the past is increasingly denser; less expanded, and finally becomes an infinitely dense point. Gravity pulls all matter together. So obviously gravity is at least trying to recreate the past. In fact the reach of gravity is infinite, so it not only tries to recreate the past locally, it is trying to pull the whole universe back together as it was in the distant past. Essentially gravity is in a battle with the expansion of the universe. If expansion ever became the weaker of the two forces then gravity would successfully recreate past-like conditions by collapsing the universe in on itself.

We can easily identify the portion of possible states that pull at our universe, trying to recreate the past. They are all the states which are more (positively) dense than our present. All the states between Alpha and our present form a group of states which are denser than our present. Each state is a possibility and collectively those states form a strong probability which pulls at the conditions of the present.

Gravity is a force produced by the set of all the possibles on the Alpha side of the present, in opposition to all the possibles on the Omega side of the present. The present is of course the natural dividing line between those two sets.

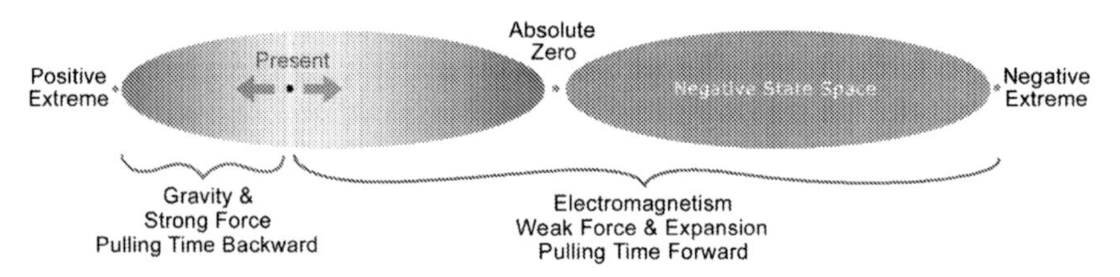

Figure 18.1: Probability arrows reflect the pull of the past and future which manifest as the four forces of nature and cosmological expansion. All the dense areas in the universe are the result of retarded time, time moving backward or not moving forward. Note that this present shown is much earlier than our present location in state space.

Gravity is time moving backwards. Taking this a step further, simply saying the same thing in another way, gravity is time in reverse. We can recognize that anyplace where gravitation is successful in increasing the density of the universe is a case of time or conditions moving backward in time. If you have a really strong desire to travel backward in time, just visit the sun. The sun is an example of how the whole universe used to be billions of years ago. In a very real sense the sun is still in a very retarded state compared to most of the universe. Even the gravity holding us to the Earth is time moving backwards. Again this is something simple and self evident. A necessary portion of the universe must travel backward to accommodate the possibilities of past-like conditions. Areas of the universe which retreat in time of course become denser and areas that are advanced in time become less dense or expanded. In hindsight it is actually very surprising that gravity has not been imagined to be the influence of past-like states since the influence of such states are also evident in Boltzmann's way of modeling all possible states.

Expansion is a force from the future. The strongest probabilistic trend of nature is not at all toward disorder, as physicists presently claim, the overall trend is for balance to increase. The most fundamental force of nature is simply the tendency for all things to balance out. And in being drawn toward the balance of zero the universe invariably expands. There are other forces pulling time elsewhere. In every direction that possibilities exist, there is a pull. But extremes balance out with the opposite extreme, and balance always ends up the winner, which is why gravity is losing the cosmic battle against expansion. Gravity is a force from the past and expansion is a force from the future. The past-like set of states which pull time backward is always smaller in comparison to the future-like states pulling time forward, at least until the two sets finally reach an equilibrium at Omega.

The expansion of the universe is time moving forward. Cosmological expansion isn't a product or consequence of some chance explosion in the past. It is a force just like gravity. Just as we can describe gravity as time moving backward, expansion is time moving forward. The why of gravity can be understood in a simple way and so also can expansion be understood in a simple way. Although the set of states producing gravity is very strong on the cosmic scale, more of the universe is moving forward in time toward balance than backward toward imbalance, so a greater portion of the universe is presently expanding and cooling, moving us slowly more forward into the future than backward into the past.

Time is not moving purely in one direction. Any gravitationally contracting area of the universe, such as a star, is an example of a group of time directions moving backward toward Alpha, while the large expanding regions of the visible universe between the galaxies reflect the majority of time directions moving forward to Omega. Two steps forward, one step back. But we have only considered time directions on the largest scale. What about time directions in our immediate environment?

Electromagnetism is time moving forward. Electromagnetism is the perfect balance of absolute zero in the future influencing our present. The great balance of the future, being the most probable state, is a great cosmic attractor of all universes and all change. The future-like conditions located between the present and zero are all more like the flatness and uniformity of zero than past-like conditions. They are less lumpy, less grouped, and more uniform than past-like conditions. The future-like states also includes the whole of all the inverse negative states, which are less positive than the present. So, as that dominant set of possibilities funnels time toward absolute zero, the cosmos must become increasingly less lumpy, less grouped, more uniform, and more neutral. Those changes are produced by cosmological expansion, the electromagnetic force, and entropy.

We can actually feel absolute zero pulling and pushing the cosmos around. We feel the balance of zero as electricity and magnetism. If we imagine a state of perfect balance, it would be perfectly smooth and uniform. It would be neutral. If positive and negative particles are moving nearer to that balance, like particles will naturally repel while opposite particles attract, because that moves them toward greater balance. Electromagnetism is that simple. Like expansion, the force of electromagnetism is the present being influenced by our ZAT future. The difference between expansion and electromagnetism is that expansion applies to the large scale universe. Expansion works on the whole while electromagnetism works in the micro-world of particles, but they are both caused by the same inevitable future.

To highlight the fact that electromagnetism is a force creating balance we can just imagine electromagnetism in reverse. Like particles would then attract and opposite particles would repel. Such a force would cause positive and negative particles to divide apart into separate groups, just like dividing up colored checkers. In fact, when like particles such as positive protons bond together, time is actually moving backward. The strong force is essentially electromagnetism in reverse.

The Strong Force is time moving backward. Of course a group of possibilities fights against the fundamental force toward balance even in the small world of particles, so in the same way that gravity battles against expansion, the strong force battles against electromagnetism, by causing positive protons which are like particles to attract at very short distances. If protons and neutrons get close enough, the repulsion of electromagnetism is overcome by the strong force, which is time moving backward, just like gravity. The strong force holds a group of protons and neutrons together to form the nucleus of atoms, just like gravity holds together particles to create stars and galaxies. The strong force is the gravity of particles which has a short range because electromagnetism is dominant on the larger scale, just as expansion is dominant on the large-scale over gravity.

The Weak Force is time moving forward. The one flaw in the strong force is the weak force, which can cause the nucleus of atoms to decay, and properly so. In the future, something has to eventually overcome the strong force, because electromagnetism cannot break down the bond between like protons by itself. So in order for electromagnetism and expansion to eventually win the battle against gravity and the strong force, at some point in the future the weak force has to break down all the complex atoms in the universe into individual protons and electrons. The weak force is very much an extension of electromagnetism. In fact we know the weak force grew out of electromagnetism in the early stages of the big bang, prior to which there was just one force called the electroweak force. We can generally recognize that electromagnetism and the weak force together are working against both gravity and the strong force. Isn't it really funny and amazing how a sensible purpose to the forces of nature can suddenly seem self-evident. The forces of nature aren't arbitrarily selected in some way by chance or design, they are simply the possible directions of time.

The Three Axes

Time is said to be like a sandy wind which erodes all that rises up against it. Time is more like a flowing river, dominantly following the path of least resistance, but not purely following one direction. The flow of a river twists and turns and has eddies which flow backward. Likewise, the cosmos doesn't move purely in one direction, or even in two directions. We live in a world where time moves backwards, forwards, and several sideways directions. Time flows freely, but like a river it obeys the contours of a landscape.

In addition to the more powerful directions of past and future, in the expanse of pattern space there are four other directions as easily recognized that pull at and shape the cosmos. They are the influence of possibilities which exist adjacent any direct linear line drawn between Alpha and Omega, so they are in a sense forces concerning the present, or present-like states.

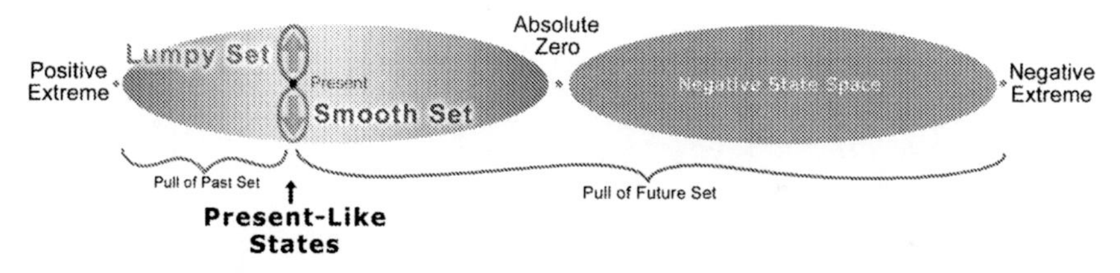

Figure 18.2: The four basic food groups for time; The Future Set (more negative), the Past Set (more positive), the Smooth Set, and the Lumpy Set.

Adjacent the location of the present, there is a force that pulls the universe toward becoming lumpy. This force created the lumpiness of the early universe, which we detect today in the microwave background radiation. Having long ago moved away from perfectly smooth conditions, those possibilities are now trying to pull us back. So an opposite force pulls the universe toward becoming smooth. Like the past and future these two forces reflect opposite directions of change, and so we naturally find the present balanced between them.

There are still two other directions that time is governed by which we have discussed indirectly but they haven't been identified clearly enough. Time is also caught between very orderly patterns and very disorderly patterns. In concert with the smooth and lumpy possibilities there are extremely cooperative patterns and extremely chaotic patterns. These two opposite directions exist at right angles adjacent the smooth and lumpy axis.

There are actually three recognizable axes along which time finds a position of balance. The first axis or "x" axis begins at the positive Alpha and stretches all the way to Omega, then like the mathematical plane continues on to the negative Alpha. This first axis is the *Density Gradient.*

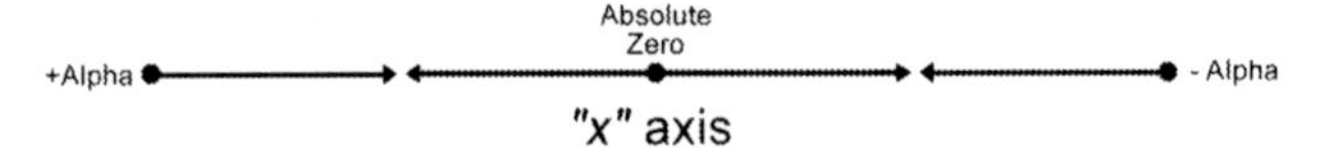

The definition of each axis requires the next axis, since each axis is in a sense pregnant with a variety of possibilities in the next axis. At each particular density in between Alpha and Omega there exists a range of possibilities between the smooth and lumpy extremes. Of course this spectrum of smooth to lumpy must exist at each point in between Alpha and Omega. These two directions form the second axis, the "y" axis. This second axis is the *Contrast Gradient.*

The influence of the smooth side causes lumpiness to spread out more evenly or uniformly. The influence of lumpiness originally produced fluctuations in the big bang that were synchronized with (naturally designed to create) the eventual distribution of stars and galaxies in the future (our present), which are still today regulated by the contrast axis.

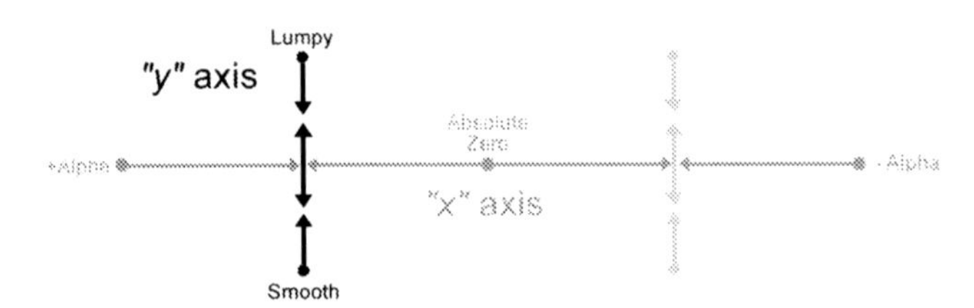

Figure 18.4: The Contrast "y" axis exists adjacent the Density "x" axis.

The "y" axis is also pregnant with possibilities. In the variety of patterns between smooth and lumpy there is a third axis. In some measure along the "y" axis the intensity of both orders can be increased or decreased simultaneously without causing a change in position along the "y" axis. So adjacent the "y" axis there necessarily exists a third axis. The third axis or "z" axis is the most interesting of all. This is the *Cooperation Gradient* or axis.

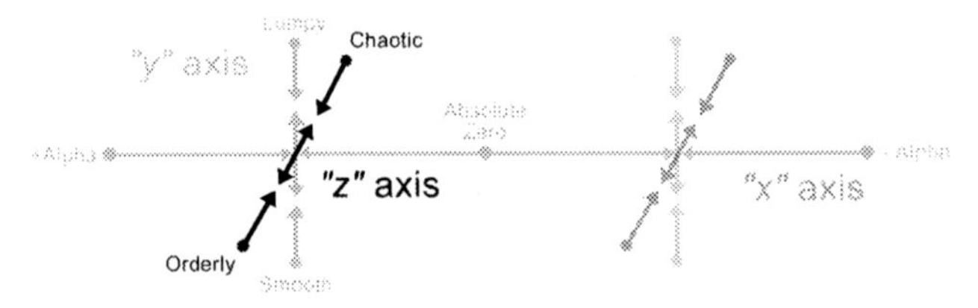

Figure 18.5: The most interesting axis, the Cooperation Gradient or "z" axis, spans between orderly and chaotic extremes.

We can notice an increasing variety of patterns possible with each axis, as well as an increasing complexity to pattern space. The "z" axis forms a range of patterns between extreme orderliness and extreme irregularity or chaos. These patterns are defined by the two orders either cooperating to create increasing orderliness, in which case the competition or tension between the two orders can be seen as high, or the two orders are uncooperative and irregular, producing disorderly and chaotic patterns, in which case the tension between the two orders is low. This axis balloons out the contrast gradient into a three dimensional mapping array of patterns.

The critically important balance between extremes in this third axes maintains an average measure of orderliness in nature. It designates improbable the extreme cooperation between two orders that would produce a rigid orderliness. And it designates improbable the extremely irregular combinations of two orders or chaos. This third axes can be compared to the order to disorder gradient as described by the second law of thermodynamics but should not be imagined as being the same, since simple order to disorder doesn't really exist.

The patterns within imaginary time shown below represent a cross section slice of pattern space. Think of this graphic as what is possible for the present excluding the influence of the past or future. With the lumpy and smooth extremes on each end, and orderly and chaotic extremes on top and bottom, paths of time are pulled to the center. The lattice patterns along the top represent high intensity of both orders. The chaotic series of patterns along the bottom result when the intensity of both orders is low, or there exists a negative tension between two orders.

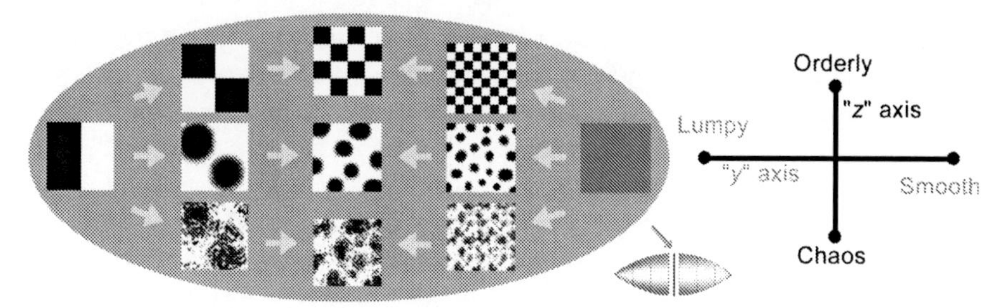

Figure 18.6: This cross-section slice of the Soaps model represents the influence of the smooth and lumpy extremes on each end, and orderly and chaotic along the top and bottom, which are the conditions time responds to adjacent the flow of time from past to future. High-cooperation is shown along the top examples, and weak cooperation is shown along the lower examples. This slice of pattern space exists at every point in between the single absolute states of Alpha and Omega.

If the intensity between the two orders is moderate, then the measure of co-operation produces the calico-like patterns viewed in figure 18.6 as the three patterns across the center. The single center pattern represents a balance between all four extremes. Sand on a beach, clouds in the sky, trees in a forest, stars in the night sky, all reflect arrangements of things which are imperfectly even. These moderately balanced distributions are the most common patterns found in nature, just as they should be.

Grouping	Symmetry	Description	Name	spatial
Low	High	Perfect symmetry – Symmetry order	Omega	x axis
High	Low	Positive-Negative split – Grouping order	Alpha	
Low	High	At each point along the gradient between Alpha and Omega there are smooth and lumpy extremes.	Smooth	y axis
High	Low		Lumpy	
Low	Low	Irregular combination of two orders	Chaotic	z axis
High	High	Regular combination of two orders (lattice)	Orderly	

Figure 18.7: Directions of Freedom in State Space

We might pause to notice here a distinction between the two fundamental dichotomies found generally in reality, one between positive and negative, in which case the opposites are identical but inverse. The other very fundamental dichotomy is between past and future or Alpha and Omega. The past-future dichotomy is not one of opposites, but rather two fundamental natures related on one hand to definite form and on the other the uniformity and implicate order of the infinite. In this second dichotomy we can recognize the complementarity of quantum mechanics which Heisenberg first described with the uncertainty principle.

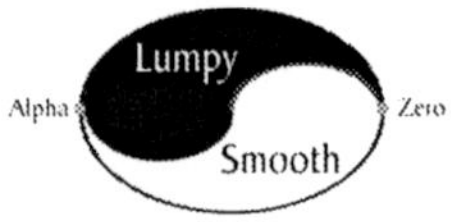

Cooperation versus Chaos

We each know a certain measure of freedom in our own lives contrasted by limitations and powers beyond our control. We each feel external and internal pressures that drive us toward orderliness. We each feel the potential of chaos and decay. We all feel social influences to fit in and be the same as others yet we simultaneously feel an internal drive to be our unique self. We shouldn't be too surprised to find that the larger cosmos feels similar forces. The cosmos is traveling through a basin of attraction that is extremely narrow in respect to the whole range of possibilities, but the basin is still wide enough to allow the flow of time to oscillate back and forth between the freedom, irregularity, and chaos that allows for diversity, contrasted by the powerful controlling forces of orderliness, sameness, and fullness.

We can try to imagine what it would be like if the cosmos traveled too near the extremes of super orderliness or radical chaos. We might look outward with our telescopes to see identical galaxies all exactly the same size spaced equal distances apart, like a giant checkerboard lattice. We might see the stars in the night sky all aligned into rigid rows and columns extending outward as far as we can see. How strange and interesting this universe would be. Such rigid control seems surrealistic. We are used to a measure of freedom to be disorderly and unique rather than so perfectly symmetrical and identical. Of course we might look out and see complete irregularity, complete chaos, no discernable order beyond our solar system. How confusing the existence of our own planet and selves would seem. Alas, instead the stars form galaxies. The overall distribution of galaxies form groups on a smaller scale, and are isotropic and balanced on the largest scale. We know there weren't simply fluctuations in the early universe. There was instead a narrow variety of fluctuations that led to the narrow variety of galaxies and the isotropic distribution of galaxies. As with chaos, when it comes to orderliness we should be grateful for moderation.

If the competition between grouping and symmetry was generally more intense, all the individual protons and electrons would be maintained in a tight square lattice, cooperating like the even distribution of squares on a checkerboard. The balance between opposites would be so controlled that the two particles wouldn't even be allowed to form pairs, so of course the chemically diverse table of elements would simply not be.

But then if we swing back too far in the opposite direction, if we widely open up the door to freedom and diversity, then the atomic bond between proton and electron would be weak, and the fluctuations of the early universe would be more varied, and the temporal universe would be far more irregular and chaotic. If pattern evolution was free and uncontrolled the universe would not be systematic and orderly enough for life to exist, or at least not with as much stability. Curiously, the balance between freedom and strictness is optimally tuned right about where we would want it.

We see the world as ruled by cause and effect. We imagine the present as evolving forward into the future due to the momentum of cause and effect, as if causes are built into the past. We imagine the present moves into the future like a boulder rolling down a hill. Yet when we roll the dice we at least hope the outcome isn't predetermined. No one wants to be a controlled robot, trapped in a single predetermined series of events. On the other hand, we need some external control over what is possible. If time was completely free, if there was no control, then anything could happen. No one could exist in the radical freedom of absolute chaos. Extremes of freedom and control are like extremes of hot and cold, we don't like extremes. Some measure of freedom is a good thing, but too much is not. Orderliness and symmetry are beautiful, but we prefer at least some measure of freedom within which to make choices and shape the world around us.

What fortunes we share in this cosmos, having freedom, the element of chance, and even randomness, balanced against the control of a predetermined final outcome. Einstein didn't like the probabilities of quantum mechanics, but probabilities are a good thing. The chance of something happening is always a probability, ranging from impossible, to low, to being equal with another event, to high, to the total inevitability of something happening, such as our zero future. All motion, all change, is inherently probabilistic. The expanding universe isn't properly described as coasting outward due to the explosion we call the big bang. There cannot even be change like that.

Of course the tiny microcosmic events are more open and free than collective macro-events. Our ability to predict the next position of an electron is very poor. In contrast the gravitational rotation of the Earth around the Sun is virtually inevitable and highly predictable. The rotation of the Sun is built up from the probabilistic travel of billions upon billions of particles, so the freedom of each is weighed against and controlled by motion as a whole. There in between the unpredictable micro-world and the predictable macro-world are all the cycles and waves of probability that govern human development.

One of the surprises in understanding our place between the two orders is how irregularity and chaos now seem to be just this small window of opportunity, one possible direction of many, where we have in the past seen disorder as vast and overwhelming, and so more probable than order. Secondly we find a potential for irregularity is required for our space-time to oscillate and become complex and diverse within the dynamic play between two orders. Grouping and lumpiness, smoothness balance and symmetry, even disorderliness and chaos, each play a critical role in a natural, yet very godly designed, governing dynamic that seems so utterly impossible in the order to disorder paradigm.

The Great Struggle of Two Forces

It could be said that the four forces of nature simplify into just two forces. Expansion, electromagnetism, the weak force, even entropy, are all ultimately a product of the great attraction of balance. They all represent the call of the future and are ultimately one force that reflects the increasing influence of symmetry order. In contrast, gravity and the strong force pull in the opposite direction, essentially from the past. They form a single force toward imbalance. Gravity influences the whole universe while the strong force influences the small, yet they are also ultimately just one force related to grouping order, a force that tries to re-create the positive imbalance of the past by grouping together protons into a nucleus, or a star, or a galaxy.

In a complete quantum cosmology all forces of nature are probabilistic and in recognizing the fundamental struggle between the past and the future, between grouping and symmetry order, we must depart altogether from a view of time as a dimension and recognize that time travels in all available directions. On the surface this would seem to eliminate the possibility of temporal paradoxes. Even if an observer could somehow manage to intrude on a past-like state, all temporal evolution from the instant of the intrusion would proceed probabilistically free from the observer's expectations of the future. Simply the presence of the traveler changes the overall dynamics of probabilities, similar to the butterfly effect.

The Super Lattice in Our Future

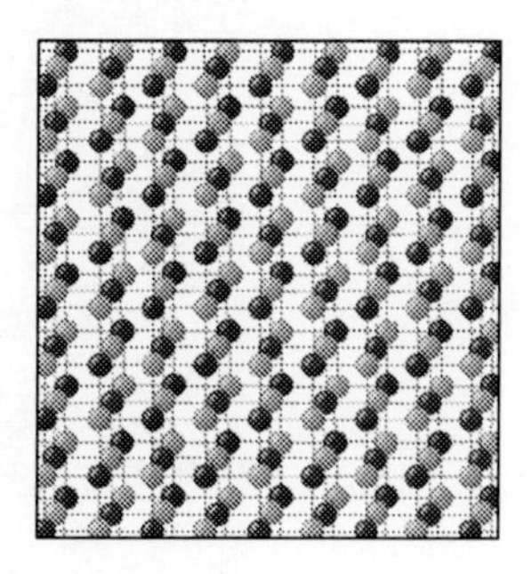

Electromagnetism is known to commonly shape atomic elements and molecules into complex lattice structures. Most of the pure materials in the periodic table of elements are crystalline, meaning the particles are finely organized into lattice patterns, built of square atomic structures. Electromagnetism creates those lattices by causing like charges to repel, and oppositely charged particles to attract.

In fact, electromagnetism is presently trying to spread all the particles in the universe into a great lattice, and if it weren't for gravity and the strong force, electromagnetism would succeed; it would spread out and distribute all the protons and electrons in the universe evenly into a perfect but simple lattice. Any physicist would agree and many have considered this fact including Julian Barbour, in his book *The End of Time*. Only here we are not just imagining a possibility that never happens.

The missing ingredient of the big rip theory proposed by Caldwell is the increasing orderliness of the distant future. The force of gravity does eventually weaken and electromagnetism will dominate the distant future. Along the way, the weak force breaks down all the complex atoms and molecules into single hydrogen atoms, so the strong force is also overcome. Near the end of time, as the cosmos forges its way into the narrowing avenue of possibilities near absolute zero, all the particles will distribute themselves into a giant lattice by lining up in columns and rows that continue outward endlessly. In the final moments near absolute zero the waves of the particles will overlap and unify. Finally accelerating expansion will stretch the remaining curvature of particles flat, and in that instant our time line will come to rest at Omega.

Infinite Labyrinths © Juergen Schwietering Mozenart.com

The concept of space as something existing objectively and independent of things belongs to pre-scientific thought, but not so the idea of the existence of an infinite number of spaces in motion relatively to each other. This latter idea is indeed logically unavoidable, but is far from having played a considerable role even in scientific thought.

Albert Einstein

~~~

That which is all that it can be is one, which comprehends and contains in its being all being. It is all that is and can be whatever other things are and can be.

Giordano Bruno
~~~

If space is nothing, then things cannot be in it. If however it is something, it will itself be in space, and so on indefinitely. But this is an absurdity. Things therefore, are not in space or in an empty void.

Parmenides

Chapter Nineteen

A Matter of Space

The Big Switch

Matter is the central focus in the sciences of physics, astronomy, geography, and chemistry. Mathematics was formulated to count material things. Skepticism plays an honorable role in science and many scientists don't want to entertain the possibility of a hidden order beneath the emptiness of space. In the past, it was concluded from Einstein's work that space does not even exist separate from mass. As many people know, Einstein went as far as to say that there is no such thing as empty space.

Today in cosmology we know that once the rate of expansion turns from decreasing to accelerating, the outer horizon of the space-time bubble begins to shrink inward relative to each observer. Eventually all other galaxies leave the event horizon, leaving only the fate of our own galaxy in question. If the acceleration overcomes gravitation and even particle forces, the outer event horizon would collapse inward on every point in space, and space-time is ended, yet it is the physical expansion of space that produces the collapse of space-time. Although direction and extension lose meaning within this final state, we have every reason to envision absolute zero relative to space-time as a pure space extending infinitely in all directions. Many if not most of the top astrophysicists concerned with the future now believe that time will reach absolute zero in either a finite or an infinite time period, which means the cosmos will physically become an empty space, or at least a seeming empty space.

It helps to realize what Einstein meant when he seemed to be rejecting the very notion of empty space. As is indicated in the quote heading this chapter, he was actually stating that the Newtonian view of space as a vessel in which things move about was abandoned by science. We are certainly not returning to such precepts here, we are actually discovering why that ideology once so logically deduced by Newton turned out to be false. What Newton envisioned as an empty space is actually properly defined as 'perfectly empty flat space', and the reason science now recognizes that objects do not move about in Newton's vessel of empty space is because a real perfectly flat space is absolutely full. Objects can only exist as distinct forms in a space where their opposites have been removed or displaced to create a half space or parallel regions. This creates the present interdependence between matter and space which led Einstein to describe space as an extension of mass.

Space is probably the ultimate challenge, and mystery, the ultimate puzzle, and conundrum, all rolled into one. It's not like we can put space under a microscope and see anything. We cannot isolate empty space to see what it would behave like without matter or time. All that we have learned about space is relative to a world of materials. And it doesn't help the problem knowing that the volume of space-time is collapsing. At zero, the point of final collapse, the expanding cosmos, existence itself, seems to suddenly disappear. For these reasons perhaps, it may be that the philosophers are ahead of the mainstream scientists in understanding space. Today there are a lot of theories about space on the internet, and many of them seem to have been sewn from a common thread. The common theme is that matter and space are two forms of the same thing.

If the cosmos can be stretched perfectly flat, then matter and space are clearly equivalent and interchangeable, but this leads to a very dramatic switch in the most basic set of values by which we evaluate the universe, whereby we see matter and particles as holes in a primary full space, rather than our present assumption that material things are arisen above and more primary than space. In this new set of values, matter is a bi-product of space. The Nobel awarded physicist Paul Dirac originally conceptualized anti-matter as holes in space. This view is very similar except that it works for matter also. Strikingly, in this model matter is seen as the absence of some measure of content rather than the presence of content.

In order for our universe to exist an identical but opposite anti-matter universe has to be removed from space. If we looked out at the cosmos with the Hubble space telescope and we could see just as many anti-matter galaxies as matter galaxies, then we could say we have an overall balance in our cosmos but there would still not be an overall symmetry. So we would still wonder, why not greater symmetry? Why is the universe out of equilibrium? Why not a perfect balance? Further still, if we could somehow see both sides of the larger whole cosmos, if we saw the exact replica of our own galaxy located like a mirror image on a replicated polarized side of the cosmos, we would then ask in amazement, "Why are we separate from our other half, why not perfect symmetry?"

If there was indeed a perfect symmetry would it destroy the two inverse worlds from existence, or would our experience of a perfect symmetry just revoke our ability to observe them? If two inverse galaxies overlap one another in the same space, they aren't cancelled from existence, they are just then being experienced as a oneness. They still exist in that oneness. Perfect symmetry doesn't create nonexistence. Perfect symmetry is just everything experienced at the same time in the same space. In fact the answer to whether opposites cancel or combine is kind of evident, simply because we are here observing half a world.

In being here (I think therefore I am), we logically expect perfect symmetry, but obviously there is something wrong with that thought pattern. Indeed, we should logically expect, even demand that there is ultimately a perfect symmetry, as much as we stubbornly refuse to accept something came from nonexistence, but along with that expectation of symmetry we have to realize that from our perspective, from our place in the world, we cannot or would not directly observe a unified symmetry. Perfect symmetry is not going to be right in front of us, but rather hidden from us. It might be in our future or in our past. But we will likely ignore it, overlook it, look through it as if it isn't there, and even refer to it as nothing, or potential, or a void, or a vacuum, or space.

We should similarly recognize that the seeming nothing of the balanced whole has to have internal content in order to be the whole. So we should recognize that we need to be here, there has to be asymmetry, in order for there to be symmetry. Otherwise we shift into a mode of expecting perfect symmetry to be nonexistence, which obviously is a radical contradiction. Nonexistence isn't. Perfect symmetry is.

The combination of our galaxy with an inverse anti-matter galaxy, the two combined, and the combination of every other pair of galaxies, all exist combined together in ordinary space. And so matter is NOT more than space, it is less than space. Matter is what happens when part of something whole is taken away, taken from the place and time we call now, and placed somewhere else in its own place and time. We know that happens simply by observing the world as such. The world we experience is out of balance, it is complex, it is chaotic, it is meaningless, only when we view a fragment of the whole. We don't observe the timeless whole because it looks like nothing to us. But we are within that nothing-ness. We are part of the whole. It wouldn't be whole without us.

It's difficult to re-normalize after such a switch, but the fullness or substantive aspect we attribute to matter is actually present in space more than in matter. Space has content; a sort of density opposite of our normal expectations. The space around us has more content than what we think of as dense matter. Dense matter is actually an absence, as represented in how the early universe is so tiny compared to the expanded universe. Still, here in this book we will most definitely maintain the normal application of the word density to apply to form, but we never the less must completely redefine the meaning of density.

Such a radical switch in perception and in one's basic sense of being might be uncomfortable at first, but the implications of symmetry order are clear (double meaning). Mass and density are both a measure of absence, a deficiency in full space, not an addition to empty space. Mass and density seems convincingly to be a value that is more than the transparent space that surrounds us, when in fact objects cannot exist unless their opposite is removed from that seeming emptiness. Things are not more than nothing, they are less than everything.

What is Positive and Negative?

In physics today we aren't required to designate density as being specifically positive, since density is by definition greater than zero density. In fact both mass and density are always given a positive value greater than zero. I think we are correct in thinking that mass is always a positive value greater than zero. Mass is a measure of substance. Mass is always seen as being greater than zero because, as its definition infers, it is a product or consequence of density. A positive or negative density always creates a positive mass. A positively dense particle and a negatively dense particle positioned next to one another will have an overall mass equal to the sum of both, while the overall density can be averaged out, such as when a particle and anti-particle collide. The basic rule in physics we are concerned with is that density multiplied by volume equals mass. For example:

$$5 \times -7 = -35$$

or

$$(+\text{density}) \times (-\text{volume}) = (-\text{mass})$$

A positive density in a negative volume would create a negative mass, which would violate the simple rule that mass is always positive (mass is a fundamental property that relates to existence itself, so a negative mass would be a negative existence or a nonexistence). In order to not violate the 'mass is always positive' rule, we must state as a law of nature that a negative density can only exist in a negative volume. Inversely, a positive density can only exist in a positive volume. This allows the mass value to always remain positive.

In a complete reversal of what we tend to assume, space is full, and matter is less than space. Matter is somewhat like bubbles in a water aquarium, tiny voids in a denser medium. Density is a positive or negative void that causes the curvature of the bubble. We could in fact imagine the whole cosmos as a big empty bubble, a void, in the solidity of Omega. In fact the existence of larger dimensions has recently become very popular with cosmologists. MIT physicist Lisa Randall has proposed the existence of a larger dimension called a *brane.*

People learning about the universe often ask, what is beyond the universe? If space is expanding, what is it expanding into? Scientists from the grouping order perspective have been claiming it isn't expanding into anything, but the universe is expanding into Omega. Omega is of course the largest dimension since by rule it is a synthesis of all possible dimensions. Our universe is a rather tiny bubble that is expanding within an infinite background.

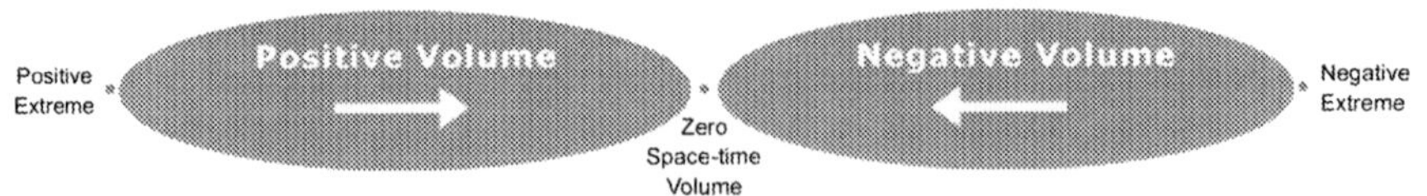

Figure 19.2: The two directions of time create opposing positive and negative volumes.

We live in a positive bubble (volume) that is actually collapsing as distant galaxies expand away from us at speeds beyond the speed of light. There is also a negative bubble and a negative volume, or an anti-time. Each has its origin from one of the two Alphas. But only a positive density can exist in our positive volume, and negative density can only exist spatially extended in a negative volume. So we detect the mass of a negative density as a point of zero volume, or what is called a point particle. The prime example of a negatively dense point particle contributing to our space-time is the electron. Electrons have essentially leaked over from the anti-time side in response to the existence of protons on our side, which is why the ratio of electrons to protons in the universe is exactly even.

One of the greater mysteries in science has been why the electron point particle has a defined mass, as opposed to an infinite mass. The law that a negative density cannot be spatially extended in a positive volume is the reason why a negative electron is a point particle with a finite mass. Mathematically, if a density is confined to a point it should have infinite mass. The reason the electron does not have infinite mass is because the density of an electron is not actually confined to a point. The mass exists in a negative volume that opens up beyond the point of the electron in our ghostlike invisible partner. Negative density also explains why protons and electrons are oppositely charged.

We are taught to imagine negative numbers as representing deficiencies or debts to be paid, but we imagine positive numbers as representing physically real things which rise above the background of space, or zero. In actuality, both positive and negative things are debts that have to be paid back to a larger neutral space. Both charges reflect deficiencies or imbalances in a spatial content. To create a particle or a star you take away from a full space, which creates a sort of indent, or curvature. Larger stars represent larger deficiencies, while black holes are the greatest deficiencies, which is why they are tiny and energetic. So opposite of our expectations, matter falling into a star or a black hole doesn't fill the deficiency, it increases it. It increases an already existing imbalance.

If we again consider Hawking's puzzle in a box, if the box is empty then we know that space is in an equilibrium state, which means all possible pieces of the puzzle, and all the possible inverse-matter pieces of the puzzle, are in the box, they just are all fit together perfectly so that we cannot see or interact with them. If so, why don't they occasionally jump out of the empty space? The reason is because a space that is already in an equilibrium state remains in an equilibrium state. So the only way for the plus and minus puzzle pieces to exist separate from one another, is if time originates from such a state, from an imbalance, and then moves toward equilibrium. There is never a case when a system in equilibrium breaks down into imbalance. Something does not come from nothing. But time can start from something, from imbalance, and evolve toward balance, or nothing, which is of course exactly what we observe happening. Our cosmos is the two half-full Alphas merging together into the full nothingness of Omega.

What I hope to be establishing here is a mode of thought where we see nature and reality accurately, and logically consistent, as opposed to the mode of thought by which we see and experience the world as individual things somehow distinct and separated from one another in their existence. It is fully ordinary for people to think about things as if they solely represent existence, and therefore imagine the absence of things or zero things as non-substantive, or nonexistence. Space may seem empty, but it can seem empty only by being full. If we could remove from that balanced fullness one part, the symmetry is broken, and the opposite part appears in the absence of that which was removed. Return the missing part and like a calmed surface of water, the whole returns to the perfect uniformity of superspace.

A Four Dimensional Space

I have myself in the past imagined the possibility that a finite universe somehow came to exist and began exploring the world of possibilities without direction or purpose, like monkeys punching keys on a typewriter. Given no restriction to the duration of time in which this finite world exists, just as the monkeys would eventually by chance alone seem to type out every book in every library, we can try to imagine this fluke universe produces every conceivable time world. With every world actualized somewhere along this timeline, it seems as if all worlds would exist simultaneously.

However, in this scenario, existence itself is evolving, and at any given point along the time line, there is always a distinct present. Time never actually becomes infinite, time is rather a continuous process, or imagined as such. I now believe that a genuinely infinite future necessitates an infinite past, and when you combine together an infinite past and future, time becomes relative. Everything suddenly exists together in one enormous moment of now, which supersedes any possibility of existential change.

In this great single moment there can of course be a seeming dimension of time in which things change, like a story in a book, but such worlds would only exist as a secondary consequence of features within the static existence. The time of clocks can be secondarily real, although semi-conscious observers within a secondary reality will likely prefer to imagine their sense of time as primary.

If any time worlds do exist imprinted within the great moment, then the time of such worlds must travel linearly from a beginning to an ending. As we have considered previously, a book must have a binding. The frames of a movie must be projected onto a screen. Individual words must be read for their meanings to fuse together. Without the continuity and binding of *linear time* each moment would exist completely separate. There would only exist a catalog of individual possibilities imprinted at different places within the one great moment, like many blocks of space. But such would not produce the illusion of change that we cherish.

We know such time blocks exist. At any given tick of the clock the world is in a particular condition or state, which we are calling *lateral time*. Each distinct moment is in some sense an individual universe. Without the distinction of each time frame there wouldn't be a present between some point called past and another called future. But this brings us to an age old mystery, the seeming incompatibility of linear and lateral time. How can distinctly separate blocks of space be simultaneously fused together into a linear progression of space? If they are fused then they are not distinct. If they are distinct, then they aren't really fused.

Suppose we try something different, and we shift our focus away from how such spaces are linked, and instead ask how such spaces are maintained in nature as separate? What separates one block of time from another? The answer is, only the definition of each time block maintains it as separate from other states. So it would seem then from this perspective that beyond any such definition, the spaces are always linked by a common existence. They are all embedded in the same great never ending moment. And importantly, they are all fragments of that whole existence. Each spatial pattern is a tiny internal part of something greater, and only the sum of all the patterns creates the whole. So we should now be able to imagine that all possible patterns are both defined and enfolded into a single superspace. Does this create the special space that we call time?

I believe that in addition to all the ordinary expected directions that construct each individual time frame, there also exists directions in space which travel across or through the multiplicity of these blocks. These directions in space are no less natural and inevitable than those which build an ordinary three dimensional block of space, except that each direction of this fourth dimension of space independently constructs the lateral component of its surroundings. Each individual direction moving through the blocks becomes a sort of parent to a lateral time world. Each direction establishes its own identity; its own four dimensional space.

Each of these directions in space are essentially free to find their way through the overall multispatiality, except that the direction of travel, and so the lateral surroundings of each linear pathway, is governed by the innate probabilities of the whole superspace, i.e., the soaps model. This replaces what would otherwise be a chaotic freedom with what is comparatively a very strict guidance system, making the surrounding environment of each linear pathway systematic and organized. The result is a fourth dimensionality of space, in which the lateral surroundings of each four dimensional parent direction are uniquely constructed strictly relative to that single parent direction. We refer to these parent spatial directions as time, but they are actually simply directions in space that exist exterior to the more easily imagined directions found in three dimensions.

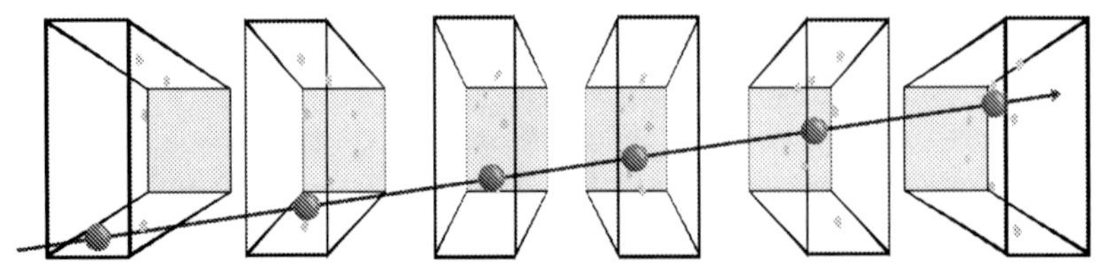

Figure 19.5: A direction of time passing through many spaces. Each block is an individual moment. The round object does not move through the spaces, rather it stays and exists frozen in each space. Only time, which is actually a special direction in space that is able to pass through multiple three dimensional spaces, is able to travel from one moment to the next.

These fourth dimensional directions through the overlapping multispace are recognizably a natural part of the inevitable existence of space. An undivided existence of all conceivable three dimensional configurations or patterns would inevitably create a fourth dimension of spatial directions that pass in between the blocks. It even follows that any semi-conscious observer passing through time would inevitably be aligned with a single direction through four dimensional space, and consequently the lateral environment of each observer would be probabilistically constructed strictly relative to that observer. What this means is that an observer will experience a lateral world which is coordinated with the probabilities of patterns space. Consequently the person will observe a history that traces backward to the most improbable state in the whole of possibilities. And they will observe a future that traces forward to the most probable state. And so they will observe an environment relative to their position between an Alpha past and an Omega future.

To whatever degree the two kinds of order dictate or allow, the observer will observe a sampling of complexity and organization in their environment in respect to pattern space, in concert with the full range of what is possible at their location. They will invariably observe a sampling of galaxies which reflects the widest range of configurations that pattern space dictates. They will observe the whole range of solar systems, planets, moons, asteroids. They will observe the whole range of geography, a whole range of possible chemistry, a whole range of other lifeforms, a whole range of personalities, all in perfect coordination with what is probable in respect to the whole of possibilities.

All that is observed, all that is experienced, is then integrated into the parent direction, inevitably causing expansion and growth. As we read the words on this page, each sentence, each paragraph adds up and synthesizes into something more than the individual words. As each person observes their world, one configuration emerges from the ethereal background in perfect synchronicity with a synthesized integration of memory and experience, all of which exists enfolded in the parent consciousness. We ourselves are the evolution of time. The cosmos exists inside of us, as real as we exist inside the cosmos.

Imagining the Higher Dimensions

At the time of writing I am collaborating with Rob Bryanton, the author of *Imagining the Tenth Dimension: A New Way of Thinking about Time and Space* who has developed a unique way of looking at the growth of spatial dimensions which blossoms into a ten-dimensional map of all possible time-worlds. Interestingly connective to my own work, Bryanton presents a ten dimensional spatial map of what I believe to be the overall multiverse. Bryanton begins his model with a point which has zero size or dimension. Two such points lead to the first dimension, a line or length, and lines that split and diverge away from one another create a second dimension, or width. The approach becomes unique and synonymous with symmetry order when the lines fold back onto themselves or converge together, which creates a third dimension. My favorite analogy from Bryanton describes our lives as four dimensional undulating snakes, with our embryonic self at one end and our old or dying self at the other, emerging from the zero point and ending at a zero point. A fifth dimensional splitting entails all our other possible future selves, as described in the many-worlds theory, and the sixth dimension once again enfolds past and future of all those selves into a single whole, returning us to a point. Then in the same way the fourth dimension joins past and future, a seventh dimensional line consists of all the timelines from past to future for all sixth dimensional points, which would be the infinity of all the many-worlds of quantum theory, or all universes that begin with a big bang. In cosmology we tend to assume that all we need for a complete many-worlds theory are four dimensions, but Bryanton seems to recognize that if time is a direction in space, then each observer is a four dimensional world, and he brilliantly shows how folding or enfolding those worlds together leads to a staircase of higher spatial dimensions.

The eighth dimensional line includes all the imaginable universes outside of the many-worlds partition, including universes with different constants and laws of nature, so the eighth dimension is what physicists today call the multiverse. All such worlds of the multiverse interconnect and enfold into a ninth dimensional super multiverse. Then finally in one last step, all conceivable paths of time, even those universes that don't begin at Alpha, the strange and weird irregular universes, the full breadth of imagination, unite with the ninth dimension, forming a single whole, beyond which there are no further possible universes. The tenth dimension returns us to the ultimate point, except now the point is understood as wonderfully full rather than empty.

Fractal Artwork: My Mind © Kerry Mitchell

Physicists like to think that all you have to do is say, these are the conditions, now what happens next?

Richard Feynman

~~~

Nature has been kinder to us than we had any right to expect.

Freeman Dyson

~~~

We feel that even when all possible scientific questions have been answered, the problems of life remain completely untouched.

Ludwig Wittgenstein

~~~

...quantum mechanics may be looking at - and mixing up - two (or more) different levels of physicality - the structure of objects, and the separately existing structure of the physical thoughts which refer to them (whether or not the objects actually exist).

Anthony Giovia

~~~

It never occurs to us that looking for the definition, origin, and nature of consciousness within the content of consciousness itself is the equivalent of searching a movie for a view of the camera man.

William Tedford

If this is a dream the whole world is inside it.

David Benioff *From the movie Stay*

Copyright © Charles Beck

Chapter Twenty

Built in from the Beginning

Scientists share a special type of humility that comes with being more fully aware of the immense universe beyond the daily display of blue and starry skies. Geoffrey A. Landis, an outstanding highly acclaimed science fiction author, winner of Nebula and Hugo awards, a poet, a Ph.D. physicist, and active researcher for NASA, expresses through a character in one of his science fiction stories the tendency in science to play down the role of our conscious experience and the role of life in awe and humility of the grander universe of matter. The paradox is revealed in the story *Winter Fire* when a character states "I am nothing and nobody; atoms that have learned to look at themselves; dirt that has learned to see the awe and the majesty of the universe."

Many scientists maintain a humble, even self-effacing attitude about life, even as they praise the magnificence of the universe. As a consequence of seeing the world from the bottom-up, many have trouble accepting the possibility that life might play a larger role in reality. In such stark contrast, now seeing the path of time traveling toward balance, there is such a major shift in perspective taking shape here. Apparently such humility is not warranted. There is an old assumption made not merely by scientists, but rather we all expect the past is responsible for creating the present. We imagine some event in the past creates the universe and natural laws, then governs it from that point on. Yet in now recognizing absolute zero as a cosmic balance point, the evolving universe that surrounds us itself has a goal, an objective, a destination. Time is not just wandering out into chaos as the second law suggests. The universe is not winding down after some fortunate but purposeless cataclysmic event in the past. An order we are not yet even accustomed to recognizing in science plays the largest role in shaping the present.

This formula of moving from imbalance to balance, from order to order, explains sensibly why we experience physical reality to be systematic and orderly. It would eliminate most multiverse scenarios with randomly selected constants that otherwise might exist, excepting the potential of a fifth, sixth, seventh spatial dimension, and so on. And there should perhaps be other cosmological systems which aren't founded on the duality of positive and negative. In other modes of reality there may be more Alphas than two, possibly triangular, rectangular, or octagonal extremes all balanced on zero. Perhaps time bounces from one to the next attractor in a spiraling spin-down to zero. Still, the overall possibilities of what infinity is like are lessened and seem almost manageable. Recognizing the role and influence of a balanced Omega has at least greatly lessened our reliance on the anthropic principle to explain why we experience this particular universe.

I hope what has been explained so far makes the orderly cosmos seem more natural. However, the two kinds of order may not allow us to eliminate the possibility of anthropic influences altogether. As many biologists and even physicists have acknowledged, our world seems optimally tuned for the existence of life. It is certainly true that the measure of expected orderliness and complexity is greatly elevated once we recognize two orders. We can now look at the stars and galaxies and know that they should exist. Should plants and animals exist? Given that there is life, we should expect to see the two orders influencing its design, a case example being all the different flowers on the Earth. Another example might be the grouping and symmetry of all the various trees. What is not plainly evident in the two orders theory is the certainty of biological life.

Left to Right, Top: © Nellie Buir, © Esther Seijmonsbergen, © Allison Choppick, © Michael Bretherton,
Lower: © Rodolfo Clix, © Nathalie Lalumiere, © Mira Pavlakovic © Elvis Santana

For myself, for some reason, understanding why the universe is orderly makes the scheme where life emerges by chance seem dubious at best. Or perhaps it is thinking in terms of timelessness, and the flow of time toward zero, where a measure of probability rigidly dictates what is able to physically exist. I have difficulty imagining life is just a thin probability having little to do with the larger cosmic arena, and just a fortunate by product. Instead, in respect for the infinite, the complexity and supernatural character of life seems more to be tied into cosmic evolution than the lives of stars and galaxies.

In addition to a more skeptical humility of some scientists, there is also a top-down humility where a person senses a deep inexplicable omniscience in nature. Einstein himself remarked, "The scientists' religious feeling takes the form of a rapturous amazement at the harmony of natural law, which reveals an intelligence of such superiority that, compared with it, all the systematic thinking and acting of human beings is an utterly insignificant reflection." How does human life relate to the ultimate sum of all? Someone rather religious once said to me, "If you were to look directly at God, you would instantly burn up". The other possibility of course is that you would freeze into a popsicle. Besides the obvious lack of knowledge, experience, and wisdom, I wonder how different we are. I suspect there is a natural consciousness to the whole of being, but we ourselves are a being that is conscious, just dramatically less conscious, speaking for myself anyway.

There would seem to be a case for at least a meager relationship between our consciousness and Omega. Simply considering that absolute zero is a product of synthesis rather than cancellation, we then are led to consider Omega to be a synthesis of all life, all experience, all thought, existent throughout the infinite expanse of galactic and planetary systems in an infinite universe. At this stage, remaining conservative is just unreasonable humility or biased skepticism. To the dismay of the skeptical, we are not here considering philosophy or religious idealism, we are rather properly applying an evident scientific theory. It follows that the emergence of life and biological diversity are probably best explained as future influences, which is classifiable as a form of intelligent design.

In fact the most reasonable conclusion is that the final state of time is by nature innately self-aware of its internal self, and thus supremely conscious (absolutely no correlation meant to any supreme consciousness as portrayed by any one religion), and that the very existence of life is directly attributable to an evolution of consciousness invariably built into the process of time reaching zero. The enfolding nature of symmetry order ultimately indicates that our intelligence and consciousness, the human desire to understand and model reality; life itself, exists relative to a predestined cosmological evolution toward the sum of all being and all life, all knowledge and all experience. As surprised as anyone, I find that I must argue that the supreme state of the universe is both shaping the universe and shaping human history toward a goal. The very same conclusion was made for similar reasons in concert with science by the physicist Frank Tipler in his Omega point theory, a conclusion in part based upon quantum theory. Tipler states:

> Quantum mechanics says that it is completely correct to say that the universe's evolution is determined not by how it started in the Big Bang, but by the final state of the universe. Every stage of universal history, including every stage of bio-

> logical and human history, is determined by the ultimate goal of the universe. And if I am correct that the universal final state is indeed God, then every stage of universal history, in particular every mutation that has ever occurred, or ever will occur in any living being, is determined by the action of God.

Of course the view suggested is not that Omega is dependent upon the summing of life, nor is the existence of life, systemization, and form, dependent upon the uniform whole for its existence. Instead there is a complementary interdependence between both realities. It is only natural and logical that there are no parts without the whole, and no whole without the parts. There is no infinite without the definite. Living systems are ultimately timeless and embedded in the whole. The summation of all life, all information, or the term I prefer is all *meaning*, forms a whole that is inseparable from its finer unfolded content. That whole is alive, by any definition of life that successfully applies to any part of the whole, all of which dramatically elevates the role of life.

I suppose it is the treasure of knowing that the infinite whole exists everywhere, in every part of the universe, which has led me to conclude that existence reduces to meaning, although in mentioning this some people are likely disturbed, as they think I am suggesting the physical world is an illusion. But that would simultaneously make meaningfulness an illusion. Instead I imagine meaning and physical reality as the same thing. Just knock on wood. That is how solidly real meanings are. The physical world is how concrete and vibrant and colorful real meanings are. So I have no trouble with the idea that everything is mindstuff, or consciousness, that is, as long as the physical cosmos isn't degraded with that association. And as long as what consciousness is made of is understood to be the same as what physical reality is made of. Consciousness arises out of the meaningfulness of being. The physical world is made of the meaningfulness of being. The physical world isn't just filled with an ethereal meaning behind objects, *it is meaning.*

Any view which separates meaning from physical reality moves us toward a division between matter and ideas or thought. Part of what can be deduced from the success of this cosmology, as well as the harmony exposed here between science and eastern philosophy, is that not only is it possible to discover good science that is based on sound reasoning, there also is a deep and fundamental relationship between physical reality and the world of ideas and meanings.

Omega isn't hollow or empty or plain or ugly or nothing at all. Not one ounce of that impression we make of zero actually applies. All the life that inevitably exists outward amongst billions of galaxies in the visible cosmos is but a pin drop of the totality of life present in the timeless matrix of Omega. Every joy, every pain, every waste, every mistake, every solution, every tear cried, every prayer, every hope, every lesson learned, every pride, every humility forms a fullness that eludes description. Omega is forever preciously omniscient simply in being itself.

There is a wonderful place of balance between science and imagination, not unrelated to the balance between thinking and feeling, that very few people even know exists. It isn't a place where a person is free to create fantasy, but rather where one is able to fully envision and appreciate the wondrous lessons that science is teaching us about the cosmos. It is really a shame that such ideas as those we are considering are so difficult to accept for those conservative and skeptical minded scientists who so successfully explore the world from the bottom-up. Physicists in particular, in tending to be practical, sensory oriented, and realistic personality types, have difficulty appreciating the two kinds of order, or I should say, they have difficulty with the enfoldment idea of symmetry order and implicate order. Sadly, it is true sometimes that those who do the work are the least capable of appreciating the rewards.

Even though religions are of the past, science in principle is not an alternative to religious beliefs. Science cannot justify skepticism toward extraordinary conclusions made of an existence that is so extraordinary. The more religious minded often either consciously and unconsciously base their spiritual beliefs and faith on how amazing existence and life are. Why should we limit our imagination with skepticism and expect that the whole of reality is no more incredible than the specific world of time and space we experience. In hindsight it is possible to see how the most brilliant minds of the past were those who were imaginative enough to unify science and religion, and the most exemplary person who comes to mind is Giordano Bruno.

Man in the Middle — A Visionary Life

At the very dawning of the modern scientific era, born in 1548, the Italian Philosopher Giordano Bruno was sixteen years older than Galileo Galilei. Bruno traveled around Europe not only arguing for the Copernicus model of the Earth orbiting the Sun, but also teaching the doctrine that there are an infinite number of other worlds amongst the stars, including infinite planets with other forms of life. Bruno wrote, "There are an infinite number of suns; an infinite number of worlds revolve around these suns, just as the seven planets revolve around our sun. These worlds are inhabited by living beings". Bruno almost certainly influenced the advent of science, although he was a top-down thinker and had no difficulty philosophically uniting the entire infinite into a single whole, which he then related to the common idea of God.

In his book, *On Cause, Principle, and Unity*, Bruno makes a statement that is eerily prophetic of both relativity and quantum mechanics, even the evolution of time as we understand it, writing, "There is no top or bottom, no absolute positioning in space. There are only positions that are relative to the others. There is an incessant change in the relative positions throughout the universe and the observer is always at the centre".

As he traveled around Europe he taught a special skill of memory, and was the guest of many dignitaries including the King of France and the Queen of England. In 1584 Giordano wrote a thesis entitled *On the Infinite Universe and Worlds* in which he argued that if a person believes it logical that even one other world likely exists, that it reasonably follows that all other worlds exist. Giordano writes:

> We are not compelled to define a number, we who say that there is an infinite number of worlds; there no distinction exists of odd or even, since these are differences of number, not of the innumerable. Nor can I think there have ever been philosophers who, in positing several worlds, did not posit them also as infinite: for would not reason, which demands something further beyond this sensible world, so also outside of and beyond whatever number of worlds is assumed, assume again another and another?

Bruno's conception of infinities is reminiscent of the transfinite mathematics of Georges Cantor, the mathematician, as Bruno describes infinities within other infinities. Bruno writes:

> Whatever is an element of the infinite must be infinite also; hence both Earths and Suns are infinite in number. But the infinity of the former, is not greater than of the latter; nor where all are inhabited, are the inhabitants in greater proportion to the infinite than the stars themselves.

Bruno's writings portray a strong sense of a unity to being, and eternity or infinite time. His view of a holistic infinite universe was fully intertwined with his belief and view of one perfect God.

> Therefore the perfect, absolutely and in itself, is one, infinite, which cannot be greater or better, and that which nothing can be greater or better. This is one, everywhere, the only God, universal nature, of which nothing can be a perfect image or reflection, but the infinite.

And yet Bruno's belief in an infinite God wasn't enough, it did not save him from the ecclesiastic members of the Christian Catholic inquisition, who considered his beautiful prophetic visions to be heresy. Bruno returned to his homeland of Italy where he was accused and imprisoned. He was kept imprisoned for six years then during a festival he was tied hanging upside down to a stake and was burned to death. This occurred early in the year 1600, a mere four hundred years ago. This murder by religious leaders was perhaps the grossest crime ever committed in order to limit what is imagined of reality.

I personally don't accept any particular religious interpretation of God. Personally, as did Bruno, I have come to see God and the infinite Universe as being the same, as something which all religions and philosophies intuitively and spiritually strive to describe and comprehend. I agree with Bruno that nothing could be godlier than the infinity of being, or as Bruno put it, "he who denies the infinite effect denies the infinite power". Today we don't fully understand or appreciate such things because that is precisely what our own evolution is all about. In my studies I have found signs of both natures, what I call human gravity and expansion, within every belief system and human environment that I have ever explored. The influence of grouping order and the past are particularly evident in religions. What religion does not in some way group themselves apart from others? Religious grouping may be the main source of conflict in the world today.

Fundamentally speaking, grouping order involves division, separation, distinction, individuality, density, pronunciation, opposition, and conflict. Grouping order has an obvious dark side, and yet these key words define the human condition, they define the splendor of the physical. Such qualities and behaviors seem quite unavoidable in human life. Opposition and conflict have certainly been a function of survival for the past evolution of each species, and unless human populations are managed somehow, in the future as land and resources are depleted we will surely experience horrible divisions and conflicts. Symmetry order involves balance, integration, combination, uniformity, homogeneity, singularity, formlessness, symmetry, and unity. These words obviously represent the ideals, ethics, principles that we manage at times and strive toward. They also represent what we often fail to accomplish.

The cosmic struggle between the two forces is the most common human theme. Grouping and symmetry orders reflect the two basic natures of the Universe. We can easily find these two forces in our own selves, in our homeland, in society, in religions, in politics. We might define one direction as better than the other if they were not so interdependent. Ideally we can use a new found clarity to learn to fully appreciate the value and necessity of both sides of each dichotomy, understanding the necessary balance between selfishness or identity and the greater sense of self that is human society, the Earth, and the Universe. If imbalanced, either force of past or future can detrimentally dominate a person's mentality, a group's culture, or a period of history.

Of course the key is balance. We can avoid extremes of selfishness or selflessness, and rather nourish fullness, meaning that we fully develop our selves, yet also develop our function and place as a member of a community, country, as human beings or simply life. And finally it is essential that we become aware of our ability to dissolve all boundaries, to know that we are ultimately one with the universe.

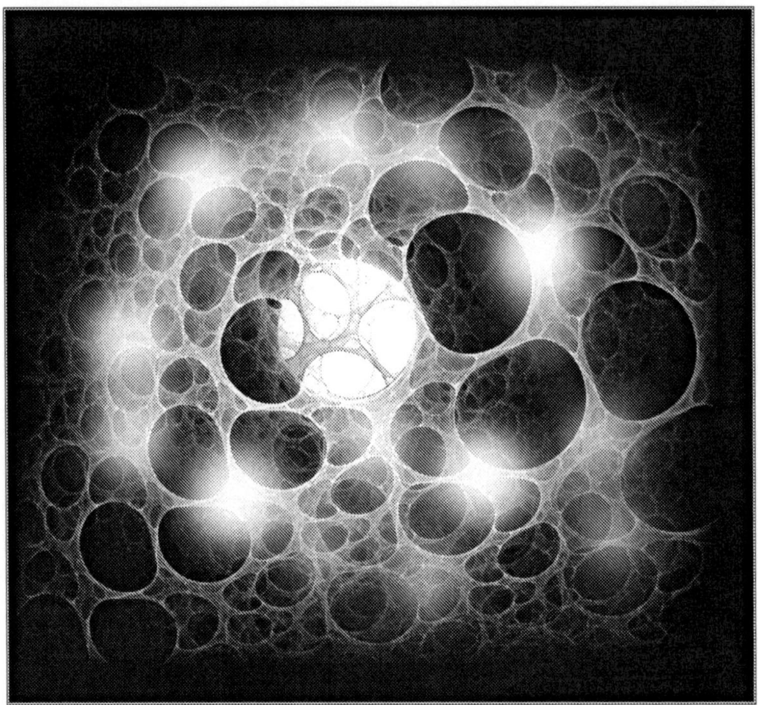

Modified from Original Photo: Digital Sponge © Yuri Hahhalev - Fotolia.com

In the final stage of egolessness there is an obscure knowledge that All is in all - that All is actually each. This is as near, I take it, as a finite mind can ever come to perceiving everything that is happening everywhere in the universe.

Aldous Huxley

~~~

All things are contained in the One, by virtue of the fact that it is one. for all multiplicity is one, and is one thing, and is in and through the One. . . The One is not distinct from all things. Therefore all things in the fullness of being are in the One by virtue of its indistinction and unity. [Sermon LW XXIX]

Meister Eckhart

~~~

The world globes itself in a drop of dew . . . . The true doctrine of omnipresence is, that God reappears with all his parts in every moss and cobweb. The value of the universe contrives to throw itself into every point. If the good is there, so the evil; if the affinity, so the repulsion; if the force, so the limitation.

Ralph Waldo Emerson

~~~

The weak overcomes the strong and the soft overcomes the hard.

Lao Tzu
~~~

I have yet to meet a single person from our culture, no matter what his or her educational background, IQ, and specific training, who had powerful transpersonal experiences and continues to subscribe to the materialistic monism of Western science.

Albert Einstein

Part Six

The Cosmic Psyche

A Universe Thinking Itself Alive

It was a sunny day, a perfect blue sky without a cloud in sight. I was somewhere near the age of fourteen, daydreaming as usual when I was supposed to be pulling weeds in the back yard. I had read something about the early universe in *National Geographic Magazine*. I think it was the theory that during the Big Bang our universe was just one single bubble within an infinite foam. I looked up at the sky and imagined our universe as a great bubble, and imagined another universe bubble, and another. In my mind I extended these bubbles outward into that perfect blue sky, trying to imagine an infinity of them, and as I tried to imagine the bubbles endlessly extended outward, something happened that I did not expect. I somehow imagined what I was trying to imagine.

I was standing there, but in the next moment I was not in my body, and not anywhere really. In any normal sense I was not myself. Instead, there was just a space that extended outward, my experience of the vastness of it seemed mostly in two opposite directions, and there is no way to explain it with words, yet it was so vividly real. There weren't any objects within the space. My experience was just of an immense distance that extended far beyond any expanse I could have imagined in the usual way. Actually, I believe for a time, I was the distance and the space, or I was not, and the distance was, but anyway, it was so much more real than even my youthful normal experience of reality, maybe because it was less fleeting than each moment of my life.

The distance was probably infinite, but of course completely beyond any other distance I could imagine, such as to the moon, which seemed minute in comparison, or at least it seemed so in the next few moments. What was next is what shook me, perhaps because anything at all came after this enormous place. Suddenly with a great shock that ran through my insides like thunder, as intense as if lightning had struck two feet from where I stood. I literally felt time begin again. Apparently, for me, time had somehow stopped because I quite literally "felt" time (or perhaps it was myself) originate once again after timelessness. It was like this timeless universe of space had somehow interrupted the moments of my life, somehow squeezing in between.

As I felt time and my self consciousness begin again, immediately but only then, did I know I had experienced the vast space. The memory of it came flooding in. As I experienced the distance I could not think, "Wow, what is happening to me." I could not think at all. Only afterward, once I felt time begin again, was I able to remember whatever it was that had happened. I was severely shaken, and I went in the house and cried a bit from nervousness, and didn't even consider going outside again till the next day. I don't remember telling anybody about it. For years the memory would just sit there inside me, an experience to learn from, and a realization that there was something more than the world I knew in everyday life.

Today most of us see the world as things, as something from nothing, as strange and improbable, as order that shouldn't exist. When we are young the physical world of form seems to vibrate and space has greater significance, but that usually fades as we grow older. It is only natural that we identify with our immediate environment, but as a consequence, the way we look at space is sort of like the cresting waves in an ocean saying to the larger wave and even the ocean below, "we are what exists, you are uniform and undefined. You are nothing. You don't really exist." It's rather silly to think we might have even the basics correct, considering how our expectations don't add up. If the universe doesn't make sense, we shouldn't blame the universe. Better to reason that our mental picture has produced expectations that we would not have if we understood reality properly. The universe not seeming logical, and instead seeming complex or even incomprehensible, are really just clues that we aren't viewing reality correctly.

Such states of mind where everything seems impossibly complex and chaotic are not uncommon in human situations. They seem to be a stage of growth and part of the evolution of each of us. We move through them at certain stages of youth before we have learned enough about ourselves and our own wishes of what we want out of life. Such states of mind exist in times of strife, injury, times of economic stress, or whenever we don't have the information we need to solve a serious problem. Whenever we are blocked from seeing the actual problem, the world seems painfully complex and chaotic. Only when there is some kind of mental or spiritual breakthrough, when we for some reason manage to look at things differently, then the clouds disappear and the sun comes out, and the world is made right again. In those rare lucid moments, the problems we are trying to solve, the purpose and meaning of life, everything, is made clear and simple, and we are forever changed.

I came away from my one out-of-body experience changed forever. I don't seem to have any memory of it left after so many years. I only remember my descriptions of it over and over again. It hardly seems real now, and a part of me wants to say it didn't happen, that it was a dream. But why did I later find my experience obeyed and fit in with physical reality? Even though I tried, I didn't experience an infinite succession of bubbles, or something totally unrelated to the known universe. Instead, just as it should have, my sense of material, my sense of form disappeared. There was just space, endless space.

Right now the volume of our space-time is collapsing inward as accelerating expansion pushes distant galaxies beyond the event horizon. Even though the cosmos is expanding outward our own time bubble collapses until expansion reaches the speed of light locally. At that instant all sense of the universe, all sense of form, is erased, except space. There will be space, endless space. I know this as the basic physics of the cosmos now, but why did my experience back then match what I know technically today? Why would someone's naive inner imaginations or metaphysical experiences ever match up with the outer reality? I think that the cosmos is as much the stuff of mind as it is matter and measurement, which likewise means the stuff of the mind is the stuff of the cosmos.

Zero is powerful because it is infinity's twin. They are equal and opposite, yin and yang. They are equally paradoxical and troubling. The biggest questions in science and religion are about nothingness and eternity, the void and the infinite, zero and infinity. The clashes over zero were the battles that shook the foundations of philosophy, of science, of mathematics, and of religion. Underneath every revolution lay a zero – and an infinity.

Charles Seife
Zero; The Biography of a Dangerous Idea

~~~

Where did the substance of the universe come from? . . If 0 equals (+ 1) + (-1), then something which is 0 might just as well become + 1 and -1. Perhaps in an infinite sea of nothingness, globs of positive and negative energy in equal-sized pairs are constantly forming, and after passing through evolutionary changes, combining once more and vanishing. We are in one of these globs between nothing and nothing and wondering about it.

Isaac Asimov

~~~

Philosophy is written in that great book which ever lies before our eyes - I mean the Universe - but we cannot understand it if we do not first learn the language and grasp the symbols in which it is written. This book is written in the mathematical language, and the symbols are triangles, circles, and other geometrical figures, without whose help it is impossible to comprehend a single word of it; without which one wanders in vain through a dark labyrinth.

Galileo Galilei

~~~

I want to know God's thoughts; the rest are details.

Albert Einstein

~~~

The universe can be best pictured, though still very imperfectly and inadequately, as consisting of pure thought, the thought of what we must describe as a mathematical thinker.

Sir James Jeans

~~~

The essence of mathematics is not to make simple things complicated, but to make complicated things simple.

Stanley Gudder

~~~

If we do not expect the unexpected, we will never find it.

Heraclitus

Opposites are not contradictory but complementary.
Niels Bohr

Winter Solstice: © Damien Jones

Chapter Twenty One

God's Math

Counting the World from the Top-Down

What is the largest number you can think of, no wait, what is the largest number of all? What is the total sum of all numbers? Of course the answer is that there isn't an answer to this question. As I am sure you know why, there is always a next greater number. But let's put it another way. What is the greatest universe of all? What if we imagine all things that exist? Can we at least find a single concept, a simple word, which includes all things combined together into one single whole universe? Is there such a word? Sure, this is easy. Even the word everything does that. Also words such as Universe, or existence, or being, can be meant to symbolize everything and anything that is remotely associated with existence.

But what word or number represents the whole in mathematics? How many numbers in mathematics symbolize an 'everything' in the number world? Is there some place on the real number plane which symbolizes the sum or the whole of all numbers? Intriguingly, the answer to this question is no. The reason is that there is always a next greater number when counting and so it isn't possible to count to a final largest number. There is just something different about the nature of the system of mathematics which makes it impossible to represent all numbers combined together within the system itself.

We could use the term positive infinity to refer to all the positive numbers combined together, but such a term would not actually represent a combined whole. Since there is always a next greater number in this group there cannot be a single definite value. A mathematical positive infinity is more a representation of a never ending process; a series of numbers, and not a number itself. Of course the same is true of the infinity of negative numbers. Like the positive side, mathematically there isn't a whole of all the negative numbers.

Most of us expect there to be some direct relationship between mathematics and reality, but there isn't a single number in ordinary math that symbolizes the 'everything' of numbers? We all are accustomed to using words such as Universe, existence, or being, and meaning them to symbolize the whole of all that exists. Why then, if it is so easy to refer to the universe as a whole, why is it impossible for a number to represent the whole of all numbers? What is so different about the nature of mathematics which makes all numbers impossible to represent?

There has never been a number to represent the whole. But what if we combine together all the positive numbers with all the negative numbers? Is there an answer to that equation? Isn't that a way of combining all numbers into a single whole. Actually, it makes sense that all positive and all negative numbers would sum to zero. It is not hard at least to imagine 'trying' to sum all numbers into a single ultimate number. If we try to combine together all the even and odd numbers, all the rational and the irrational numbers, then the equation would look something like this equation shown below, here using just the integers. This equation, as designated by the three dots at the end, continues infinitely so it includes every integer in the mathematical realm.

$$(0) + (1 + (-1)) + (2 + (-2)) + (3 + (-3)) + \ldots = 0$$

This equation appears to be the same as:

$$(0) + (0) + (0) + (0) + \ldots = 0$$

Wouldn't that be strange if the sum of all numbers somehow equaled zero. We could then say that the number zero represents the 'everything' of math, couldn't we. Except what about the other meaning of zero? Zero normally represents "no things". So zero as everything would not really make sense overall, because the meaning of zero is also very much related to the word nothing. How can "no things" and "every thing" be the same? That obviously doesn't make logical sense in terms of the real physical world of things, which is what math is supposed to deal with.

The equation shown above certainly does make it seem like zero is the sum total of all real numbers. We can reason there is always a negative value for every positive value. However, there is a problem, because it is possible to sum all numbers several different ways, and the sum does not always equal zero. Several equations sum all real numbers yet each yields a different result. As you can see for yourself below, this next equation adds up all integers also, but it adds up to an infinity of ones, so it has a different sum:

$$(1 + 0) + (2 + (-1)) + (3 + (-2)) + (4 + (-3)) + \ldots = ?$$

Same as:

$$(1) + (1) + (1) + (1) + \ldots = ?$$

Furthermore, there is another way we can rearrange the numbers where we have a different sum when we add up all the integers. This next equation adds up an infinity of negative ones and doesn't sum to zero either:

$$0 + (-1) + ((-2) + 1) + ((-3) + 2) + ((-4) + 3) + \dots = ?$$

Same as:

$$(-1) + (-1) + (-1) + (-1) + \dots = ?$$

Notice that this second and third way of summing all numbers does include all the numbers that the first equation included. All three sum every integer. All that we have done is place the same numbers in three different arrangements. So logically when we sum each equation they should produce the same sum. And if they don't, if they produce three different answers, then we have a problem. We have a logical inconsistency. Consequently it is said in mathematics that the sum of all real numbers is undefined. You may initially want to object to this but it really kind of makes sense. Otherwise, zero would be a mathematical nothing, a positive and negative everything, and a unified everything simultaneously. So the final conclusion mathematicians have made is essentially that zero represents nothing and there is no ultimate number that represents all numbers.

Zero cannot with logical consistency represent both nothing and everything in the same mathematical system of values, and as long as we understand that and accept it, if we appreciate that logic, then we can go on to discover a new mathematical system, one that is very similar to ordinary math, and yet very different, because in this different system, zero represents a mathematical everything. What's more, just as there is no number in ordinary math to represent everything, in this new system there is no number to represent nothing.

It is said that the sum of all real numbers is undefined but logicians and mathematicians made a critical mistake in formulating the rules concerning zero. We tested the hypothesis that all numbers might sum to zero, using a mathematical system where the value of zero is axiomatically pre-set and defined to mean nothing. In ordinary math, all values are relative to zero as nothing, or no-things, so of course we would discover that all real numbers do not add up to a nothing, or no-numbers. If it were not so, the logical consistency of mathematics would be destroyed.

There is no reason to wonder at why no one has ever truly considered zero to be the sum of all numbers, although it can be done. It just can't be done half way. As the saying goes, it's all or nothing. The proper test has to be a genuine reconsideration of zero. If we test zero as the sum of all numbers we must allow its usual value of nothing to change to a value equal to the summation of all numbers. So naturally its value would have to be given a value greater than all other numbers. Do you see what I am saying? It's a bit radical. If we combine all numbers into zero rather than cancel all numbers into zero, we alter the entire value system, and suddenly you have what appears at first to be nonsensical, because if zero is the greatest value; the sum of all numbers, what then is the value of number one, or number two? Which is greater, one or two, if zero is now greater than both?

See the problem! How do we resolve this? Is it just nonsense? Or could we be at the threshold of discovering something new and important? Naturally in order to find out we must explore some unfamiliar terrain. However, keep in mind, that we are not considering here a change to ordinary mathematics. The mathematical system developed since the dawn of human reasoning functions in relation to the definitive world of things that we observe each day. It is a valid system evidenced by its application to the physical universe. And yet it is noteworthy, even important, that we notice how that system cannot describe the universe as a whole. As we count a world of things we count upward into an endless abyss of numbers. If we wish to understand and describe the universe with a mathematical system that is able to represent the universe as a whole, then we have to make a switch and see the world in an entirely different way. What I mean is that we ordinarily see the world as if everything is more than nothing. What follows is a way of seeing the world mathematically as if all that we know as physical is merely less than everything, rather than more than nothing. In our ordinary mathematical system nothing is a foundational axiom. In the mathematical system I am about to explain the idea of nothing has no place or meaning.

$$(\quad 0 \quad) + (1 + (-1)) + (2 + (-2)) + (3 + (-3)) + \ldots = 0$$

We begin by looking at the original equation which seemed to sum up all integers in a simple and straightforward way. For a moment we will imagine that the correct sum of all numbers does equal zero. This means that we switch the value of zero away from nothing and make zero the largest value in the mathematical system. What I mean is that zero has now become a number that contains all other numbers. Every positive and every negative number on the real number plane is summing or combining together and forming an ultimate number of absolute value. This is certainly not math as we know it.

What effect then does changing the value of zero have on the value of other numbers? If we are going about this bravely, we perhaps could expect that the value of other numbers would now be different, transformed in the same shift that we have taken with zero. Ordinarily the nothing of zero is a foundational axiom. Our foundation has shifted. What now is the value of one or two?

If zero is seen to contain all other numbers, then logically all other numbers must have a lesser value than that of zero. If zero is the largest value, the only way there can be lesser values is if we remove some measure of value from the whole of zero. For example, suppose that we take away a (-1) from zero. What remains? Zero is now no longer an absolute value containing all other numbers.

Something has been removed from it. But what value does zero transform into to show that loss?

The answer is simply that zero has become the value 1. If zero contains all numbers within it, and we take away a value, zero then contains all numbers except the removed value. If we remove a negative one from zero the value of zero records that loss by becoming a positive one. And if we treat this as the logical rule we can now discover the values of all other numbers in this system. For example, one is the sum of all numbers, so it contains within it all numbers, except (-1) is removed. Notice the content of the number one is less than but nearly as great as the value of zero. The number two is the sum of all numbers except (-2) is removed, so it is also near zero but its content is less than zero and less than one. And so on, and so on. The value of three is less than two, the value of four is less than three. So 1 billion is a much smaller value than 1, since -1 billion is removed)

Everything has turned around, although the transformation in content is not simply an inverse reversal of ordinary math, but a hidden content of numbers in this system decreases as we count toward greater numbers. Of course there is content in ordinary math also. The number one contains two halves. The number two contains two ones. The number three contains three ones, and so on and so on. In this new system we are just summing up in a different way. We have created an axiom that sums up every real number into a whole.

In order to notice something, we need to switch to the negative side of symmetry math. The number (-1) is a combination of all numbers except that a positive 1 is removed, which would otherwise create the balance of zero. And in removing a positive two the whole shows that loss by becoming a negative two (-2). Dramatically unlike ordinary math, the numbers (-1) and (-2) are very large values in this system, in fact the content of (-1) is equal but inverse to the content of (+1). This is notably identical to the physics of matter and anti-matter, which are equally substantive, yet inverse in form and structure. In ordinary mathematics numbers 1 and -1 are not logically symmetrical, since 1 represents a physical thing and -1 represents a debt of one thing.

Of course this new math feels odd to anyone at first exposure, and to a mathematician who is learned and naturally entrenched in ordinary math, all this likely seems absurd and useless. Ideally however everyone is interested in logical consistency and maybe wise enough to not expect to immediately see how a set of ideas can be applied for some practical purpose. Keep in mind that we are no longer counting finite things, so it is certainly not being meant here that two things are less than one thing. None of this applies in any way similar to how ordinary mathematics is applied to the world in which we live. In what I shall now refer to as Symmetry Mathematics, zero is a complete and infinite value. As we remove a part from the whole, we create other number values which are somehow both infinite and definitive. As we shall see, the whole infinity of possible values in this system is absolutely definitive, and not merely part of an endless series or a process.

2| equals the set of all real numbers except (-2), or (-1)+(-1). The symmetry value of 2 can be drawn on a number line as shown below:

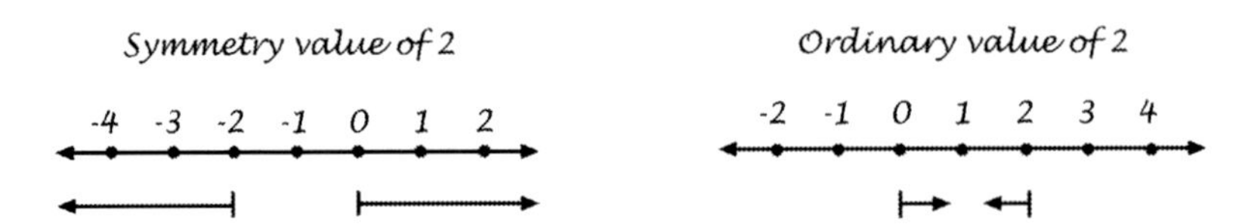

It should be clearly noted here that we are not merely reversing the general value system of mathematics, we have changed the very nature of the system. This is most evidenced by the fact that there isn't any number in this system that represents nothing. There is no basic duality of something/nothing like that which exists in ordinary mathematics. In ordinary math 1 represents one thing. In symmetry math zero represents one whole. It is as if one's whole focus changes (another accidental double meaning).

In the same way that there are two distinct forms of order in nature, there also exists two entirely different ways of seeing zero and all other numerical values. It is a whole other type of value system, a system as valid as the one we presently use, one of no use within the abstract world of individual things, yet immeasurably valuable in cosmology where a mathematical value for the universe as a whole is of critical importance in any attempt to understand for example, the implications of the Many-Worlds Theory, or how the big bang past relates to an absolute zero future, or the whole of possibilities, or the whole of existence. Someone wrote to me once and remarked that this is God's math, because of the way the system sees and enfolds everything.

One shouldn't assume this unique value system threatens our normal value system in any way. Each is built upon a perspective. Two apples are still more apples than one. We can still divide up and see the world from a finite perspective, in which case the infinite can be seen only as an indefinite process. What this system does, is allow a radical shift of perspective, so that we can also see the universe as an undivided whole, where two apples are part a single universe. In symmetry math, one of the conclusions we can draw is that for there to be a positive two apples (matter), there must be a negative two apples (anti-matter) removed from the pattern that we observe. All finite form requires that the two positive apples are less than the whole of the four apples combined. Such ideas are really very simple and even sensible once one is accustomed to switching from one perspective to the other.

Modified from original © Paha L fotolia.com

A new scientific truth does not triumph by convincing its opponents and making them see the light, but rather because its opponents eventually die, and a new generation grows up that is familiar with the idea from the beginning.

Max Plank

~~~

Science is simply wonder that gets pursued with some kind of discipline.

Brig Klyce

~~~

Infinitary mathematics is a fantasy world in which we fantasize about the completions of processes which, realistically, we can only begin.

Matthew Donald

~~~

He thought of the incomprehensible sequence of changes and chances that make up a life, all the beauties and horrors and absurdities whose conjunctions create the uninterpretable and yet divinely significant pattern of human destiny.

Aldous Huxley

~~~

...one could say that through the human being, the universe is making a mirror to observe itself.

David Bohm and Basil Hiley

~~~

I am at two with nature.

Woody Allen
~~~

This All is universal power, of infinite extent and infinite in potency, a god so great that all his parts are infinite. Name any place, and he is already there.

Plotinus

Chapter Twenty Two

Proto and Elea

The Infinite is Finite

Infinity as a real physical thing is still often treated with skepticism. The nature of infinity is very much an unresolved mystery in both physics and mathematics, but there are a few tolerated infinities, such as electrons and black holes. And there are infinite series equations which mathematicians say are defined because they express Convergence. For example, .999... is said to be equal to 1, because the value converges to the number one. At first inspection .999… seems ever slightly smaller than one. Some mathematicians treat .999… as one because they consider the difference between one and .999… so infinitesimally small that it has no relevant value. A slightly more advanced understanding explains that .999… would equal one in an infinite period of time. However, even the universe's fastest computer would never be able to appreciate the .999…as a single definite value.

Other examples of convergence include:

The sequence: 1/1, 1/2, 1/3, 1/4 ... converges to 0.

The equation: 4 - 2 - 1 - 1/2 - 1/4 - 1/8 - ... converges to 0.

The equation: 1/2 + 1/4 + 1/8 + 1/16 ... converges to 1.

And the equation: 1 + 1/2 + 1/4 + 1/8 + 1/16 + ... converges to 2.

Obviously (1 + 1 + 1 + ...) is not a convergent series, it has an increasing value, so there is never a completed sum. There is no convergence. In ordinary math we think of the set of positive or negative numbers as continuous and indefinite. If we add (1 + 1 + 1 + ...), there is always a next greater number for the sum to equal and there is never a last number to the series. The value expands yet it is never nearer to an ultimate end. It is hard to imagine how an indefinite infinite series would even exist in timelessness. Perhaps there is an abstract nature to properties of ordinary math which only have meaning in time.

On the other hand, symmetry math works well in timelessness. There is one ultimate value which contains all other numbers and on either side of the great

Omega Zero other values decrease. As we count into ever larger numerals, the symmetry value of ever greater or lesser numbers diminishes, decreasing toward an infinitely small value. Consequently, the values of larger numerals converge toward two outer extreme points. This clearly visible convergence allows positive infinity and negative infinity to have definite values. So they can be represented as two numbers, rather than as an indefinite series. This might feel like relief away from the anxiety that the indefiniteness of ordinary math creates.

Unlike our present finite system of values, which definitely does not consider positive or negative infinity to be a number, in symmetry math we can define a final number to an infinite series. In symmetry math, if we add ones endlessly (1+1+1+ ...), there is still always a next greater numeral, but the value of the sum is decreasing and converging toward a point of infinitely small value. That point is identifiable as a last number. I have named the convergent sum for positive numbers Proto, which means first in time. Proto is a numeric representing positive infinity, here written +∞. I call the ultimate negative number Elea, which all negative numbers sum in value to. Obviously I derived the names from the proton and electron.

In treating a positive and a negative infinity as numbers, we then can write this simple equation:

$$+\infty + (-\infty) = \infty$$

All of symmetry math is defined and whole, bounded by extremes, just as pattern space is defined and whole. The symmetry plane is infinite but bounded. As a result, we can begin from Proto or Elea and count toward zero.

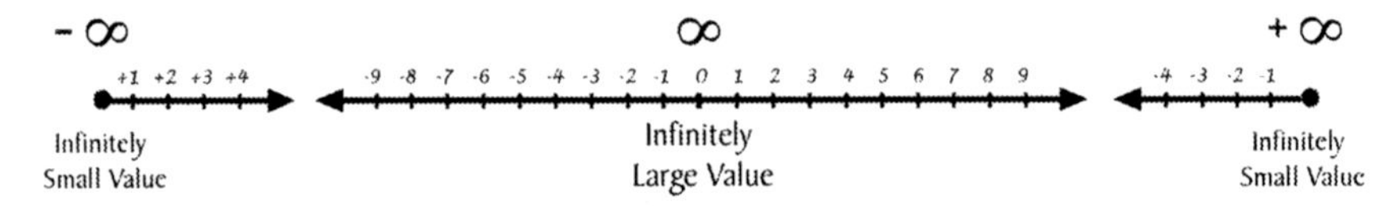

Figure 22.1: In symmetry math the whole mathematical number line is a spectrum of infinities extending out from a whole infinity toward two half infinities.

In ordinary math the nothing of zero cannot relate to positive or negative infinity in the way convergence allows the extreme values of this symmetry mathematical plane to relate. Unlike ordinary math, in symmetry math we can say both sides of the plane are perfectly balanced around zero and not balanced around any other number, simply because Proto and Elea are definite and absolute values. They are definable as numbers, where in ordinary math there aren't ultimate positive and negative numbers, since we can never reach the end value of a non-converging series.

Three Different Answers

I can try now to address the issue of there being three incompatible answers when summing all reals or integers, as mentioned above. Does that same problem exist in symmetry math? There appears to be a solution in symmetry math since symmetry math is a complete and definite system which is symmetrical, perfectly balanced on zero, which at present mathematicians agree cannot be said of ordinary math. If we keep in mind that in symmetry math we have boundaries, zero on one end, and Proto and Elea on opposite ends, we then again consider the equation:

$$(1 + 0) + (2 + (-1)) + (3 + (-2)) + (4 + (-3)) + \ldots = ?$$

As written this equation is an asymmetric arrangement. Since positive infinity is now a number, we can consider the other end of this equation, and realize that the equation is necessarily made symmetrical before it ends. The equation ends with the same displacement that it began with, and since the equation by rule includes all numbers, the equation ends by also summing the number Elea into the final sum, making the full equation equal zero, as shown below:

$$\ldots((+\infty -2) + (-\infty + 3)) + ((+\infty - 1) + (-\infty + 2)) + ((+\infty) + (-\infty + 1)) + (-\infty) = 0$$

same as...

$$\ldots((\quad +1 \quad)) + ((\quad +1 \quad)) + ((\quad +1 \quad)) + (-\infty) = 0$$

or

$$(+\infty) + (-\infty) = 0 \quad \text{or} \quad \text{(infinity)}$$

Figure 22.2: Since the equation begins with the number plane shifted, the same shift exists at the converging end of the equation, producing a remainder of negative infinity to cancel the positive infinity of ones.

Note that Proto can be subtracted from or reduced, but not added to. Elea can be added to but not subtracted from. Also note that in symmetry math the addition of $-\infty$ and $+\infty$ is actually a combining of both sets into an 'everything', rather than a cancellation of positive and negative which creates nothing. Now with the tools at hand, we can use symmetry math to better comprehend the cosmos.

Math and Order

One of the surprising and elegant features of symmetry math is that although the two smallest numbers, Proto and Elea, are points of infinitely small value, each number still represents half of the whole. What do I mean by half of the whole? Where the symmetry value of the number one includes all numbers except (-1), the symmetry value of Proto equals all numbers except all the negative numbers. All the negative numbers must be made separate, just as grouping order separates the game pieces of checkers. All we can do is remove all the negative numbers, or all the positive numbers, from the fullness of zero, which makes Proto and Elea the two smallest possible values in this system. Consequently we don't have a value of nothing in this system. Proto and Elea are as small as any values can become. So even though Proto contains the infinity of all positive numbers, relative to all other values, it is infinitely small.

Obviously the infinitely small value of Proto (and Elea) can be related to the infinitely small size of the Alpha state of the big bang. Just as space-time collapses in our past to a smallest possible size, Proto and Elea are equivocally the smallest possible values in symmetry math. Then just as the Omega zero of symmetry math is infinitely large and contains all numbers, the perfectly flat space of ZAT is most correctly known as an absolute fullness that extends infinitely in all directions. Symmetry math is essentially a mathematical model of the two kinds of order and pattern space.

Thus symmetry math provides insight into the properties of the Alpha state. Relative to itself, the mathematical Proto is all positive numbers unified into a positive singularity. It's a positive oneness, a single color. I believe this is why some big bang theorists have mistakenly identified Alpha as a type of nothing or a vacuum. Some of Alpha's properties such as uniformity are similar to Omega's properties. The greatest difference is energy content. Since the very nature of being positive or negative is imbalance, the extreme positive of Alpha is naturally a very energetic state, a probabilistic energy all aimed at moving toward balance. Once time arrives at Omega there is no longer an imbalance, so there is no longer any usable energy.

Hopefully it is obvious that symmetry math relates so much more gracefully to the large-scale evolution of our spacetime than does ordinary math, particularly so since the WMAP anisotropy probe soundly determined the large-scale cosmos is geometrically flat. Most astrophysicists recognize that if the overall cosmos is geometrically flat presently, then even the Alpha state had to have been flat, which means that it extended infinitely in all directions. A cosmos that is flat in its present is always flat in its present state, even at the very beginning of time. This seems to lead to a paradox, since we see Alpha as a point, and a singularity, but it is simply a matter of perspective. Everything is relative, meaning everything is intimately related and interdependent. Even though we view Alpha as an infinitely small point at our beginning of our time, we can see from symmetry math that Alpha is still half of the infinite whole. Just as Proto is an infinity of numbers, the positive density of our Alpha is an infinity of positive space, it is just not the larger whole infinity of Omega. Viewed internally positive Alpha or Proto would extend infinitely in all directions as a universe within itself. It would be flat and infinitely extended because it has infinite content (I will now sometimes refer the Alpha states as Proto and Elea).

In comparison to other states, Proto is a very simple state that is not further defined than itself, much like the word something. It is all the positives divided apart from all the negatives, and that creates a seamless union of all positives. Semantically Proto relates to the basic idea conveyed by the word 'somethingness'. Alternatively we necessarily perceive anything beyond that somethingness to be a 'nothingness', when it is actually the flip side of the coin, or the negative Alpha or Elea. The other side of the balance doesn't seem as real or valid to us, because it isn't here. We can't see it. We can't detect it.

We also sometimes conceptually deny the reality of Proto in our past. It is said the laws of physics break down at Alpha. We deny Proto because we can't imagine how time could have begun from a single point in time, or from an extremely ordered state. We only view Proto from the outside. Because we view Proto relative to the much larger expanded space of our own cosmos, Proto appears to us to be an infinitely small point. In fact our sense of dimension or size is ultimately relative to both Proto and Omega.

Our most basic sense of scale originates with Proto perceived as the smallest physical size possible in nature, half of the whole, and ends with Omega Zero as the largest physical size. It seems a little strange but we ourselves can seem larger than the infinite extension of Proto. Keep in mind that in science we already know or imagine our physical bodies to be larger than the tiny point of the big bang, which has been described as being as small as a pea, smaller than a proton, small enough to fit under one's fingernail. We don't relate to the internal dimensions of Proto, we are of the larger dualistic world, where positives and negatives combine into something larger, so naturally our world is comparatively larger in comparison to the vast inner Proto universe.

Finally, the size or volume of Proto and our own cosmos are both viewed relative to the much larger absolute fullness of Omega, which is one of two reasons why the expansion of the universe is presently accelerating. We see a ZAT future as a hyper expanded space, with every spatial direction traveling away from every other spatial direction, therein creating the most extreme state of spatial inflation imaginable; which also happens to be the largest Universe imaginable, and naturally so. Due to the basic principle of relativity, we are naturally physically connected to both the bottom-up and the top-down of reality. We are invariably connected to the big picture.

Conclusions

Everyone unconsciously thinks in accordance with a logical framework. That framework is greatly supported by a highly successful mathematical system, one primarily based on the gain and debt of countable things. We see the world either as a thing, or as an accumulation of things, and can't help but imagine the many parts or the whole to be arisen above the backdrop of nothingness. This standard view of zero as nothing (meaning no distinct things or objects) is a valid system of understanding. Our ordinary math is based upon a perspective derivative of grouping order, or thingness. But a view where "the total summation of things equals zero" is also a valid system of understanding. There are two logical frameworks to utilize in order to understand nature, the bottom-up and the top-down. They are not independent. They are complementary. This new and unique mathematical system is derived from symmetry order, with its foundational axiom reflecting the innate singleness and wholeness of existence. As a perspective it doesn't see isolated or individual things. All its values are definitive but infinite, so it cannot divide up or count the world of objects. It sees geometric fields, patterns, curvatures in space, that extend outward infinitely, that always remain enfolded in the whole.

In ordinary math we count upward into an endless staircase of numbers, with no finality or boundary, and thus reality modeled by such a system has no ultimate sum. In that ordinary math fundamentally counts things, there is naturally a number that represents no things, while no number can represent everything. If we instead switch into a mathematical mode that is able to represent everything as a whole, then naturally we find that the system represents reality in an entirely different way. We haven't merely reversed values, we have changed the very nature of our system of understanding.

In symmetry math, infinity is no longer constrained to a never ending process, but rather the infinity of mathematical values is whole, bounded only by infinite extremes. Engulfing the finite, the entire symmetry mathematical plane is real, complete, and consequently quiescent and timeless. In symmetry math, zero represents everything, and because the smallest values of this system still represent half of the whole, we no longer confuse the nothings in this system with non-existence. Nothings in this system are singularities. In a purely philosophical study of the three fundamental states, Omega is of course denoted as everything while interestingly the positive and negative outer poles can be related to ultimate illusory concepts of something and nothing; two singularities annotated the simplest of any two meanings, concepts which are certainly real, but which eventually break down in an ultimate perception of reality.

In considering the new axioms of this system, we would not expect the values of the symmetry plane to be derived from an elementary first thing somehow emerging from nothing or an empty set, as is imagined of ordinary math. All is not magically arisen above nothing. In fact there is no axiom of nothing in symmetry math to imply a non-existence, from which we question the existence of the rest. Not surprisingly I hope, this new system feels more like the primary system, while the axiomatic structure of ordinary math now seems more to be an abstract representation of reality. We just need to keep in mind the underlying lesson we have just learned from this math, that there cannot be a greater infinite without the finite, and there cannot be a finite world of things without the infinite whole.

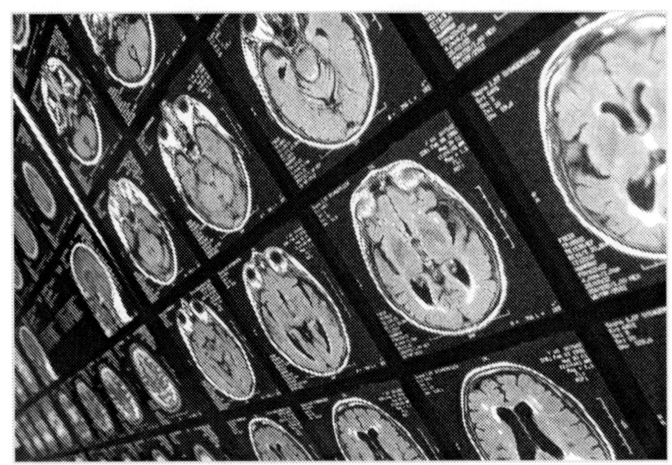

Photo: X-Rays © Jason Woodcock

I am as I am not.

Heraclitus

~~~

Anyone who wants to know the human psyche will learn next to nothing from experimental psychology. He would be better advised to abandon exact science, put away his scholar's gown, bid farewell to his study, and wander with human heart throughout the world. There in the horrors of prisons, lunatic asylums and hospitals, in drab suburban pubs, in brothels and gambling-hells, in the salons of the elegant, the Stock Exchanges, socialist meetings, churches, revivalist gatherings and ecstatic sects, through love and hate, through the experience of passion in every form in his own body, he would reap richer stores of knowledge than text-books a foot thick could give him, and he will know how to doctor the sick with a real knowledge of the human soul.

Carl Jung

~~~

They that see the Real in the midst of the unreal,
They that behold Life in the midst of death,
They that know the One in all the changing
Manifestation of the Universe,
Unto them belongs eternal peace,
Unto none else, unto none else.

Upanishads

~~~

You know the trouble with real life is, there's no chamber music.

Jim Carrey
~~~

There is much to be gained by appreciating differences, and much to be lost by ignoring them or condemning them. But the first step toward seeing others as distinct from yourself is to become better acquainted with your own traits of character.

David Keirsey

Chapter Twenty Three

Our Basic Natures

Defined By What We Are Not

Oneness naturally destroys individuality and uniqueness. Uniformity erases the pronounced differences between groups. When many things unify we end up with the white world of no things. We can try to toss all human traits into a soup pot and cook them all together into a single personality, but our strengths define our weaknesses and vice versa. We may be defined more by what we are not than by what we are. The Swiss physician and psychiatrist Carl Jung defined the opposing sides of basic personality traits which he considered to be archetypes. Archetypes are like attractors, they are the natural possibilities of what we can be in terms of human personality, or rather in terms of life. Like lumpy and smooth, they are out there pulling at us to be one trait or another.

Katharine Briggs and her daughter Isabel Meyers advanced Jung's work with a test that defines sixteen distinct personality types or temperaments. These types were further advanced and popularized in the insightful book *Please Understand Me*, written by David Keirsey. The book outlines four axes, with each archetype balanced around a center, out from which our personalities grow and are defined, as shown:

Extrovert -----|----- Introvert
Sensory -----|----- Intuition
Thinking -----|----- Feeling
Judging -----|----- Perceiving

The first divide describes the range of traits between Extrovert and Introvert. The Extrovert type is not merely more social than the introvert, they simply have a preference for the external world, which not only includes people but places, objects, nature, and activities. The Introvert type is by no means correctly defined as being socially shy, but rather they are defined by a preference for the internal or subjective world of thought, creativity, ideas, feelings, and knowledge. Where the extrovert is energized by their external interactions with people and activities, the introvert is energized by internal activities of decisions, planning, organizing, deciding ethical standards or laws, and exploring spirituality. An introvert needs peaceful alone time to find their center. In contrast, the extreme extrovert needs interaction. They are more comfortable relating to the surface of people while the introverts are more comfortable with the inner depths of people. Extroverts tend to have many friends but tend to be generally less personal or intimate, while the introvert has fewer friends but they are more bonded with the few friends they have. The more extreme introvert might be shy and feel like they don't fit into social situations but many introverts are very comfortable in social situations, especially when those situations are oriented around the inner world of ideas or mental activity. Oppositely, extroverts can be uncomfortable and even phobic about profound ideas or when forced to be alone with their internal world. Three out of four people are extroverts, or seventy five percent of the population.

Next, the Sensory and Intuitive divide describes how a person takes in the world; as how they read or see things. The Sensory Type is oriented more firmly in their five senses, so they read the world and trust what they read in reference to the physical. This makes them healthfully attuned to the here and now, and reliant on the practical. Sensory types typically have crisp memories and they are good at deduction, so they make very practical decisions and accurate calculations. Three out of four people are sensory types and so seventy five percent of the population.

Oppositely, in addition to having the recognized five senses the Intuitive type utilizes what can seem like a sixth sense, as they see the world as an integration of all they have felt, thought, and experienced throughout their life. The intuitive utilizes the vast unconscious mind in order to sense and more deeply appreciate the world. Consequently the intuitive mind is better at seeing the less obvious. They perceive impressions and notice subtle relationships that others miss. They have a very active unconscious that sees from the future, sometimes seeing possibilities rather than actualities. The more they strengthen their skills however, the more they lose touch with the distinct abilities of the left brain. It is only natural that the more crisp and definitive senses get in the way of sensing the aggregate big picture.

Intuitives have excellent but vague impressionistic type memories, and are sensitive to the larger meaning of things. They walk around in a world of greater synthesis and oneness. Jung himself, in finding common symbols in myth and dreams in all cultures, believed in a collective unconscious that is shared by all. This collective is a storage of all human experience, and Jung believed the intuitive type is more sensitive to this background reservoir of knowledge and information. Just one in four people are more gifted intuitively than sensory, meaning they have developed their intuitive voice. I personally remember consciously developing this skill after I realized that I could ask myself a question and days later return to the question only to find that I knew the answer.

The next division is between thinking and feeling. We all know the differences between thinkers and feelers. Thinking types are logical and rational. They consider the effect of actions and acknowledge consequences. They tend to make sound decisions based upon an objective reality, although being thinkers they often under appreciate the more emotional and subjective experience of others. Consequently, thinkers can seem cold or cruel to a feeling type, who themselves make decisions more often based upon their own and other's emotions. Oppositely, feeling types tend toward defending ethical standards and values, and consider how an action, a situation, or a consequence will make them or others feel. The feeling types experience a rising up of emotions in situations and they make decisions in reference to those feelings, which can seem foolish and illogical to the thinker types. The middle ground between thinking and feeling is probably the most favorable balance point of the four axes, since one side without the other tends to produce poor decisions incomprehensible to the other. The number of these two types in the population is even, or fifty percent each, although two thirds of men are thinkers and two thirds of women are feelers.

The judging and perceiving divide doesn't correspond as well to the words chosen. The Judging types prefer life to be planned and organized, they prefer things finished and closed, determined, and settled. Judging types like regularity and sameness. They want a plan of action. A better word to describe this tendency would be organized. Oppositely, the Perceiving types prefer life to be unplanned and irregular. They like to be spontaneous and moody. They prefer to follow the flow of things, to follow the changing rhythms and currents of time, and their own internal flow. A better word to describe this tendency would be spontaneous. In extreme the perceiving types are beautifully harmonious and natural, or free, but also flighty and undependable. Oppositely, the judging types in extreme are inhibited and inflexible. Half of the population is more judging and half is more perceiving.

It is possible to notice there is a common thread running through four types on each division. You may have noticed that the extrovert versus introvert divide is noticeably similar to the sensory and intuition divide. The sensory versus intuition divide is especially similar to the thinking versus feeling divide. A clue to how they relate exists in how the extrovert, the sensory, the thinking, and the judging archetypes are always presented together on the left side, while the introvert, intuition, feeling, and perceiving archetypes are always presented together on the right side. How do all the archetypes on the left relate? How do those on the right relate?

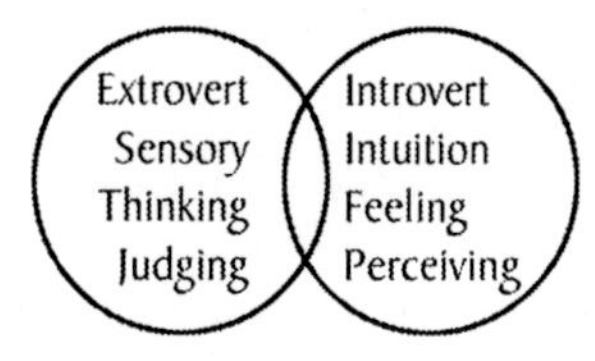

Figure 24.1: Which types relate to the form and definition of grouping order? Which types relate to the integration and enfoldment of symmetry order?

People who are dominated by their left brain are known to be more logical, rational, analytical, and objective. They tend to be focused on the physical. They think in a linear fashion, have crisp memories, and are procedural and systematic. They are verbal, mathematical, practical, and in extreme can seem emotionally sterile. Left brained types see the world as many distinct things, and analyze from the bottom-up. The left brain seems to have a temperament of its own that is extroverted, sensory, thinking, and judging (estj). Interestingly, all these traits can be related to the definition and distinction of grouping order. The left brain is essentially what results of grouping order manifesting as a living breathing personality.

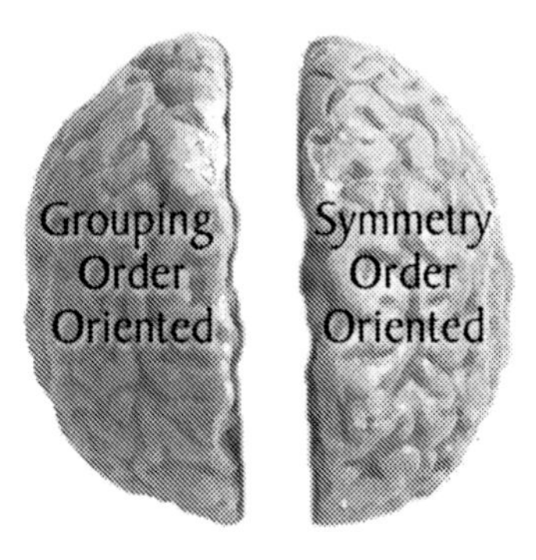

In contrast, people who are dominated by their right brains are known to be more creative, imaginative, intuitive, as well as spontaneous and random. They are more personable and intimate. They think in a more holistic fashion and their memories are more feeling-like, integrated, and even extrasensory. They see hidden relationships, meaningful symbols, and are empathic, artistic, and occasionally even mystical or psychic. Right brained types see the world holistically, more as an undivided whole, and they assume a top-down perspective due to the way they sum up the world. The right brain also has an identifiable temperament all its own that is introverted, intuitive, feeling, and perceiving (infp). The right brain has a personality recognizably aligned with symmetry order, and a nature more in tune with enfoldment and the future.

Personality Space

We could reasonably conclude that the reason we have developed two brains is purely because there are two kinds of order in nature, creating two ways of seeing the world; two modes of perception. The structure of the Universe is equally the structure of ourselves. We are finding that in the same way the cosmos is influenced by attractors in state space, our personalities are influenced by attractors in a personality space. Suppose next we try to map this personality space. How closely is this map going to relate to the larger pattern space of our cosmos? Let's try to determine how distinctly each dichotomy relates to the three axes of two orders.

Sensory characteristics relate easily to the definitive nature of grouping order, they are grounded, distinguishing, focused upon the distinction and form that exists at the surface of the world, while intuitive characteristics relate easily to an enfolded order, they are more in tune with an integration and synthesis of form underneath it all. This sensory and intuitive divide seems to be a fundamental divide, a primary axis of personality that seems to align with the great divide in between Alpha and Omega. Sensory relates to Alpha while intuition relates to Omega. Jung noticed how sensory and intuitive types have difficulty relating to one another more than the other types.

The thinking and feeling divide is more difficult to assign, but as we discover this dichotomy it slowly seems to acceptably fit with the smooth and lumpy axis. We might start by imagining emotions as water waves with a smooth surface, like slow waves on an ocean, while thoughts are sharp and distinct oscillating wave surfaces or a choppy surface of water. Like Morse Code, or computer data, the choppiness conveys distinct information. Thinking and feeling are at least very related to a base common to both. They both come along with an underlying content that isn't definite but with care and effort can be understood or expressed and so made definite. Thoughts and feelings can be understood and translated into one another. Feelings can become thoughts. Thoughts can become feelings. Perhaps the accuracy of truth exists when the crest or surface of thinking accurately represents the entire underlying content of feeling.

It certainly could be said that feelings are smooth and enfolded while thinking is lumpy and sharp. But thinking and feeling seem also to have a multidimensional nature that cannot be understood simply as a linear axis, although this is true of pattern space in general, with each axis pregnant with the next axis.

At first, thinking seems more definitive and less enfolded than feeling however in being related to knowledge and truth our thinking produces human growth forward, while feelings and emotions are youthful and often in error and yet in contrast they also have another side. Feelings can be a guide to knowledge and would seem to be the fruit of both understanding and an emotional unifying with the reality of the world around us. Obviously thinking and feeling both produce states of mind that are advanced in some ways and retarded in other ways in terms of growth, wisdom, and enlightenment. The strengths of one often tend to negate the strengths of the other.

Also in the thinking versus feeling dichotomy there is an obvious masculine and feminine element to the sides which is first diametrically opposite but also past-future. I think in the same way the yin and yang natures are defined both by diametric and contrast opposites, the thinking and feeling dichotomy is both diametric and contrasting, with thinking relating to distinct ideas and concepts, while feelings relate more to a convoluted synthesis of thoughts and perceptions. I once noticed that in the sudden rising up of an intense emotion there also were what seemed to be almost instantaneous thoughts, particularly noticeable in fear. We sometimes forget that every word or idea requires a mountain of other ideas to support it and give it meaning. Words and ideas are like the definitive crest of an underlying wave of meaning. Feelings are like waves without a distinct crest.

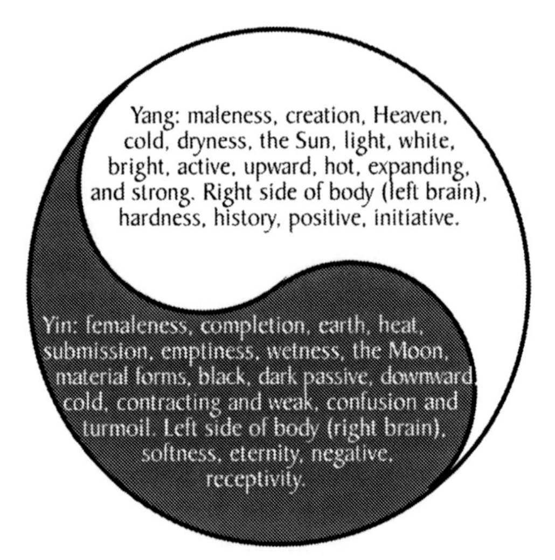

Figure 24.3: In some ways Yin and Yang are diametrically opposite and in other ways they are contrastingly opposite (past-future). The Yin nature is conservative and the heavenly creator of form (grouping order), while the Yang nature creates possibilities and yet is simultaneously destructive or disintegrating of form (symmetry order).

So then, can we reasonably place the thinking and feeling divide at right angles adjacent the primary axis of sensory and intuition? Does a thinking-feeling axis relate well to descriptions of smooth and lumpy? I do believe thinking and feeling effectively relate to smooth and lumpy. Thoughts are distinct ideas, and feelings are less distinct ideas. It also makes sense to see a thinking and feeling axis as an expression and outgrowth of the more primary axis of sensory and intuition, with thinking relating best to the senses and feeling relating best to intuition.

This smooth-feeling connection versus the lumpy-thinking connection actually helps us to further appreciate a deeper level of meaning to the smooth and lumpy extremes of pattern space, exposing them as being more than dry meaningless patterns, instead suggesting an underlying substance and meaning which a more scientific interpretation tends to overlook. It is after all correct to imagine that beneath the surface of the physical world there exists a supportive world of deeper meaning, containing rhythms and interconnections that we aren't ordinarily conscious of.

Finally, note that the even distribution of thinking and feeling types in the population lends support to relating thinking and feeling to the pregnant smooth and lumpy axis. The average distribution of types in the population should be equal, of an axis that exists at right angles to the primary axis. Let's continue on and see how far we get with this.

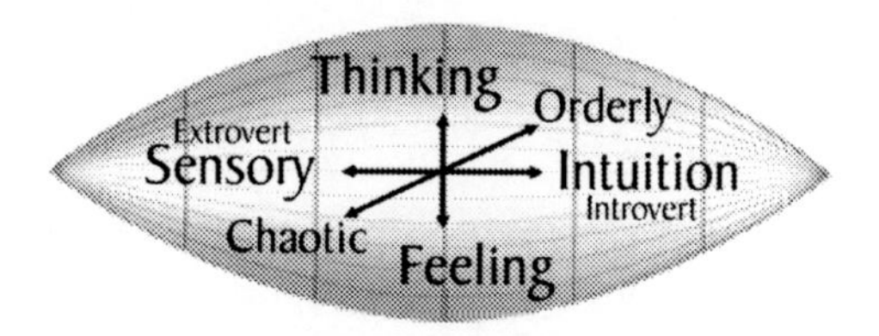

Where correlating the thinking and feeling divide to smooth and lumpy is a challenge, the judiciary and perceiving divide is easily correlated to the orderly and chaotic axis. I can make the comparison without any effort at all. The judiciary type is clearly more comfortable in an orderly environment where the tension between the two orders is high, while the perceiving type prefers the openness and freedom of low tension which in extreme leads to chaos and unpredictability. In fact the terms orderly and chaotic would have been the better choices for naming these two psychological archetypes. Furthermore, we can easily imagine the more complex judiciary and perceiving divide growing out of the more basic thinking and feeling axis. Thinking certainly relates to planning and organizing while feelings relate to the perceiving type's spontaneity and preference for going with the flow. Again, as we would expect of an axis adjacent the primary axis, there are an equal number of these four types in the population, which suggests that what is possible in personality space maintains a balance in the population.

We should further notice that with this second and third divide we don't observe a single sharp balance point, like a zero point, and this is what we should expect of an axis at right angles adjacent the primary axis. An integrated balance of thinking and feeling produces a wide region of sensibility and level headedness, while the medium between judiciary and perceiving creates a very diverse and productive region of personality, such as that which results when a person combines discipline and creativity.

Finally, if we don't imagine the extrovert and introvert divide as social and shy, and we instead recognize the extrovert's preference for activities related to external thingness, and the introverts preference for the internal and a need for periods of solitude, it appears then that the extrovert and introvert divide also is contrasting and runs between past and future in exact alignment with the first axis. This parallel is supported by how sensory types and extrovert types are both represented by seventy five percent of the population.

However, there might be more to this split. If a fourth axis of even greater complexity arises within the third, we would expect that it might directly relate to, or simply be life as we know it, and this divide seems to generally cover the two directions that life can explore, the inner and the outer universes. Perhaps the introvert and extrovert axis relates better to time and life existing along a fourth dimensional axis.

Past and Future Personalities

This model of personality space actually relates to pattern space and the soaps model very well, in my mind suggesting a very deep relationship between life and physics. The model also has become a very revealing model for explaining the various characteristics and attitudes of two more general archetypes, a conservative type and a liberal type. Taking this modeling of personality one step further, I will suggest there are two fundamental directions in society which relate to the past and future. These two opposite groups move in opposite directions, and are more generally prevalent in society than any other dichotomy.

In the global west the definitive side of reality, meaning the accumulation of material things, wealth, status and power, is said to be more celebrated, and more glorified, than in the global east where fundamental opposites of yin and yang are seen as the two underlying natures that animate all of existence. However, these differences if they exist may be global realities of availability, ownership, or power over resources, both natural and technological. In the same region, either in the east or west, people with fewer resources are more likely to idolize living in balance without nonessentials, worshiping the indistinct middle ground as a fullness and source of spirituality and enlightenment, while those more fortunate ridicule living without extravagance and luxury, seeing the alternative as weak but dangerous, as a vacuum that threatens their cornucopia. The struggle between those who have and those who have-not is as old as bananas, and we can only expect a more intensifying struggle as populations grow and resources dwindle in the near future.

There have always been great differences in political systems. We can certainly see the divisiveness of grouping order in centralized forms of government such as a dictatorship or control by one person, a monarchy or control by a family, a plutocracy which is government control by the wealthy, and a theocracy or government control by one religion. Such governments, even a dictatorship, are naturally dependent upon power attained and established by a group of people, however, in every form of government a basic struggle exists between the power and demands existent in the whole of society and the distinct ruling powers or group. There is an innate power to any whole group of people and any government must in some way extract a measure of that power.

We can plainly see a more uniform control and distribution of power in the ideals of democracy or government by the people, socialism where wealth, resources, and production are owned and controlled collectively, and communism which is an extended ideal of socialism where there are supposed to be no classes or divisions between peoples. Unfortunately in each application of uniform government divisions develop, power ends up concentrated in some area, and there in the division between those who have power and those who don't, we find the same basic struggle that exists in a monarchy, i.e., the struggle between individuals in power and the whole.

In these examples we can recognize a series of levels that range between extremes of all the power existing focused upon one position or person, such as a king or dictator, and the opposite extreme where power is evenly distributed throughout society. In the struggle between the privileged elite and the masses, between powerful corporations and the masses, between religious members and the masses, between military forces and the masses, in each we can see the fundamental struggle between definition and wholeness; between the distinctness of a group or individual and the many or the whole. This is the same struggle that exists cosmologically between gravity and expansion. It is the primordial opposition between grouping order and symmetry order. And what we are really talking about is the fundamental difference between past and future; the finite and the whole.

Generally in all politics, whether it be world governments or the work place, there are two fundamental forces at work which we plainly recognize in the division between conservative and progressive worldviews. Present in every country on Earth, the conservative mindset either creates or accepts the divisions between those more successful and wealthy and those less so, between those of particular races, between particular religions, or between social classes. The nature of conservatives is toward grouping order, and like the innate natural biological law of "survival of the fittest", the conservative mindset is comfortable in allowing the successful to succeed and others to fail. It does not believe a political system should burden the successful wealthy with taxes that work to equalize wealth and opportunity, rather the theory is that the wealth of the successful will benefit those who contribute to the survival and strength of the system.

In the book *The Great Divide, Retro versus Metro America*, the for-profit education Billionaire John Sperling portrays the differences between blue-state and red-state America. The book explains that republicans or "Retro America represents 'Old America' with a traditional economy, high levels of social and economic class disparities, and political power firmly based in Caucasian communities." Democrats or "Metro America represents 'New America,' with the New Economy and political power more equitably distributed among ethnic groups."

Everyone knows that the conservative nature is defined by words such as past-like, traditional, old fashioned, self or group-preservation, and self- or group-interest. Indeed the conservative mindset relates to the past, like a human gravity, it wants to recreate the past, it idealizes the way things used to be, and like gravity that pull backward serves to positively maintain the present. Conservative voices are bold, loud, and defined, they are willing to utilize and support power, they know which part of America they stand for, God, flag, and family, which in America still translates to mean Christianity, chiefly United States' interests or (white American) self interests (in 2004 less than two percent of Republicans serving in Congress and state legislatures were minorities), and conservative nuclear family values.

The progressive or liberal mindset is less pronounced and more difficult to define, as might be expected. The progressive worldview looks to the future, it has concern for all. It wants equality, freedom, and a fair and equal opportunity base for all. In extreme the progressive forces want to even out the wealth, like Robin Hood, liberals want to take from the rich and give to the poor. The progressive mindset believes that people are equally gifted and will respond to opportunity positively if given the chance, with an end result that irons out wealth and success, and integrates races. Further still, the more liberal progressives even want to equalize differences between the United States and other countries. Where conservatives want to strengthen borders and restrict immigration, or design trade and immigration to benefit American corporations, the liberal progressives want to open borders and increase free trade in a way that equalizes wealth between nations and doesn't exploit the workers and environments of other countries.

Progressive forces are pro-science, thoughtful of the future, and environmentally concerned, where conservatives, contrary to their title, are more willing to allow businesses to exploit the environment. Conservatives think more in terms of present gains for what they identify with and who they identify themselves as. Conservatives positively champion the self, one's region, one's country, and all levels of concern for the self. They represent the force of becoming defined and strong. Yet in extreme this political force becomes increasingly divisive, naturally moves toward the centralization of power, even to the fascism of tyrannical governments, and consequently sacrifices diversity, freedom, individual rights and equality, for a brutal crushing order related to gravity.

On the flip side, in extreme the liberal progressives, like an emotional child resistant to thought, disregard the brutal facts of reality. They disregard the consequences of limited resources, overpopulation, overtaxing, and the impact of unrestricted immigration. They unrealistically assume all problems are solvable, or simply fail to acknowledge looming threats. They have trouble defining a clear and distinct political goal. In extreme this political force rips apart and expansively tears away the definition and pronounced structure of an individual or nation, even as it increases sameness and unity, all very much like the expansion of the universe closing in on a galaxy or a star.

Threatening individual freedom, personality, and identity, we can sense the dull thoughtless conformity that exists in a future of increasing sameness. We know this threat and the related struggle as the great battle between self and ominous powers that be; big brother, big government, political correctness, social conformity, all threatening even if in becoming reasonably similar and moderate we are ever more able to harmonize and live together peacefully.

This basic conflict between the centralization of power and extreme equalization of power is a common theme in literature and television. Anyone who watches Star Trek is familiar with the Borg. The Borg are not simply one group of aliens, they are a collection of alien races from all over the galaxy who have been assimilated and linked to a collective mind. Each species is forced to become part of one great functioning system linked by computers. Motivated toward a goal of perfection, the Borg assimilate the individuals and technologies of each planet, yet the collective has no respect or appreciation whatsoever for individuality. The result is that everyone assimilated becomes a mindless drone, a slave to the collective will of the whole.

But of course the Borg as villain are never able to threaten the planet Earth directly, rather they loom in the distance, in other more remote regions of the galaxy, where they are growing in uncompromising power. They loom only in the distance because symbolically the Borg are a threat that exists in the future, not the present. The Borg represent the threat of ugly symmetry. Even the Borg's spaceships are giant symmetrical squares.

The Borg are of course we ourselves in the future. Escalating computer and medical technology will undoubtedly lead to implants that dramatically augment and expand our mental and communicative abilities. Cells phones will evolve into cybernetic communication systems interfaced with the brain. The internet will evolve into a pandemic virtual reality, brimming with every imaginable branching of education, entertainment, and commerce.

The future pulls us forward. The past pulls us backward. The pull of the past is noisy and pronounced. The progressive voice is comparatively quiet and not nearly as focused and bold because such is the nature of symmetry order, which is the hidden force behind progressive forces, just as grouping order is the hidden force behind the conservative forces. In the same way that the order of one is the disorder of the other, from our perspective the good of one is the evil of the other.

Each side, each force, is an imbalance for the human world, each is less than the synthesis of both conservative and progressive traits, just as thinking and feeling, orderliness and chaos, are imbalances. Yet if imbalances are what everything we know is constructed out of, if we ourselves, if life is the product of imbalances, then imbalances probably should be placed in high regard, certainly not valued above integration and fullness, but perhaps the definition of form and imbalance should be valued equally to the balance and oneness of the whole. We should and can easily prize the paradox of there being definitive form within the great singularity of existence.

Like gravity and expansion, these two great uncompromising forces shape the global human world. We should expect that beyond the nearly equal but less controlled balance of the two forces in each country, there exists a more constant balance across the planet. If the political conditions in one country swing to the left or right, somewhere on the planet there is a swing in the opposite direction. A centralization of power in one region, unifies other regions against it. It is likely that our individual propensity for one mindset or the other is governed globally as strictly as gravity and expansion hold each other in check.

The Whole Self

In *Please Understand Me* Keirsey stresses the fact that our strengths define our weaknesses. He points out that people are gifted in intelligence, emotion, social skills, and creativity, in various ways, and each person's make-up is dependent upon their particular nature. Each archetype is like a separate dimension that a person is exploring, which by nature tends to exclude their ability to explore the opposite dimension. Along with this model of personality space we need to remember that since enlightened knowledge of the whole involves actually being all form, and so increasingly becoming the model itself, and not simply being the uniformity of zero or attempting to not be anything, it is therefore essential that we work to develop and appreciate each archetype. We actually grow and move forward by developing and integrating both sides of each dichotomy into our personality, in order to become our more full whole self.

But we might wonder then what happens when someone's personality is balanced and resides in the middle ground between the two sides. Many people when tested score even in one or two of the dichotomies and so are balanced in that region of their personality. Logically, such persons can only be either consistent in both traits, or they oscillate regularly between being one or the other. If they are consistent, then a person might be neither one or the other trait, or they might be both traits intensely at the same time. In other words, they might be empty of both traits, or full of both traits. And so we see the challenge before us.

When time ends there is no here or there, no past or future. The definitive world is there, it is everywhere, and yet nowhere in particular. All places exist in the same place, all within a fullness that is everywhere. All meanings exist in the one meaning that is everywhere. The definitive patterns we have explored in this book, the patterns of each moment that make the present feel physically real are very much like words. The patterns are not separate, nor are they separable from the full tapestry which includes all greater depth and even life. The same patterns that shape the cosmos shape our personalities. The same attractors that influence time, on a deeper level influence our lives. We know the pull of the definitive past and our sensory side, we know the pull of the future and our intuitive side, we know of our more definitive thinking side and our more intuitive feeling side, we know of the pull of orderliness and structure as we know of the pull toward both freedom and chaos. Whichever direction influences us most shapes and gives meaning to our lives, and those lives add up to create the collective being that is first humanity, then all life on this planet, and finally all the life of many worlds. The reality of enfoldment means that each is a real evolving body, and each can be recognized as simply greater levels of ourselves.

If matter, separate objects, time, nothingness, are all secondary or abstract aspects of reality, if they are indeed ultimately a type of illusion, not a sensory illusion but more a sort of false impression or a limited truth within the single truth of unity and oneness, if we are moving away from a semi-consciousness defined by this limited truth of separation and grouping, moving toward knowing the reality of the unified whole, then the source of the grand illusion can easily be traced to the past. Like the expansion of the cosmos it traces backward to Proto and Elea, to the original positive and negative division of grouping order, as in, all the white checkers on one side of the board and all the black checkers on the other, and our knowing and becoming traces forward to Omega, to what we define in physics as absolute zero.

I suspect a far deeper level of consciousness than what we can observe in physics, a level where time is simply life. I believe it's reasonable to look at the entire evolution of time as a developing awareness, something Steven Kaufman has referred to as "a 'where'ness". I believe from the very first moment of the big bang a semi-consciousness originates as two mental statements, "I am", and "I am not" present at the very first moment of time. Such are the opposite natures of Proto and Elea, as seen from our side of the great divide. The two are inseparable from the real Alpha state science sees in our past. Each nature is a product of boundaries, even though such are really just assumed boundaries, existent within the oneness of Omega, or pieces of a whole pie. And so the cosmos is not a dry lifeless evolution of patterns that is dying, it is a grander living spiritual evolution away from a semi-consciousness that innately exists permanently in being, and it is evolving toward a fully-aware consciousness, finally becoming the ocean itself.

In each individual life, a person's initial development, the unfolding stage of becoming defined, is analogous to the emergence of a virtual particle or the formation of a sun or planet. It is analogous to some region of the universe gravitationally moving backward in time toward the density of Proto, while the formation and a person's later development, the enfolding stage, is analogous to the expansion of the cosmos or the spiral of a galaxy, with crystals or diamonds, with winter, with electromagnetism, and finally entropy. Of course the final stage of reaching full consciousness is analogous to the physical cosmos reaching zero, although in that age we would no longer be ourselves but perfectly at one with an evolved Proto and Elea.

All the tissue and blood cells, the separate organs, the two distinct halves of a brain further divided into layers and regions, the hands, the nose, the mouth, the ears, the eyes, all cooperate to create a unified whole that hardly notices the distinctions of its parts. In the end all that we view as physically real ends up seeming like a beautiful grand illusion, as all the distinctions we make of form and substance are at heart a belief in, or experience of, boundaries between self and the other. It seems strange to imagine that all of what we know of reality is formed by imaginary almost dreamlike abstractions of the whole. Such boundaries are also the base of selfishness and selflessness, both of which deny one side of a boundary. The self and non-self can be defined as not having full consciousness of the other, and not having full consciousness of oneself. I honestly believe we evolve spiritually and pass through each life eventually traveling away from these definitive states along a staircase of growth and unity toward the greater awareness of "I am, you are – everything". I don't think we can do anything else.

Although technically it matters not, I do suspect there is some sort of bridge between the progression and learning of each life. There appears to be a dreamlike stage of consciousness that bridges our life to some measure of the larger whole. The near death and out-of-body experiences people report appear to be simultaneously very real and very dreamlike. Is reality just a dream that can be dreamed differently if one realizes they can?

The profound out-of-body experience in my youth which contributed greatly to my later understanding of the cosmos and space, made me conscious of the greater depths of reality that seem in comparison to be foreign to daily life. So I have been able to appreciate the experiences of family members and friends who have had out-of-body and near-death experiences. I also have found my own day to day experience of reality to be organized and dreamlike at times, due to experiencing a measure of synchronicity, often linked to the writing of this book, which has been astonishing to witness. Michael Talbot defines synchronicity as "coincidences that are so unusual and so psychologically meaningful they don't seem to be the result of chance alone". There have been so many coincidences in recent years, it has seemed as if the universe was directly teaching me or at least making me think. It has created a truly wonderful feeling of relationship with the Universe, albeit one of questioning and reservation.

In my early twenties, in a sleeping dream, I once found myself lucidly floating above the small ocean town where I lived. I could see the local rivers and lakes and the pattern of streets below me, and strangely I knew the names of each.

Feeling salty winds blowing against my face, I could see or at least vibrantly sense below me an intensity to each name and a deeper meaning to each. I was deeply aware of the interconnectedness of each river and lake and street, which I experienced primarily as a tapestry of meaning related to the names. I have had many lucid dreams since, but I have never remembered a dream near as detailed or intense. As I awoke it felt as if the experience was beyond my sensory abilities, as if it was not simply a dream within my self, but more similar to the out-of-body experience of my youth. I have always believed that it was a real experience of the depths that exist beneath the surface of form, which I suppose would make it an out-of-dream experience...unless it's all the same dream.

Cooperation and competition are the very foundation stones of social life. It is difficult to think of any form of social behavior, whether primitive or sophisticated, simple or complex, that is neither cooperative nor competitive at its most basic level. This is because a person's or an animal's behavior, if it affects others, is generally calculated either to promote the common interest or to give an advantage to the individual at the expense of others.

Andrew M. Colman

~~~

The mortals lay down and decided well to name two forms, out of which it is necessary not to make one, and in this they are led astray.

Parmenides

~~~

Battle not with monsters lest ye become a monster, and if you gaze into the abyss, the abyss gazes also into you.

Friedrich Nietzsche

~~~

Man did not weave the web of life, he is merely a strand in it. Whatever he does to the web, he does to himself.

Chief Seattle
~~~

Nothing is right in my left brain, nothing is left in my right brain.
Unknown

Chapter Twenty Four

Cosmic Lovers

The Psychology of the Fourth Dimension

I sometimes imagine the two directions of time as two great cosmic lovers evolving opposite one another. My own names for them are Proto and Elea but I'd say there are many other names for them in stories and myth. I believe we are each remotely linked to them on a spiritual level. Some would call them one level of our higher selves. The Swiss psychiatrist Carl Jung referred to a human collective unconscious which all human beings share. I believe, on our side of the great divide, there is both a masculine and a feminine collective. As absurd as this may sound to some, this belief stems from the rational conclusion that life is simply change and evolution. If we properly appreciate the physical oneness of the universe we have to conclude that not only are we alive, but the Universe, even being itself, is alive. That follows logically simply from our own existence. Therefore, just as we are parts of an evolving cosmos, we are also like the cells of the evolving bodies of Proto and Elea. I suspect self-identities exist at many levels working upward, similar in the way that countries of people have such distinct and unique identities.

Proto and Elea, the two sides of the evolving collective, relate directly to the proton and electron. I believe they originate at the beginning of time from the big bang unknowing and unconscious of who and what they are. It is a necessity of their natures, the natures of imbalance which they originate as, to originally be ignorant of the larger whole of reality. I believe this takes on two very minimal states of consciousness, one state where Proto is self aware but blind to everything else, and one state where Elea is aware of Proto but blind to everything else, or herself. Then slowly as the universe evolves they learn and grow and wake up in synchronicity with one another. They are as diverse and incomplete as human society as they reflect our present stage of human development.

I believe Proto and Elea have two distinct personalities which originate very pure and simple but then complexify with time. The two basic personalities are manifest in us as well. They can be thought of as states of mind and are easily explained.

We commonly refer to existence generally with two words, something and nothing. These are the two most general terms we use in language. Often we place value on others or ourselves either as being something significant or meaningful or valuable, or we place value on others or ourselves as being nothing significant or meaningful or valuable. The dichotomy of 'something versus nothing' is very basic to how we see the world. Usually, being something important is thought to be a very good state to be in, and not being something important, or being nothing, is a very bad state to be in. At other times self sacrifice is valued and we give ourselves over to something else which we see as more important.

There are four basic mental statements we make constantly as we relate to our own being and the existence of others. These statements are:

1. I am (something).
2. I am nothing.
3. You are something (important, purposeful, of value).
4. You are nothing / or you are not (relevant, important, existent).

People say and think these simple statements constantly in their everyday interactions and conversations with others. In stating we or "I am something" a person is defining a boundary between themselves and what is not. In stating "I am nothing relevant or purposeful" a person is again creating a boundary between themselves and what is. In looking down on someone or a group we are saying "I am something, you are not". In being submissive to another or a group, the person is saying "I am not important, you are". Very few people remain in one statement consistently in every situation and with all the people they interact with, but rather their statements change constantly in every interaction. With one person we feel like we don't match up, like we don't matter, with another we feel ourselves to be superior to the other. We can discover that the myriad of ways that each person defines these statements in their attitudes about themselves and others creates the core and complex tapestry of each person's personality and identity.

First we can note how these four states are very similar to the attitudes identified in Transactional Analysis, popularized in the best selling book *I'm Okay, You're Okay*, written by Thomas Harris. In transactional analysis there are the four basic attitudes a person can maintain, of which the first two are the most common:

I'm not Ok, you're Ok —— (I am nothing, you are something)

I'm Ok, you're not Ok —— (I am something, you are nothing)

I'm not Ok, you're not Ok —— (I am nothing, you are nothing)

I'm Ok, you're Ok —— (I am something, you are something, the same, undivided, wholeness)

Notice how the first two statements of Transactional Analysis differentiate and thus define a boundary between the self and the other. Competitions and even wars exist where groups make statements of "We are, you are not." In such cases one group attempts to expand their personal boundary, and so their definition of self, over the other. These basic statements exist hand in hand with our attitudes and opinions about everything. They largely define where we are in life. But further still, they are the most basic mental statements of all possible mental statements. In forming any kind of identity we are forced to make them. You either are (something) or you are not (something). If you are something then you are defined, and you are pronounced apart from other things, and you are arisen above being nothing. If you are nothing you are held back, irrelevant, below or apart from the world of form.

As a next step, surprisingly and a bit disturbingly, it is possible to recognize that these attitudes intimately relate to what we consider to be masculine and feminine natures. The following mental statement is considered masculine:

I am (something, that is relevant: that which experiences, knows, feels, sees, the one, the best, in extreme: I am all that matters, and even the delusion that I am all that exists).

Imagined Boundary

You are not, or you don't matter (not relevant, not important, not independent, not feeling, in radical extreme: unreal or non-existent).

The following mental statement is considered feminine:

I am not important (not relevant, not worthy, not strong, unable to think for myself, unreal, in extreme: non-existent).

Imagined Boundary

You are everything/something that matters (that which is relevant: important, safety, that which thinks clearly, food and comfort source, creator, i.e., mate, family, boss, job, friends, groups).

It is rather easy to appreciate how the masculine identity places self above others while the feminine identity places others above self. Of course neither of these statements are even remotely exclusive to male or female. We each switch back and forth constantly, and harbor layers of one attitude then the other, such as the deep feelings men feel toward their parents or spouse or family as everything that matters in life, and the identification women sometimes feel (I am) being child bearers or creators of life, to mother nature, God, the Universe.

The 'feminine' nature places relevance outside of self and the direction of energy is out-going. To some degree the feminine state denies their own emotions, attitudes, knowledge, strength, power, or beliefs, thus denying its own reality. A somewhat mild representation of the feminine nature is visible in someone saying, "I am not important, my spouse, my children, my church, my God, is important." This familiar mindset is what we often respect and honor in others and ourselves. It is what we appreciate of our feminine side, as it fosters child care-taking, family commitment, and devotion to miscellaneous groups such as employers, religions, sport teams, communities, governments, and countries, fostering survival of the greater whole beyond self interest.

The direction of energy of the masculine state is in-going. In absorbing energy, to some degree the masculine denies the reality of another. A common representation of the masculine nature is visible in someone saying, "I am the better, stronger, smarter", "I am the best", "I am right", "I am what matters". Beyond caring more for oneself, the masculine nature tends to assume ownership and thinks of spouse, children, group, community, more as an extension of themselves than as a separate and independent entity. Consequently, there is always some degree to which the masculine denies the separate existence, experiences, emotions, attitudes, knowledge, power, and beliefs of the other. This denial is particularly exercised toward those considered to be lower in status or less fortunate, meaning those who are perceived as smaller, weaker, less attractive, poor, less skilled, less adept, or less powerful.

It is a bit shocking to realize how much time and mental activity we invest in these attitudes of identity instead of being fully conscious of the reality of all. Again, we each ordinarily move in and out of these attitudes in different situations, so it is easiest to relate these attitudes to people who are extremes on the personality scale and seem to constantly remain in one state or the other. An obvious example of the masculine "I am" is someone who is arrogant, egotistical, driven; someone who recognizably needs to be dominant and have power over others, or needs to seem more important than them. Naturally what creates the "I am" drive is the fear of being the negative other, or of being what they perceive to be a nothing, or someone unimportant. At some level the "I am" extreme harbors a great fear of the "I am not" statement. This kind of overlapping of mental states is common in the full tapestry of our personalities.

Some people are stuck in exercising the "I am not" statement and they identify their existence instead with others or something outside of themselves. They might identify with another person, their family, their job or employer, or some large club or group. We all to some degree at different times identify with something outside of ourselves. We all know and appreciate the attitude of the wife who considers herself unimportant in comparison to the well being of her children or her husband. We all know the mentality expected of the foot soldier who fights for their country. At some level someone stuck in this nature harbors a disdain for selfishness and aggrandizement. Of course various ways of mildly suspending the self are necessary in any cooperative venture, such as raising children or facilitating a business. The self is normally lessened in the workplace, given over to business success. People often find self-identity through being associated with a club, a political party, a sport team, a famous person, or a geological region such as a state or country. Sports are almost always a competition over obtaining the rights to the statement “I am the best” and yet interestingly the cooperation necessary of a team requires individuals give themselves over to the team.

I have known about these basic states for many years, it is easy to identify them in human situations, but I still have only begun to see the incredible dynamics within each person, myself included, as we alternate back and forth between each attitude and the next in different situations. I find it amazing how behind one attitude there exists layers of opposite attitudes somehow woven together into a cohesive way of thought. The husband stands tall claiming “I am the man”, yet in the next moment all his work and accomplishments are seen to be for his wife and children. We all oscillate back and forth between similar attitudes. Fortunately, as sure as the direction of the universe is toward zero, it is cosmic law that we all eventually end up at the "I am, you are" state, and so finally we end up in an "I am the Universe" stage of awareness where there are no boundaries.

The Ugly Extreme of “I Am Nothing”

It is always helpful to expose extremes although we should appreciate them as such. In extreme the “I am nothing, you are everything” attitude sees themselves as being absolutely nothing relevant. To function such persons typically become followers who commit to some person, religion, group, or a cause, and sacrifice all their own choices, powers, and perceptions. The extremes of this personality are marked by a seeming inability to think original or independent thoughts, due to their own avoidance of such thoughts. By nature, they internally restrict objective interpretations of reality. The person actually shrinks toward becoming a point surrounded by all else. The person even feels as if they exist within the point and is only vaguely aware of the real self with a sense of withdrawal and contraction.

One example helpful in understanding the effects of the extreme “I am nothing” neurosis is someone who has destroyed their own ability to think rationally, a common example being where a person has given themselves over to religious dogma. In cases of schizophrenia, the person minimizes themselves to the degree that the conscious mind is increasingly disabled from having any influence upon its own content, the result being any cyclic fear or irrational thought can permeate consciousness, leaving the body helpless against paranoia and hallucination. At the core a person maintains an internal perception that one's self is nothing, leaving the person unable to act or control their own mind based upon their own self interest or preservation.

"I Am Everything" Ugly Extreme

Of course the extreme result of "I am, you are not" is not much better, since instead of denying the self, here the person denies a large part of the outside world. This person is naturally isolated within the boundaries of self. They have continual difficulty making the outside world cooperate with their aggrandizement of self, so eventually in extreme they remove themselves from the persistent leakage between their self and the greater reality. They break away from an outer reality and form an inner reality. They resort to imagining their own attitudes and self to be all that exists or matters. In this case the internal chaos is self designed.

In most cases events in a person's life pull them back from the inward spiral toward either extreme attitude, away from positive or negative directions toward a more reasonable balance, in which case the person's world begins to expand as the boundaries between self and others dissolve. However, it is interesting that with those who are not pulled back, in both cases, even the person trying to expand their borders outward, instead finds the result to be an ever shrinking sense of control and power, neurotically resolved only by further denial of the other, denial not only of the outer world but eventually one's own physical senses and raw perceptions. In psychosis the boundary of self ends up shrinking inward toward a single point from which the person finds no solace. It is no coincidence how this direction toward a collapse of self directly correlates with the shrinking of a star into a black hole, or the shrinking of the cosmos backward toward the big bang, as well as the shrinking values of symmetry math toward Proto and Elea.

Expanding and Contracting Boundaries

Our less extreme identities are constantly changing. In the natural quest to be significant, an ordinary healthy person tries to expand the self, or what they see as oneself, outward, and ends up doing this either creatively or destructively. A person might expand their sense of self to include others either by becoming dominant or controlling of them, or by submission to the control or mindset of another person or group. We might imagine that each individual or group has a circle (boundary) around it which can expand outward. Like borders between two states or countries, we create these mental statement boundaries, so naturally we can move them around.

Often we pretend our influence on others expands our own personal identity and makes us greater than we are as an individual person, such as when we become the person in charge, or when we own property or a business. Oppositely the boundary circle can also contract inward so that what exists in the outside world to some measure overruns the real self or the body of the person, which allows the person to identify with that other. We expand our boundaries creatively with art, with writing, with music, with education. But we can also influence the world around us destructively with vandalism of property and disruption of improvements. People can lie and gossip and distort the truth intentionally in order to influence the world around them toward their own ignorance or chaotic internal environment. Or they can teach or stand for truth and knowledge in order to make the world a better place.

The "I am", and "I am not", are often described by psychologists in the study of human behavior and personality. In a must read book, On Disobedience, Erich Fromm describes what he called basic dominative and submissive attitudes. Fromm writes:

> Man is torn away from the primary union with nature which characterizes animal existence. Having the same time reason and imagination, he is aware of his aloneness and separateness, of his powerlessness and ignorance, of the accidental-ness of his birth and death. He could not face this state of being for a second if he could not find new ties with his fellow man which replace the old ones regulated by instincts.
>
> There are several ways in which this union can be sought and achieved. Man can attempt to become one with the world by submission to a person, to a group, to an institution, to God. In this way he transcends the separateness of his individual existence by becoming part of somebody or something bigger than himself and experiences his identity in connection with the power to which he has submitted.
>
> Another possibility of overcoming separateness lies in the opposite direction: man can try to unite himself with the world by having power over it, by making others a part of himself, and thus transcending his individual existence by domination.

In being a creator a person moves toward unifying with the environment, expanding their sense of self outward by positively shaping the world in their own image. In being a destroyer or vandal a person expands their circle as well, they expand their sense of void or personal chaos, in order to destroy the positive, to destroy that which makes them seem less relevant or valuable. In destructively expanding one's own state they become big enough to encompass the problem they feel they cannot solve any other way, which is primarily the identification of oneself as negative derived from their relationship with others, not necessarily others who are positive, but who define themselves as positive or better in relation to them.

Typically in being destructive a person has not yet learned to be creative, to be functional, or to be cooperative, which once learned a person tends to utilize such skills and refuses to abandon them. Also the person hasn't learned to form a stable identity irrespective of the opinions and influence of others. Cooperation is also a general way that we positively transcend the imaginary boundary of self toward a union of self with the rest of the world.

A full union with the world is clearly visible in the "I am, you are" statement. "I am, you are, the same existence. We are both the universe." Such conclusions are ultimately inescapable. All individual identities are merely secondary aspects of a unified reality. As much as we seem to be separate and individual, ultimately we are not.

Physically we are each made of the same atoms which exist in other people and the objects in our environment. Each person is merely a different pattern of those objects. We each constantly trade atoms with our environment as we breathe, as we take in food and exfoliate. We also constantly absorb and emit heat and energy, just as we constantly absorb and emit all kinds of information through our senses and muscles. We each are a past and become a future. We are absolutely dependent on the universe around us because we are inseparable from it. The separation of objects and boundaries we assume are ultimately illusions. There are no existentially independent objects and no disconnecting boundaries. We are merely a Universe even as the definition of the world and the self is real.

Through the course of time we are each naturally evolving away from identities and boundaries toward the statement of "I am the universe, you are the universe...We are the same, we are one being." It is a very long journey, but when those boundaries do finally break down, seeing the wholeness and unity of the universe is not merely an attitude, it is more a quite different reality and very unique way of seeing existence. The difficulty is of course simultaneously becoming all that the self is capable of becoming which is absolutely necessary of the final transcendence. The enlightenment of "I am and everything else is" requires we actualize the self and not abandon the self while we fully know and appreciate all else.

The Transparency of Good and Evil

Conflicts are naturally disturbing, and thus the boundary between the self and everything else is disturbing. Since we assume that something and nothing contradict one another, and essentially imagine each as an individual reality, we often attempt to resolve the conflict between one state and the other by making one overcome the other. Describing this, Fromm writes.

> Man can create life, by giving birth to a child....by planting seeds, by producing material objects, by creating art, by creating ideas, by loving one another. How then does man solve the problem of transcending himself if he is not capable of creating, if he cannot love? There is another answer to this need for transcendence; if I cannot create life, I can destroy it.

We all can appreciate the necessity and importance of self identity, self maintenance, self control, self exploration, and most of all, self interest. Competition for the basic necessities of life is a reality. A measure of self concern and self importance exists in each individual, in each group, in each city, and in each country. Self interest is necessary for self survival, and then beyond survival, individual desires and self expression drive us to create and actualize the potential of each person. And yet existent within desire, within those very same essential drives that make us alive we find the potential for the extremes of selfishness and evil. What is evil if not the myriad of ways that people deny the reality of others for self benefit?

The world is not simply divided up between good and evil and it is easy to misjudge the good or evil of individuals and groups, in fact this is commonly done based upon self-interest and self-identity. What I personally find most evident is the deep hypocrisy in human beings, where those who have experienced a predominantly pleasant life define what is right or wrong, good or bad, merely from the base of their own life experience, without having any respect for the real conditions and experiences of others who have had very few pleasantries and instead destructive experiences in their lives.

I strongly believe it is generally the denial of others (or self) that is evil, and that there is a deeper measurement of what is good or bad, because the rich and beautiful are not the good people and poor and misfortunate people are not the bad people. The more healthy people I have known have experienced lots of pain and yet still had lots of love and good things in their life, so it all balances out for them and they learn from both sides. The bad experiences give them compassion and awareness for others and the good gives them strength. But are even these people better than others? I don't think so at all, they are simply the most fortunate. They aren't blinded by having just good, and they aren't disabled by having just bad. I don't really believe in bad and good people. There is just life. There are just fortunate and unfortunate people, or rather, fortunate and unfortunate periods for what we each are; the life force passing through time, caught between the two orders, existing as a fragment, as less than, working our way home, but often times deeply lost.

There are so many different settings to people's individual lives and each setting establishes a different perspective on reality. None of them can be correct since each is incomplete, and the incorrectness of both sides is by nature quite extreme at this semi-conscious stage of human evolution. What is most certainly always unhealthy and can reach the extreme of being evil is the extreme attitudes of "I am, you are not", or "I am not, you are". Both are extreme imbalances and both are related to the cosmic past. Both result either in damage done to others or in damage done to self. The evil of being destructive to self is often overlooked, even though damaging the self is technically the same as damaging another when we recognize everything as a oneness. A great deal of confusion exists due to how we simultaneously idealize and chastise self importance and self love. We chastise the self in part do to the harm done by extreme selfishness but more so it has to do with the fact that the fundamental direction of self is toward becoming defined and pronounced, so it is a direction related to grouping order, a backward direction toward the past. But we idealize the self literally for the same reason, as self is the rising out of uniformity and sameness.

Long ago Heraclitus said, "I am as I am not." The definition of self is quite a paradox. As any increased definition or form moves away from balance and unity and fullness, it moves toward distinction, individualization, uniqueness, and diversity, all highly appreciable qualities. It is critically important to appreciate the direction of self development and actualization apart from the fallacies of selfishness, and yet they are intimately related. The realization of the self is extremely important and healthy. It is what we are, definition, form, even if all such form relies on imbalances and parts being absent from the whole. Know thyself, even if ultimately, we are defined by what we are not, rather than what we are.

We must always keep in mind that the unified whole would be a non-existent nothing without finite form. There cannot be a great cosmic balance without imbalance. There is no infinite whole without the finite. There is no God without life. The innate objective of life and growth is to become whole by becoming full, rather than becoming one by becoming empty. So actualizing our individuality and being what we are is a healthy and required part of that process.

As we explore each part of ourselves we become fuller. That path first unfolds toward definition and form, then enfolds back toward unity. It is a twofold path. So our evolution is very much a sort of exploration of imbalances and individuality, a road that often leads away from unity even if the same direction eventually produces transcendence past illusory divisions and ends at unity.

"I Am Moving in A Direction"

Life moves in directions. Life is change and life is always moving away from one place in some other direction, but like time, like the stars and galaxies, life doesn't just evolve forward. Sometimes life and consciousness moves backward, or sideways. Another psychologist, Karen Horney, who I personally believe was the most brilliant and important theorist of all the great psychiatrists, describes what she identified as basic conflicting attitudes toward others as well as three major attempts at solution. These solutions are essentially unhealthy directions of motion or movement in respect to the external world, directions in respect to other people. Horney identifies her three basic personalities as:

1. Moving against people — *(I am, you are not)*
(domination; expanding one's self boundary)

2. Moving toward people — *(I am not, you are)*
(submission; collapsing one's self boundary)

3. Moving away from people — *(I am not, you are not)*
(resignation, destructiveness, nothingness)

Since Horney is defining neurotic attempts to solve conflicts and problems, she doesn't identify an "I am okay, you are okay" or an "I am, you are one" type of statement as one of her directions, since it is a healthy state of mind. The excess of the "I am, you are" statement merely results in a greater sense of unity and oneness, greater cooperation and awareness, and not neurosis. The other directions in excess become problematic and generally define neurotic behavior. In one of her early books explaining her model, entitled *Our Inner Conflicts*, Horney describes her first recognition of a system behind her patient's neurotic trends:

> I could see that a neurotic need for affection, compulsive modesty, and the need for a "partner" [I am not, you are] belonged together. What I failed to see (yet) was that together they represented a basic attitude toward others and self, and a particular philosophy toward life. These trends are the nuclei of what I have now drawn together as a "moving toward people. I saw too, that a compulsive craving for power and prestige and neurotic ambition [I am, you are not] had something in common. They constitute roughly the factors involved in what I shall call "moving against people."

The *moving toward people* and *moving against people* types can be understood as inversely or diametrically opposite. Existentially the two types are interdependent. They could also be described as *moving over people* and *moving under people*, and they easily relate to the basic states of "I am" and "I am not". The *Moving away from people* type is a contrast opposite to both. The moving away type is marked by resignation and withdrawal from human relationships and the conflicts associated with life in general. This type responds to emotional stress by attempting to outgrow personal relationships, and attempting to become totally self-sufficient. In discussing an extreme case of *Moving away from* Horney writes:

> Instead of moving away from others, the neurotic moved away from himself. His whole actual self became somewhat unreal to him [I am not, you are not] and he created in its place an idealized image of himself in which conflicting parts were so transfigured that they no longer appeared as conflicts but as various aspects of a rich personality.

I included this part of her writing in part because of the way she describes how the false self is like a separate reality existing apart from the actual self. Each type develops a false image of oneself which ends up overrunning the genuine self. I believe the false selves are very much like computer programs, they are separate realities programmed within the actual reality of the self. In order to solve conflicts and deal with complex challenges, we unconsciously program them into our selves as a way to adapt. What makes them neurotic is how the attempt at solution ends up creating a problem, so the program locks us into a vicious circle of applying a wrong solution to a problem that is created by applying the wrong solution.

Horney's overall theory of personality explained in *Neurosis and Human Growth* is so compelling and applicable to every person that one finally realizes that neurosis is inescapable, it is a part of being human. It comes along with having a mind and the development of thinking. It is a part of the evolution of semi-consciousness to consciousness. We naturally inherit neurotic solutions due simply to not knowing how to solve every problem. Although Horney does not often focus on what is healthy in her writings, she does at least explain that growth is an evolution away from neurosis toward the 'whole actual self'. Wouldn't you know it. It is always easier to describe the more empty definitive imbalanced selves than the whole actual self.

An Enlightenment of Emptiness or Fullness?

I have tried to explain that our ideas of somethingness and nothingness are realities within themselves, and although they are real, they are only real to that extent. The separateness of things and seeming nothingness of space are both only secondary aspects, abstractions, half truths, of the reality of wholeness. Yet in the same way that the Alpha of the big bang is an individual and an absolute state, something and nothing are both complete universes within themselves. I believe the two concepts of something and nothing, the most elementary divide in all meaningfulness, form the Proto and Elea personalities at the beginning of time. "I am" is the nature of Proto while "I am not" is the nature of Elea. Those original states, the neurotic selves Horney identifies, and the illusory "I am" statements create the grand evolution of time and consciousness.

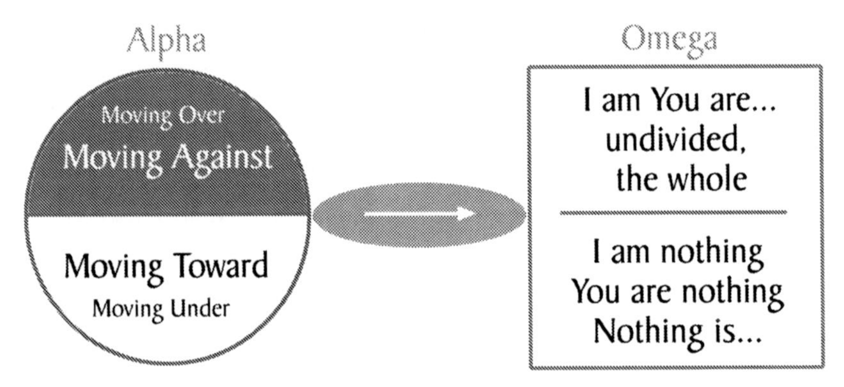

Figure 23.1: This diagram portrays how the moving toward and moving against types relate to Proto and Elea of symmetry math, and we recognize them also as the high contrast polar opposites of grouping order and Alpha. The neurotic states of semi-consciousness are intimately connected to the evolution of the universe.

We can identify the moving toward and moving against types as inverse or diametric opposites. Both types place relevance on one side of a boundary and devalue the other side of the boundary, be that side self or other than self, as in

"I am relevant, you're not", and "I am not relevant, you are". Note however there is an asymmetry here that we should be conscious of. Both statements exist on our side of the great divide, like the proton and electron. The Proto mentality is the positive side denying the reality of the negative side. The Elea mentality is here, like the electron, it is a part of our system, but it relates with the other side of the boundary and it sees that side as nothing, just as we ourselves are blind to the inverse universe. Both impressions are in a sense correct. The singularity of either Alpha can be seen as the ultimate somethingness or a real nothingness. The only difference is one of perception. In fact on the other side of the divide, what Elea sees as nothing is actually identical to Proto.

Fortunately there is the 'whole self' type in the future and also the 'moving away' type, which are related in an odd sort of way, as they both can describe Omega. The statement of the whole self type is "I am, you are", which translates into "we are one, we are everything, we are the infinite whole". The statement of the moving away extreme is "I am not, you are not" which translates into a rather nihilistic attitude of "I am nothing, you are nothing, everything is nothing". Interestingly, when compared, the 'moving away' type and the 'whole actual self' type are diametric opposites. One statement sees Omega as not relevant or as a 'nothing', and the other statement sees Omega as being very relevant and as an 'everything'. Again both are measurably true. Omega is a real nothing and a real everything simultaneously.

In eastern philosophy there are two angles on the path toward enlightenment, one which focuses on a type of nothingness and emptiness, and moves consciousness toward an enlightenment that claims to be indescribable in terms of being, which would seem then paradoxically to be non-existent. This enlightenment is often described in both ancient and modern writings using the term emptiness, or beyond emptiness. Similarly if not identically, the moving away neurotic personality type in extreme is the "I am not, you are not" state of mind, or "all is nothing". When described, usually poetically (which is quite paradoxical) this state of enlightenment seems almost to see being from a distance, as expressed in this verse of Jalal al-Din Rumi:

> Not Christian or Jew or Muslim, not Hindu, Buddhist, Sufi, or Zen. Not any religion or cultural system. I am not from the East or the West, not out of the ocean or up from the ground, not natural or ethereal, not composed of elements at all. I do not exist, am not an entity in this world or the next, did not descend from Adam and Eve or any origin story. My place is placeless, a trace of the traceless. Neither body or soul. I belong to the beloved, have seen the two worlds as one and that one call to and know, first, last, outer, inner, only that breath breathing human being.

The idea of experiencing nothingness as sometimes advocated by eastern philosophy has always seemed to me to be a strange goal, although the climax of existing in such a state certainly produces an interesting and attractive state of mind. I personally expect the peak experience is of 'nothing except existence' and thus no things. I think attempting to experience a non-existence type of nothing simply bounces someone back to clearly appreciate the innateness of being. In a sense, you can't get there from here, but you do at least touch upon the surface of an innate existence. Many if not most ancient eastern writings seem to toggle between describing the ultimate enlightenment experience as emptiness and fullness.

I think the more direct path toward enlightenment in eastern philosophy moves consciousness toward an awareness of the totality of being all unified into a physical oneness. Those returning from this enlightened state, as well as those having spiritual awakenings or rare out-of-body experiences, all describe an integrated totality, a fullness, with the person ultimately seeing one's self as everything. This state of mind equates to the "I am, you are" statement. I believe expectations and intuitions of this state in the east and west lead people to generally expect the existence of a God.

Likewise there are two angles in world religions. Some writings portray God as separate from the physical existence of the universe, as if God somehow isn't a part of being and is unrelated to existence. Others portray all life and form to be part of one existing God. Descriptions of God seem to toggle between all things being separate from an incomprehensible God, and God being all, omniscient, omnipotent, and omnifarious, which would arguably place everything within God. In this same dichotomy, conservative science has tended toward seeing empty space and absolute zero as being non-existent or impossible, while the more idealistic or intuitive scientists and philosophers have expected a perfect symmetry or infinite whole.

We can relate these two perspectives to whether a person views the surface or the depths of Omega. These two perspectives are two different views on the personality of God, of existence, and of the Universe. One view sees God or the Universe as indifferent, as not having a personality that relates to life forms, where the other view sees an omniscient personality in the integration of all life and being. Which is correct, empty or full? As I am sure you know by now, I personally am soundly rooted in a belief of fullness, and I also somewhat religiously believe an eternal enlightenment no one ever returns from is out there destined for us all, because all life shares one reality.

They flutter behind you
your possible pasts
Some bright eyed and crazy,
some frightened and lost.

Roger Waters

Leafless Willows: © Sergey Pristyazhnyuk

Part Seven

A Spiritual Science

Adjusting to timelessness and Omega

Not many people see their life as a statement of reality, or live their life accordingly. Most people recognize their lives to be meaningful only as they live in the passing of the moment. We sense the present as change, not eternity. We feel the solidity of each moment only in a sense of the immediate past and a faith in the immediate future. So the question arises, would people behave any differently as they approach the future if they knew that their decisions and actions define themselves forever in timelessness?

The reality we don't sense is that each moment has existed forever and each will continue to exist forever. Forever has already happened. The moments are unalterable because ultimately there is no form of time in which to change what exists. Such would be like trying to change what can be imagined, or what has already happened. The real pattern space and the imaginable big picture are both just the way existence is. Why I wonder have we ever considered that we might be able to go back in time and rewrite what exists, if not for just the entertainment value. All that happens will always exist forever. And it isn't that we arrive in time at a place that already exists. We were there at the beginning of time, and are there at the end. We are just there. We have always been there.

How might people react to learning they cannot change who they are? I certainly don't mean to suggest that we are not defining or creating ourselves as we flow through time. The extremely complex theme of timelessness reduces to a very simple principle that we can't change the timeless Universe, and yet I make a decision and take action and a particular future unfolds in recourse. All the decisions have already been made. And yet we still have a measure of power to determine where we find ourselves in the future. We have a power to shape what part of timelessness we experience, even if events are also guiding us toward a future goal in a way that is beyond our control.

Intelligence, cleverness, personality, are all valued for their ability to select and create the future of greatest benefit to ourselves and others. We envy and desire strength and beauty because it seems to create a future of greater benefit for the individual. We idealize goodness and self sacrifice because it selects a future of benefit to others. How might people react to learning they are perfectly at one with the whole of the Universe? What if people were given the chance to appreciate the fact that they are purely a personality riding along time like the turbulent surface of a wave in an ocean?

In timelessness there is a much stronger sense of being at one with the flow. There is a stronger sense of being part of a collective. What kind of person will embrace such knowledge? What kind of person will deny the simple reality of it with disdain and contempt? We all occasionally exercise divisiveness over unity. It seems to go with being human.

It is wonderful to know that we ourselves aren't an anomaly and that we should exist, we are as perennial as the trees and the stars. Nor are we beyond just the appearance of such, we are not a separate materialistic thing existing physically apart from the rest of the objects in the universe. There is no gap filled with nothingness dividing our physical reality apart from the reality of other things. It is as if there are invisible strings connecting all things together. And yet the reality of our defined selves is a powerful force that actually feels threatened by the thought of such oneness. Many do not want to know of how interdependent and inseparable we are. Many do not want there to be an alternative way of viewing reality.

People fear change. They fear being wrong. I think Eckhart Tolle pretty much hit the nail on the head by explaining that the way people identify with their minds causes us to fear being wrong as critically as we fear death, since being wrong is essentially a form of mental death. The psychologist Samuel Bois argued that we each live in our own reality, evidenced by how we each interpret and react differently to words. Speaking of our individual realities, Bois writes:

> It is the world of whys and becauses, the world that makes sense to us, the world of our logic, preferences, and principles. It is the semantic world that we project upon the empirical world, forcing both of them into a whole that becomes for each of us the only real world. If you threaten to shake loose the structure of my world by inviting or forcing a change in my key notions, my all-pervading values, or my well-established preferences, you encounter a resistance that mobilizes all the resources of my mind, soul, and physical strength. I cannot let my world crumble. Its collapse would mean for me the end of the world!

If we ever manage a collective view of the big picture on this planet, it would have to be derived from actually understanding the infinite. What else would be truly universal? But how will that comprehension influence us? How different is such a view than how we view things today? I personally can't help believing that our learning two orders and applying it to the realm of possibilities will one day create a sophisticated mathematical and rational science, even a unified theory of nature. There will be one single comprehension that far outperforms any other attempt at representation. That prospect is wonderful and yet kind of scary. How will we respond to actually knowing something about the mind of God?

I do not feel obliged to believe that the same God who has endowed us with sense, reason, and intellect has intended us to forgo their use.

Galileo Galilei

~~~

I think of consciousness as a point, an "eye," that moves about in a sort of mental space. All thoughts are already there in this multi-dimensional space, which we might as well call the Mindscape. Our bodies move about in the physical space called the Universe; our consciousnesses move about in the mental space called the Mindscape...Just as rock is already in the Universe, whether or not someone is handling it, an idea is already in the mindscape, whether or not someone is thinking it.

Rudy Rucker *Infinity and the Mind*

~~~

The observer appears, as a necessary part of the whole structure, and in his full capacity as a conscious being. The separation of the world into an 'objective outside reality' and 'us,' the self-conscious onlookers, can no longer be maintained. Object and subject have become inseparable from each other.

Fritjof Capra

~~~

The will arises from the pool of all consciousness - a pool formed by small contributions of each without spatial or temporal bounds. This collective will has the power to bring about events in the physical world that transcend the physical limits of information transfer or kinetic events, suggestive of (but more complicated than) the ideas of omnipresence, omniscience, and omnipotence.

Evan Harris Walker

~~~

Death is not an event in life: we do not live to experience death. If we take eternity to mean not infinite temporal duration but timelessness, then eternal life belongs to those who live in the present.

Ludwig Wittgenstein

~~~

When one sees Eternity in things that pass away
and Infinity in finite things, then one has pure knowledge.

Bhagavad Gita

~~~

In ultimate analysis, the universe can be nothing less than the progressive manifestation of God.

J.B.S. Haldane

~~~

He felt that his whole life was some kind of dream and he sometimes wondered whose it was and whether they were enjoying it.

Douglas Adams
~~~

You are a child of the universe,
no less than the trees and the stars;
you have a right to be here.
And whether or not it is clear to you,
no doubt the universe is unfolding as it should.
Max Ehrmann, The Desiderata

No one was ever wise ahead of time.
Tom Gillette

Sunset Dreaming © Iwan Beijes

Chapter Twenty Five

Becoming Aware

Growing Forward into an Inevitable Future

My favorite analogy for describing the timeless big picture is a simple book with pages that tell a changing story. As we follow the story and turn the pages, naturally the rest of the book's pages continue to exist. The story has a beginning and ending, but only the story actually begins and ends. The book, each page, each word, is always telling the story. The one great book of all stories has always existed and will always exist, and all the pages of the one book collectively shape the stories within, especially the common end for all the stories. But what about each person? Are we each a storyline moving from cover to cover, is each life a chapter, or just a few words on a single page?

I don't know if we evolve through many lives, but if we do exist in some form prior to this life and then somehow cross into this world from some other place, we seem to lose all trace of memory of ever having been someone else. Perhaps past memories are forgotten during years of infancy. Perhaps the human life we become is much more than what we were, so that the past seems vague and unimportant. Perhaps we just forget for a while. I only know that it really doesn't matter, because Omega necessarily includes every life each of us can possibly live. So it doesn't matter if the identity of each person in some way consciously transitions from one life to another. I think the more interesting question concerns the enfolded side of ourselves. In some way our lives are enfolded into a higher self who, somewhat like a guardian angel, must move along with us through each life.

What would it be like to know this higher self, the enfolded side of ourselves, to suddenly remember an ensemble of many lives, and in those memories find one life had been very difficult with abuse and not much love, so that we never gained much trust or love for others, and became hateful and abusive ourselves, while another life had been filled with kindness and positive support, so that we found it so much easier to be happy and maintain a sense of self respect. In this way it seems the enfolded whole might be considerably more than the sum of its parts.

What would it be like to remember the lives of the great philosophers like Plato or Socrates, but also remember a lifetime mentally challenged in some way. What would be gained in comparing the experiences of one who was rich and one who was poor, someone lost in the currents of life, another whose life is spent in tune with one's inner bliss, perhaps exploring remote countries and peoples. To have all such memories come flooding in, we would remember being a King of some past century, and yet also a life as a peasant working in the King's fields. Eventually we would know lifetimes of blindness, abuse, and sickness. We would remember being quiet by nature, or outstandingly giving to others. We would know life as a daydreamer, as a creative artist, as a socialite, and as a hermit. And in each remembered life we would know a unique body, living as a woman, or living as a man, and in each varied degrees of coordination and attractiveness. We would remember having beliefs in every religion through the eyes of many generations of all different races. To know so many lives would allow one to understand humanity as a whole. Such a being would understand each person and to hate and reject any person would be to hate and reject the existential situation common to all life. We all are just being a part of a whole that must be. There are no exceptions.

Imagine speaking with a thousand voices. What wisdom would result of the collective experiences of a hundred people's lives, a thousand, a million, a billion, even every person who has lived on the earth throughout history? Imagine human lives combined together with the grumbling roamings of a bear, with the graceful movements of a deer. Imagine all the oceanic lives spent underwater, the whales that break the surface to see the sun and stars, the salmon that charge through the waterfall. I think of the big cats, the dolphins, the wild horses, the apes, the dinosaurs. What compiled wisdom would exist in the sum of all life that has ever lived on Earth?

We see pictures of galaxies from the Hubble space telescope but there are other planets in other galaxies where a second nearby galaxy such as Andromeda is displayed across the entire night sky. Imagine the beauty of seeing a full galaxy spread across the night sky. How would that influence the development of science and spirituality for intelligent life on that planet? What would it be like to read from the light of a galaxy? I think the spark of seeing the profound beauty of a giant galaxy each night would profoundly influence life in ways that would surprise us. With all the possible life out there as certain as we are here, what would result of knowing that life and the cosmos in such a broader context, knowing all the diversity of life that exists in all the various environments everywhere throughout the infinite galaxies of Many-Worlds?

With the wisdom of many lives, knowledge would replace ignorance, peace would replace fear, compassion would replace hate, and acceptance would replace judgment. Such a being would know the world and understand every little cause of evil and every reason for good. It would recognize the limitations of the physical body and know the way the body can shape the personality in sickness and in health. It would understand the nature of limitations, of circumstance, of ignorance, of deprivation, to cause what we think of as evil. It would have respect and love for all life and all people regardless of individual imperfections and weaknesses. It would see into time, and know the struggle against nature and feel empathy as all living creatures wrestled with their trials and their failures. It would passionately love and take delight in human life because it would know that we exist at a stage of growth that is very difficult but beautiful, a place between a past of innocent unknowing and a future of becoming increasingly aware.

Those of the bottom-up perspective may eventually acknowledge absolute zero as perfect symmetry, and even as an enfolded order, yet reasonably object to the suggestion that Omega has feelings or thoughts. But what is implied here doesn't need to be of thought or decision. The nature of Omega being suggested here is the natural result of the sum of lives, the result of an integration of the smaller parts; the smaller experiences, just as we are individually the sum of our parts. If all life is but fragments of the whole, how could we imagine a whole that is lifeless? Science needs to learn to appreciate people's intuitional sense of such things. Such intuitional expectations have been around for a long time.

Religion and Oneness

The word universe (not capitalized) is best defined as the Earth and the vast sea of galaxies and stars that surround us. The word cosmos is the whole dynamic system including the inverse-matter direction of time. The word Universe (capitalized) is best defined as everything that exists. The distinction is never more apparent than when we consider if it is reasonable to say "God created the Universe". The claim that God created the Universe suggests that God created all ideas, all maths, all logic, all sensibility, all patterns, all knowledge, all meaning, all life; literally everything that has existed, exists now, and ever will exist. On the other hand the claim that God created the universe (of space-time) indicates that God created the starry heavens and the Earth. One claim somehow places God above existence and so it elevates and so separates God from the whole of everything, and makes God by our standards absolutely indefinable, even nonexistent. The other allows God to be either a part of the Universe, or God is defined to be the Universe itself, and so God can be the everything, the sum of all knowledge, all patterns, and all life.

One idea is quite sensible and rational. It fits together. The universe of stars and galaxies we live in is certainly created by the greater single whole. The universe exists and the whole exists, and the whole is infinite. The Alpha, the beginning, and the Omega, the end of time, both shape the universe. But going so far as to claim that God created existence is just one of those hypothetical ideas that doesn't fend very well under even mild scrutiny, because as a statement it denies all else that makes sense. In the same way that nonexistence can't create a Universe, a "something else other than existence" type of God cannot create a Universe either. It really is recognizably absurd to imagine that God created the universe only thousands of years ago and made it look billions of years old. We are each free to make the claim, but the world makes sense otherwise. That, if nothing else, is the point I am trying to make in this book. We can actually make sense of things, of the world we live in, we can even make sense of reality.

The Universe is everywhere we go and everywhere we see. There is no place where the Universe is not. Yet we still imagine in a vague way an alternative, as if the Universe could stop existing, as if we ourselves might not exist. And so we spend most of our lives unconsciously thinking the physical world shouldn't exist, not realizing how inevitable and innate we are. That is the true magic inside us, not the miracle that we are, but the reality that we are inevitable and ultimately timeless. Our most direct connection to God is that timelessness. It is a wonderful realization really, to be aware that the whole Universe always has been, and shall always be.

We all need something to look up to, something greater than ourselves. We really don't need a God to be the creator of all existence. We need God to exist as real as we exist. We certainly do not need a God that is powerful or that alters reality. We essentially need God to be, at the very least nearly perfect at understanding, at love, at compassion, and honesty. We know those are the most advanced parts of ourselves. So it is fitting that people who have had near-death experiences describe a warmth and love beyond anything they have known in this life. The most common report is of a light or image that speaks without judgment, who sometimes asks them what they learned in their lifetime, often placing emphasis upon the importance of love.

If instead we heard reports of a fearsome god who threatened pain and torture if we were not good and obedient, we would feel a distortion to what we know within ourselves to be just, and such reports would create a deep sense of futility in mankind. We do not respect threat and power in our hearts, only love and understanding. So God has to be virtuous and compassionate near infinite degrees, because there are many people right here at this level who are very good at love. They don't think of themselves when faced with someone in need. And they forgive and understand others by seeing both the faults and strengths of others within themselves. And if such wisdom arises in our earthly world when such a person is only humanly intelligent and vaguely aware of the unity flowing through the world, how could the true nature of the whole be any less loving?

Some people think God has to be powerful and strict like a parent to a child. They believe that if people do not fear God's judgment and punishment they will not obey the rules, meaning, they will not abide by certain ethical standards that most of us agree upon. They expect that without fear people will do only what they desire and not care about others. Such concern is really just fear of the ignorant and selfish side of ourselves, which we must remember is simultaneously the source of our passions and strivings. Stifling our self with fear doesn't actually accomplish the goal we wish for. It actually makes the problems worse. We see so little unity in the world around us, we feel cut-off and disconnected from others and from our oneness, and as a result we make decisions based upon self-interest. Such is the reality of our present world. If people felt more connected to one another the world would be a very different place.

The divisions we assume, the divisions we fear, are not ultimately real. They are just the whole pie divided up into pieces. So nothing disturbs me more than the claim that God would torture a human being eternally. There is nothing more contradictory to the relationship between God and love. Love is about being conscious of the reality of others. There are certainly people who have to pay for their crimes against others, there are people who in being a fragment of what they are create a debt, and in the requirement that we spiritually evolve there exists cosmic laws that govern and produce justice, but not with those who are hurt by others becoming the punisher. Such is not justice but rather two sides of a coin.

There is no escaping the reality of others. It is always there regardless of how much we pretend or reason otherwise. Only someone denying the reality of another would even imagine causing endless suffering to them. Certainly a God would not cause endless suffering, not any God that is whole. Simply causing someone to cease to exist would be more reasonably just, especially considering God defined as such apparently created the same people originally. Even we humans, average people, regardless of religion, in civilized societies, find it unjust or unethical to torture people who have committed even horrible crimes. The worst criminals are put to a painless death for the simple reason that to torture them would make us no different than them. Why then do we imagine that the most advanced stage of awareness leads a supreme being to torture, as if God can escape what it would become in causing endless pain and suffering?

There is a lot of symbolism in religion and I have considered the symbolism of the fallen angel, who wanted to be God but couldn't, so he falls into an illusion of being a God. The same mistake exists in us. It is the great mistake of mankind. I think the fallen angel represents the self who doesn't want to see itself as one with the universe or God, and prefers to be individual or definitive. Identity is all about boundaries. No definitive form or part of the universe can be God, because God is the whole. One can't be both self and God, both fragment and whole. In becoming the whole one loses the definition, the identity, and the freedom of self to be defined. Man doesn't want to give up selfness, and for rather good reasons. From our present perspective in seeing things as more than nothing, the whole is to us a uniform, indistinct, nothing. In seeing things as arisen above nothing, we in some ways see ourselves as the supreme accomplishments of nature or God. We don't want to lose that miracle. Oddly enough, being more than nothing can seem better than being less than the infinite. So in a way it is a choice to be incomplete. Or is it a choice? Are there any choices?

There is no way to resolve the issue of a God inflicting pain and suffering, even if free will is a reality. Each person's life is different, the entire range of experiences being dramatically different, and the assumption that everyone has an equal measure of free will somehow totally denies the reality of each life. If there is free will at all, there are varied measures of free will, degrees that change at different times of a person's life. In reality, people are mainly products of their environment, their genes, their education, their intelligence. Occasionally, perhaps we manage free will, in times of insight, lucidity, or great strength, we understand our choices, and we know what we have to do, except that then, there is no choice, because in actually knowing the whole truth, people naturally do what is best both for themselves and the greater good. Isn't the underlying objective of society's governing rules and criminal punishment to make people conscious of the greater good, almost to wake people up?

In those lucid times, the boundaries between ourselves and others break down, we know and appreciate the reality of others and are not blinded by the selfishness, or the sense of separateness, that haunts our experience otherwise. We know then the oneness of humanity, of life, with the universe, and ultimately with the fruition of all life that is God. There is great power in choices. Choice is an aspect of everything that we are; our knowledge; our level of awareness and development. Yet we have no choice in being who we are in each moment that we exist. And we can't change the timeless universe. Our choices only change the neighborhood that we live in.

There is an ancient Zen proverb, *To know is to do*. In other words, we can only act upon what we understand. Whatever we are fortunate to truly understand or know in life defines our actions. It defines our internal selves. It defines the person others interact with. There are two extreme sides to the experience of life that we all know are there, one where a person knows strength and happiness, and another where they know pain and suffering. In each life we experience some measure of each side but we are strictly defined, albeit in a variety of ways, by which side we know best, the point here being that lacking either experience takes away from the fullness of our potential. In not knowing both sides of life we are harmed. Without some pain and suffering in a person's life they invariably lack compassion and empathy. And empathy, knowing the reality of others, is a very valuable talent as it helps us to feel connected to others. Knowing the value of suffering is why once we have made it past a difficult time we don't wish to give up the experience. What doesn't kill us makes us stronger. On the other hand, in knowing very little happiness or love or trust a person cannot make decisions that produce a collective good which they have very little experience of. It is only when we know how to love and care for others that we refuse to give up the self respect we gain, in trade for anything.

I have told a story sometimes to make this point, of a thin and scraggly homeless man living on the street of a city who shuffles up to another dressed nicely in a suit. The scraggly man asks for money and in response the homeless man is pushed and falls to the ground hitting his head on the wall. The other in his clean suit swears at him contemptuously, "get away from me you filthy bum" and quickly walks away toward a nice car and warm apartment, disappearing into the crowd as the thin older man crawls up from the ground. He is dirty and worn down, but he swears to himself, "I wouldn't want to be that jackass for nothin." At the end of this story I like to ask people who they would rather be: the homeless man at the bottom of his world, with no money, no safe place to sleep, and hungry, or would they rather be the wealthy man in the nice suit, and it surprises most people that they have to stop and think about it. In his suffering, the old man has a heart, he knows what it is like to suffer, and he knows that he shouldn't trade that for everything the other man has. But the man in the suit knows to defend and take care of himself, and doesn't wish to give up anything that he has. The ideal of course is to integrate both worlds. The truly advantageous in life have been on both sides of the fence, as they know the fullness of having experienced both fortune and misfortune.

We all have wondered why a powerful God would have created such a world, where there is such extreme misfortune, where there is so much pain for those who aren't deserving. No child born deserves a life without basic nourishment, without love and self respect, so why are so many without. There must surely be a deep hidden conflict in those who see the world as selected by God, although few would admit it, but deep inside they must reason that God is unfair to have fashioned a world where children starve and the innocent suffer. And there is certainly nothing sensible about a God that creates the same imperfect world, then sits in judgment, and punishes that very thing which it has created. I hope that all can eventually realize that this is not the mentality of a knowing God, but rather just a reflection of our same imperfect selves.

It is the nature of the unknown to create fear, and our human nature to feel anxiety over fear, and often times religion seems to hold a license to use fear against us. We tolerantly accept all the problems of religion because it is part of our culture, when what we crucially need to do in this period of history is just read and study and explore so much, that we each learn how to think for ourselves, and thus our thoughts evolve and we gain a truer sense of what is good, right and important. Our religions have always made the mistake of not freely acknowledging what we genuinely don't know, about the cosmos, about God, so that we have more room to explore. Religions have held humanity back in ways, naturally so, because each religion includes the human manifestations of gravity and expansion. Each contains time moving backward and forward. Each religion is a grouping of people who define boundaries and divide themselves from others. Of course religions include both grouping and symmetry.

We all know the conservative side of human politics and personality, which is pronounced, which divides us apart, which tends toward divisions of us and them, of rich and poor, strong and weak, right and wrong, moral and immoral, good and evil. Naturally the creation of such boundaries tends toward conflict. It leads toward a growth and centralization of power. Just like gravity, the conservative nature prefers how things were in the past, and that force positively serves to maintain the present by trying to recreate the past. It holds us back but it also serves to define the present and keeps us from moving too fast into the future. We all know the slow progression of symmetry order forward, toward integration, toward equality, toward sameness, toward unity, which naturally tends toward peace and a decentralization of power. The dichotomy of two orders exists plainly in religion, in politics and government, in clubs and organizations. It exists plainly in science, in physics and biology. It is visible in every human venture.

It exists especially plainly in all great myth and storytelling. In the story of J.R.R. Tolkien's *Lord of the Rings* it is easy to see that the battle of middle Earth is between the past and future. The varied peoples of middle Earth fight against the barbarians that live underground in caves, in the denseness of rock. They are united only by a remote menacing selfish thirst for power and an identity that wants to overrun all. Otherwise the selfishness of each Orc or Troll would turn against themselves. The threat, the power, like the past, is centered at one place, Mordor, which is a hot volcanic region. There are many diverse peoples of middle Earth but just one enemy. And what gives that enemy power? A gold ring. What might gold represent?

Interestingly, in the movie film *The Matrix* the same dichotomy exists except that the villain is the future and not the past. In the Matrix the enemy is a computer which isn't in one place, it is everywhere, it covers the entire Earth, and the force of humanity, of life, of form, of self interest, must rise up against the control of machines. Of course the machines represent our own technological future. But note how the good in this story, those who have managed to free themselves from the control of the futuristic machines are all living in a central location deep underground, in caves, in a place they call Zion. In the Matrix the good guys represent the passions and desires of grouping order, the flame that drives us toward individuality, uniqueness, distinctiveness, and survival.

So of course we can see the symbols of past and future in religion too, with hell as a hot and dense cavernous place underground, obviously characteristic of our hot and dense past, and with the cloudy white laden heaven invisibly above us in the sky which relates easily to the features of infinity, eternity and perfect oneness of Omega. In religion we have groups of people in all religions claiming they are the special ones that get to go to heaven, divided apart from others that don't, others that somehow don't matter, even in the eyes of God.

In all religions, as well as eastern philosophies, sin is recognized as selfishness and carelessness (I am, you are not), while people are generally encouraged to be loving, acknowledging, and respectful of all others (we are all important and the same, we are one).

I have come to believe that universal symbolism doesn't simply arise from people being spiritually insightful in a common way, synchronicity is built into the way the world unfolds. How the two orders theory is interpreted and applied to the world religions is bound to be of great debate, but what I am certain of is that no single religion is the one accurate religion. Religions somehow must learn to grow because if they don't then they become the past holding us back. As we should expect of any clear comprehension of nature, of spirituality, of God, the two kinds of order requires us to re-evaluate how we conceive of God, the Universe, reality, nature, space, time, and the infinite. We could even expect that an accurate understanding of the Universe removes the seeming disparity between these separate concepts.

Unfortunately the conservative side of religion too often messes up our forward growth toward unity by causing divisiveness, by creating an image that man is by nature sinful and evil, while at the same time emphasizing our ignorance, our faults, and encouraging shame, when man is by nature only unknowing, and there is a big difference between the two. A long time ago every religion developed from the work of people. Why did that growth have to stop? Isn't religion a part of us and our attempt to understand, and so if religion holds on to the past too tightly and doesn't try to move forward it faces the same problems we would face in doing so! Hopefully we will realize soon how related science and religion are. Somehow there has to be a peace based upon respect forged between the two, which means that both have to compromise.

Sometimes learning makes life more difficult for a while, because it pushes us into the next stage of a hierarchy of growth, and we find a whole new set of things unknown that we are unable to deal with. And so emotionally it might seem better to stay unaware, until the world crumbles around us and we don't know how to fix it. And sometimes the problems become more complicated as we grow, as modern society is discovering.

Religion

Religion is embedded in our society and the expectations of others always have a strange overwhelming power over us. I know what I went through before finally giving myself the freedom to search freely and learn, and then trust my own thoughts about things, even things which I thoroughly understood. So in some ways religions makes me angry. I do not like to see others use natural fears of the unknown to have power over others. I do not like portrayals which distort how young minds perceive physical reality. I do not like any philosophy which implies that we should not trust our ability to reason. I don't like that religion works against the disciplines of science that open our minds to the reality that is beyond our usual self involvement. And I don't like that religion has altered the definition of faith, pretending it makes sense to say we believe in something that we have no real certainty about, as if faith is something we choose instead of something which happens to us as we educate ourselves and learn to think. I believe people who use religion to gain power over others are just the 'moving over' types.

Inside me there is a love for large parts of religion. It is an expression of humanity, our faults as well as our strengths. It is a union of people in an attempt to move closer to what is in the highest sense right and good, but we can never delude ourselves into thinking we can attain right and good simply by following. I believe there are lessons and great truths conveyed in each religion, as well as fallacy and intentional deception. I find this sense of shame that seems to be built into humanity, which makes us vulnerable, to be damaging, but it is also a strong testament to how life is naturally caring and accepting of responsibility. I know the problem is not really with religion, nor is it even with us. It is just that life is complicated. We face so many really complex problems, so many ethical issues. Taking a step back to look at those problems I see our search for answers and how much we have accomplished through religion, and government, and science, to be indescribably beautiful and admirable, and any anger then seems foolish.

Sometimes we want a God that is less quiet, that makes life easier, but we need to learn to think and understand on our own, and if God changed the world in ways we wish seeming to make things better it would not necessarily help us in the long run. I hope we soon recognize our place and role in existence and nature as more than a chance fluke or the whim of a powerful god, so we can feel innate and at home, and we can take ourselves more seriously. Unless we learn to love ourselves, regardless of our imperfections, it is human nature to create a much greater punishment than any compassionate god would ever cause us. I am not saying here that I believe nothing that happens is our individual or collective responsibility. What I mean is that ignorance and error comes with existing here in the process of life, it is the very nature of being here, and what is important is learning from mistakes and growing out of them. Growth beyond ignorance is the only purpose for shame or punishment.

I once realized that after crossing into another world where we realize that there really is more to reality than the Earth, and there really is a perfect and loving God, that many people upon dying might not let that God ever have a chance to judge them or show forgiveness, because those who feel ashamed, or know they have hurt others in this life, might run from the light and wander in gray shadows, until nature pulls them back into life again. I deftly fear the way that we are taught to feel ashamed of our imperfectness, instead of inspired by how much we have become, and excited by the adventure before us. From a grouping order perspective it might seem like we ourselves, the evolving cosmos, or the glass of water, is half empty, but ultimately the truth is that each is half full.

I know the following is a truth that is hard to appreciate, but we pay a price for our crimes against others because they are crimes against ourselves. Long ago Plato wrote, "The greatest penalty of evil-doing is to grow into the likeness of a bad man". What we seem to be missing in the human equation is that people always pay heavily for their sins as they move through life. We pay for our sins in the person that we become. So as odd as it may seem, someone lost, someone without empathy and concern for others, someone without the knowledge of balanced experiences, deserves our compassion for the situation they are in. They deserve compassion for the person they are, just as much and perhaps even more sometimes than those they hurt.

Like most people, I occasionally get angry at the people who have done me wrong or have been abusive to others, but I also know that I should feel sorry for them, because for them to do those things, means that they have not known what I have known. They have not seen what I have been fortunate enough to learn. They have not known the kind of love and caring that I have known, and out of that knowing found their own. We all begin from the same place. We all end at the same place. At every stage there exists a part of ourselves; a part of the whole. Such is so simply true it really cannot be denied. So we shouldn't only try to identify with God, or the future. We are also of the past. In every step forward it is so wonderful to know one's heart, to care about others, to be responsible to others, while simultaneously caring about and caring for oneself.

I don't know if things get easier, but I know we are in a difficult stage of life. We started somewhere knowing nothing, clamoring up out of the dark, and we are slowly wonderfully becoming aware, but at this point we are still unknowing of so much, and weak because of it. I can't say the path of history for mankind that we share in this one world won't end up like the dinosaurs some day. Perhaps there is a big asteroid out there with our name on it. I only know it doesn't matter within the big picture. And I know my own life often seems more dreamlike than it is supposed to be. Ultimately even ideas and the physical universe are seen to be one and the same. We see our selves as being a physical body with borders between our self and the world, but we are really part the great mindscape. We are each a single pathway through the infinite mind, finding our way as best we can, as we must.

I fill my days with lifetimes.

Jennifer Ebenroth

~~~

Be in truth eternal, beyond earthly opposites.

Krishna's advice to Arjun in the Bhagavad Gita

~~~

It moves, It moves not.
It is far, and It is near.
It is within all this,
And It is outside of all this.

The Upanishads

~~~

Thought is time. Thought is born of experience and knowledge which are inseparable from time and the past. Time is the psychological enemy of man. Our action is based on knowledge and therefore time, so man is always a slave to the past.

Jiddu Krishnamurti

~~~

Thinking and consciousness are not synonymous. Thinking is only a small aspect of consciousness. Thought cannot exist without consciousness, but consciousness does not need thought.

Eckhart Tolle

~~~

Human mental experience seems to be of two kinds – an experience of facts, memories, emotions, body states – a thoroughly classical kind of knowing which we might call "computer consciousness," which takes place against a peculiar background of "raw awareness" – that uncanny yet familiar feeling we relinquish when we go to sleep and awaken into every morning. Some have called this second kind of experience "consciousness without an object."

Nick Herbert

~~~

It is one great dream dreamed by a single Being, but in such a way that all the dream characters dream too.

Arthur Schopenhauer

~~~

God is infinite in his simplicity and simple in his infinity.

Meister Eckhart

~~~

Transcendence is the only real alternative to extinction.

Vaclav Havel

I am the Alpha and the Omega, who is, and who was, and who is to come
Revelation 1:8

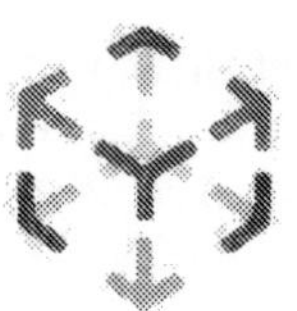

Chapter Twenty Six

The White World

From Beginning to End

Imagine a universe of white that extends outward in every direction. This single white is endless and beginningless. It is everywhere. Wherever the universe is, there is a total and complete whiteness. And so also imagine you are the white, because the white is simply all there is. This is not a place that can be viewed from elsewhere. There is no other place. There is nothing else. You and the white are inseparably one.

In this world you are like a living child in the womb before knowing differentiation of any kind. All that you know is the unbounded oneness of your being. The whiteness of your consciousness, which is being itself, is everywhere and everything. Without beginning or ending, for you time is one enormous moment of now. So you know nothing of change. You know nothing of time. You simply exist. You have always existed. Your consciousness and your being are the same thing.

The white world is all that you are and have ever known. And although your being is powerfully real you are absolutely blind to that same whiteness, because in an undifferentiated world there is no way for you to appreciate the single most fundamental property of yourself. You are blind to white because there isn't an absence of white to differentiate your being against. You have nothing to reference the white with. Consequently, you feel yourself to be both everything and nothing simultaneously. You are a something that is everything, and an everything that is nothing. Such is the only broken symmetry in your perfect whiteness, and in that broken symmetry, in the confusion of your true identity, exists the birth of difference.

So it has been forever that you are the white, and it certainly seems as if it will always be so, except suddenly, for some unknown reason, as if a door opened that you didn't know even existed, suddenly in a flash the timeless uniformity of white shatters. Almost immediately the whiteness is replaced by a rainbow of colors, so that there is red and blue and yellow, and green and orange and purple, which now illuminate the diversity of your being. Within this change or differentiation, or what we might call time, the broken symmetry splits you into two states of consciousness. In each a world begins. In each an expanded sense of existence or consciousness begins, and in each you begin to awaken from your deep silent sleep.

In one of two worlds you originate only seeing yourself as everything, different from another world where you only see yourself as nothing, and although you exist forever in each moment you are also beginning to wake up in each world. This sudden awakening gives birth to colors which interact in an initially inexplicable dance to illuminate a surprising number of properties and attributes that can be formed of each side. In this differentiation the oneness of each side divides and you become many. All this differentiation, all this form is overwhelming. You never knew your whiteness so you can't help but begin to identify with all this form, as now you have something to identify with. The variety of form is so great that soon you know yourself only as color and form. Differences become your reality. Also since there is now change, there is a sense of difference between one moment and the next, and so a sense of time, and consequently that time takes the place of your timeless existence as if you've never experienced it.

Initially it seems an odd relationship which these parts have with each other, within time. In one sense the colors combine together to create the white, and yet the white divides apart to create the colors. There is a sort of complementarity, which is to say these two aspects of your reality are interdependent and inseparably connected, which is why when one of the primary colors is taken away from the others, what is left behind in its place produces a green, or an orange, or a purple. Eventually it becomes clear to you that in the variety of colors you are only experiencing lesser parts of yourself, parts that are fragments of the whole. The parts create the whole. The whole creates the parts. No part of you is really separate at all.

In immense measures of time you are given the opportunity to discover yourself, and slowly and meticulously as you discover the more unified colors of orange, purple, and green, and you begin to see the flow of various forms, after a while you gradually begin to realize the relationship all these colors and forms have. You begin to realize they all exist within an original white, so you begin to realize you are the white, and that you have always been the white, and you realize all the colors and forms were there in the beginning, they were within you all along. And each side of you is beginning to experience this same lesson, which makes your two selves more and more alike. And finally you realize you are becoming the white world again.

As your two sides merge, you begin to see yourself as the white world refracted into a rainbow of colors. You begin to hear yourself as a perfect silence that is divided into a symphony of sound. You begin to feel the ensemble of all possible universes within you enfolding into the eternal white world. Eventually you begin to see clearly the two great forces within yourself. One force that divides your self into parts, creating real and yet illusory material things, as well as all the numbers and all the meaningful ideas, that create the finite side of yourself. Then there is another force that works to combine together all those many parts, the spectrum of colors, and the diversity of infinite many lives and experiences, into the single unified white world. There is still no other color than white, because all the colors are internal, all form is merely fragments of the great timeless white world.

In the end we all see ourselves as the great dance that begins with the colors divided apart and ends with the colors combined together, and then we will also know the dance only seems to have a beginning and an ending, a birth and a death, when in fact this story and the sense of time it portrays are only shifts in perspective between the timelessness of the mother and the birth and experience of the child.

As you become ever more fully aware of your whole self, you realize the being of the whole is actualized in being each part. You realize the 'everything' you thought you were is just a something part of a greater Universe. You realize the 'nothing' you thought you were, is as relevant as the 'everything' you thought you were. Eventually all the boundaries of separation erode, and as you become the sum, all your pronounced experiences in time are balanced and given meaning by the fullness of the whole. You even feel yourself in the future shaping those experiences in the past to become the great completeness. In a final step you move from semi-consciousness to full consciousness. And you become the beginning, the evolution of all life, and the end, existing all in the one enormous moment of now, where you know yourself for the first time.

Come Together © Kerry Mitchell

It is all one to me.

Parmenides

~~~

Look well to this one day, for it and it alone is life. In the brief course of this day lie all the verities and realities of your existence; the pride of growth, the glory of action, the splendor of beauty. Yesterday is only a dream, and tomorrow is but a vision. Yet each day, well lived, makes every yesterday a dream of happiness and each tomorrow, a vision of hope. Look well, therefore, to this one day, for it and it alone is life.

Sanskrit 4500 years ago.

~~~

It is good to have an end to journey toward; but it is the journey that matters, in the end.

Ursula K. Le Guin

If we assume some one presiding over the infinite number of agents, we must ascend above all or descend down to the center of all, to the absolute being, present above all and within all...more intimate to all things than each is to itself, not more distant from one than from another, for it is equally the nearest to all.

Giordano Bruno

Chapter Twenty Seven

God, Infinity, and Nature as One

Is Omega Alive or just a Godly existence?

We all have felt the mixture of beauty and mystery as the evening sun sets and the window of night opens to the outer universe. Like a reminder, the stars glisten within the realness of infinite space. The reality of what is out there is diminished by any scientific or mathematical description. The best way to sense the vast contents of interstellar space may be just quietly gazing upward on a clear night, just letting go, perhaps the smallest of effort in reaching out with the mind to touch and feel what is there.

Most of what we learn about the Universe is awfully hard to appreciate. Who can fathom the expanse of our Milky Way galaxy, let alone the enormous distances between the galaxies or the immense size of the cosmos? Who appreciates the diligent cells at work in one's own body or the tiny of each atom? We can now add to the list of unimaginable things, the infinite worlds that exist all around us, just beyond the surface of this particular world. We can know they are there, but appreciating them is another story. Perhaps one day we will discover ways of communicating with or traveling into parallel dimensions, which may turn out to be more feasible even than intergalactic travel. One day we may visit parallel neighbors like the crew of the starship enterprise explores distant stars and planets in the adventures of Star Trek.

Yet we may be inseparable from this cosmos and forbidden to leave. We each have a strong sense of being a single self, a body separate from other bodies and other things in our environment. We consider ourselves independent and believe we are free to physically move about, free to make choices, free to think, free to explore. Yet it is not easy to define specifically where this independent self begins and ends. We are made of the same atoms as everything else. We constantly exchange heat with our environment. We ingest and exfoliate food and water which initially isn't imagined as part of us, then it is, then it isn't. We are each first a collective of many, a cooperation of about 75 trillion individual cells that rival the galactic empire in complexity. There are many distinct parts of our body, distinct organs. Then we have to acknowledge how much of what we are mentally is defined by what we perceive of our environment. We are primarily the sum of all that we have experienced since before birth, possibly even before conception. Our exterior environment shapes our intellect and mind, even if that influence is in continuous competition with the three billion letters of code in our genetic programming, code which is also hard to define ownership of. And of course our minds cannot work without ideas and concepts, which are certainly not of our own creation.

Even more dominating is our reliance on cosmological physics. Behind the scenes electromagnetism, the strong force, the weak force, delicately manage each and every atom while gravity controls the conglomerate enough to produce the great masses of Sun and Earth. Whatever holds the universe together, whatever creates it, manages it, unfolds it, whatever is natural, whatever the big picture is, lays claim to every ounce of what we are. We may seem free to play within this reality, but how could we escape it? No man is an island, or at least, no island can separate from the sea. We are inalienably at one with the universe. We are in the truest sense a thinking and breathing universe, and no matter how insulting that may feel to the ego, the truth hurts good. All sense of a separate self is ultimately an illusion of being something immeasurably tiny and fragmented, like one piece of the whole pie.

Once we let go of seeing ourselves as distinct objects existing apart from other objects, once we understand we are instead merely the definition of something greater, we then have a chance to know ourselves. The greater something is nothing less than all the experiences of an infinite array of living creatures of varying appearance and intelligence that exist throughout time and space in an infinite expanse of galaxies. Symmetry order sums all physicality and thought, and in that fullness there is all living diversity, all the knowledge, all the math, all the artistry, all the nature so greatly beyond imagination, so breathtakingly profound, all exist as crisp and real as a walk on an ocean beach on a sunny day.

Some initial insight toward understanding the unfolding and enfolding of life comes from Bohm when he was asked about what happens after death, about five years before Bohm died unexpectedly in 1992. In an interview Bohm described the following thoughts:

> Death must be connected with questions of time and identity. When you die, everything on which your identity depends is or going. All things in your memory will go. Your whole definition of what you are will go. The whole sense of being separate from anything will go because that's part of your identity. Your whole sense of time must go. Is there anything that will exist beyond death? That is the question everybody has always asked. It doesn't make sense to say something goes on in time. Rather I would say everything sinks into the implicate order, where there is no time. But suppose we say that right now, when I'm alive, the same thing is happening. The implicate order is unfolding to be me again and again each moment. And the past me is gone.
>
> Anything I know about "me" is in the past. The present "me" is the unknown. We say there is only one implicate order, only one present. But it projects itself as a

whole series of moments. Ultimately, all moments are really one. Therefore now is eternity. In one sense, everything, including me, is dying every moment into eternity and being born again, so all that will happen at death is that from a certain moment certain features will not be born again [into this life]. But our whole thought process causes us to confront this with great fear in an attempt to preserve identity. One of my interests at this stage of life is looking at that fear.

Actually after the evolution of two orders comes into focus it makes perfect sense to expect that life continues in time after the death of each life. An extension of viewing life as unfolding out of and enfolding into the implicate order would be to see life as what is actually evolving all the way from Proto and Elea to Omega. Life either plays a major role in the definition and characterization of Omega, or life itself is what is becoming the whole. Keep in mind that although Bohm began to clearly recognize the two orders, he did not clearly recognize the whole cosmos, the big bang and expansion, as an evolution of two orders. He did not recognize how an implicate future shapes the present.

The ultimate lesson of two types of order is essentially that we are supremely all one; one life, one universe, one existence, one infinite whole, sometimes called God. We are each individually one level of the big consciousness. You could say I believe in reincarnation, except with a timeless twist. I believe there are higher levels of ourselves all the way up, as well as lower, say, from mammals to dinosaurs to amoeba. Interacting spiritually with the universe is essentially interacting with ourselves, which of course in this view is pretty hard not to do.

Is Omega understood best as a totality of universes, or the totality of life? Actually the physical universe is considerably simple in comparison to the complexities of living beings. Material objects are one or two or three dimensional in terms of identity, while life is extremely complex and multidimensional. Imagine how simple the Earth would be without plant or animal life. Life has identities of personality, friendship, family, love, career, food, music, just to name a few of the hierarchies within hierarchies. Especially considering what we know of the role of the observer from quantum mechanics, once one understands the enfolding principle of symmetry order, one can't help thinking that time is more discernable as an arena for the evolution of life and consciousness rather than for an objective material universe.

This conclusion can actually be made based solely upon quantum mechanics. Matthew J. Donald, the Cambridge physicist and author of the many-minds interpretation of quantum theory states, "...the distinction between worlds should be made at the level of the mind of the individual observer." The conclusion one draws from quantum mechanics is that the definition of the physical universe ends at the edge of the observations of each observer, which suggests we are each an evolving universe. It would seem then we could never escape our own past, which is what leaping into another universe would involve. Donald writes:

> Everett's work [the Many-Worlds Theory] largely leaves open the questions of characterizing individual possibilities and our limited natures. My own approach to these questions has involved the analysis of minds as finite systems processing finite information. As minds, we appear to live inside physical reality, our brains apparently being direct physical representations of the structure of our minds. But, if minds are fundamental, then what we think of as "physical reality" is just a mental representation. What is "external" to mind, nevertheless, is the physical law which determines the probabilities of the possible futures of a given mental structure. Although this picture is radical, it is, in my opinion, both logically consistent and consistent with empirical evidence. Moreover, it has considerable explanatory power. For example, mind is placed at the heart of reality, rather than being just an embarrassment as it would seem to be for materialists. Because time becomes an aspect of our individual structure as observers, the idea of a fixed observer-independent background spacetime becomes unnecessary.

The timeless twist on living other lives is sort of an application of quantum physics, related to all lives that exist as parts of the infinite whole. In quantum theory we recognize that neither the past or future are defined until we have observed them. There is a wide spectrum of possibilities that form the wave until we peer in and cause it to collapse. Just as this is true in defining our present life, so is it true with past or future lives. Until we actually look into the deeper past beyond birth in some way, perhaps through hypnotization, and observe past lives, then our possible pasts (lives) are still in wave form and thus unified with all our possible past lives.

One of the questions I haven't been able to resolve clearly enough in my own thoughts is whether or not Omega is mentally conscious. Is Omega alive in that way? I haven't any difficulty admitting to or appreciating how the inherent properties of Omega resemble the classic idea of an omniscient God, in some ways identically so. Although I had not assumed a God exists prior to understanding symmetry order, from a very young age I always suspected an infinite whole exists and I always felt that other people's intuition of the same led to a hopeful belief in God. I think that science as a religion assumes the idea of God is based solely upon wishful thinking, which is really just the opposite position, made in reaction against wishful thinking, which is then only skeptical thinking. I personally prefer just to think.

In physically being the infinite whole, I think Omega is necessarily defined as being supremely conscious and all knowing. Such is inherent to the reality of being itself. One thing we should try to remember, the experience of the great now is being itself. God doesn't actively think in time at the level of being everything. God's thoughts are being itself. Omega is everything we might imagine of a God, except Omega at the ultimate stage is timeless and unchanging, even though all time occurs within it. At first that seems to indicate that Omega would not be able to interact with living beings, until we consider that the convergence of time toward Omega and the influence of the properties of

Omega precisely define what exists in time. Electromagnetism is a form of the influences of Omega. So why would there also not be larger macrocosmic influences shaping life in general? In being all, Omega knows all, so in order for the evolution of time to become Omega, Omega must organize life events that teach temporality to collectively become as conscious and omniscient as the final state is. As profound as profound becomes, all life must become all being.

Of course as we question whether or not Omega is alive, we have to ask are we alive? We are ultimately timeless as well. If we are really just part of something else, part of the whole, if we really exist purely in timelessness, and our sense that existence is changing is ultimately an illusion, then are we actually alive or are we like some computer program in the matrix? The information of the program might be definitively real, but we wouldn't ordinarily imagine that data to be enough to make us alive.

In the bottom-up perspective we are each a construction of smaller parts. We know we are many individual organs forming a larger body. We are many cells forming organs. We are complex genetic codes. We are chemicals and molecules and atoms. Is life something independent and unique from the cooperation of smaller parts? Viewed from the top-down perspective, all the definitive parts of the body and the cosmos are parts of something larger. The intermediate whole is as relevant as the parts. Are we actually alive? Of course we are, but perhaps there are components in our usual definition of life that are assumed, for example, we assume that life is something that changes during its existence. We assume life is a consciousness that is dynamic, which changes along with a dynamic existence. If existence doesn't really change, then perhaps life is more aptly defined as a consciousness that results of simply being. In other words, the very being of definitive things creates a measure of consciousness, or it is the stuff of consciousness. Certainly consciousness is derived of being, and therefore any definitive thing is semi-conscious in the meaningful act of being itself. What else could consciousness be other than the work of meaning?

Why do we assume the necessity of differentiating between living and inanimate matter anyway? We make so many of those kinds of distinctions. Are the quarks in a rock different than those in my hand? We rip everything apart without ever sewing it back together. Typically the bottom-up view characterizes space, the void, the background, as if it is some special other kind of being different than the practical tangible reality of physical things, for example, science now defines the quantum wave as mere probability. The sometimes religious top-down view usually sees the godly ultimate as an incomprehensible magnitude, somehow unrelated to definitive form. In appreciating the whole we don't have to sacrifice our appreciation for the fragmentation within that whole. The broader view of all perspectives encourages us to see the whole as infinitely finite.

My original study of infinity became a study of timelessness, and my study of timelessness became a study of the two kinds of order, and my study of two orders led me to deeply understand infinity and timelessness. The reason fully comprehending the two kinds of order changes us and is so powerful is without question because it allows us to factually know that all things and all times are unified into a universal oneness. It is a rather important lesson. Indeed it is surprising that we don't already clearly appreciate this, or even recognize the two orders, but I think our lack of awareness is necessarily built into our own evolution, and I suspect this explains a great deal about history, such as the absurdity of past mentalities. In hindsight we will realize more clearly the negative consequences of missing out on a quality of the universe so basic to nature, but such was merely an inevitable stage of cosmic development.

If we are not, existentially speaking, remote objects floating in a background of nothingness, if we are instead parts of an inseparable oneness, then if anything is alive it is Omega, it is the Universe. This shouldn't be considered an ideal, or wishful thinking, but rather fact, an ultimate truth that we need to acknowledge as such. We are not separate from the Universe. Our existence is not separate from anything else. Ultimately there is just a oneness that is alive, even if the living activity of Omega is all internal. We are wonderfully the internal biology of the infinite.

As so many people deeply sense, there is really a purpose and a goal to cosmic evolution, even at the scale of stars and galaxies. There is really something worthy of being called God. There is something going on here; a profound education and growth. And I am not suggesting this hypothetically or idealistically. Although two orders extends far beyond science into philosophy and spirituality, even politics, economics, and psychology, it is first and foremost basic physics, for it is an explanation of the most fundamental way in which all patterns evolve in nature. If the two orders indicate there is a universal God, then there really is something definable as a God. I say this with reservation, with concern for the way religions minimize and manipulate the idea of God, both over glorifying and over humanizing the truer nature of omniscience.

All the same, as it should have turned out all along, the place in time in which the universe is whole, a unification of all life, all experience, all knowledge, something we can call God, is actually an integral part of science, it has just been so far misidentified. This universal God is so real that we can scientifically know that God exists. In summary it all kind of seems funny, especially since the way we have responded to Omega is designed into the evolution of our species and our consciousness. But an omniscient God has been a part of science all along. Mother Nature has been a part of religion all along. The enfolded balance, the implicate order, perfect symmetry, the timeless infinite, the unity of existence, are all masquerading as the ordinary empty space we know so well.

References

Chapter One

[1] Einstein, Albert, Relativity; *The Special and the General Theory.* Crown Trade Paperbacks (1961).
[2] Einstein, Albert, *Letter to Michele Besso's Family.* Ref. Bernstein, Jeremy., *A Critic at Large: Besso.* The New Yorker (1989).
[3] Feynman, Richard, P., *Quantum Mechanics and Path Integrals.* Mcgraw Hill, New York (1965).
[4] Hawking, Stephen W., *Cosmology from the Top Down.* [arxiv.org/abs/astro-ph/0305562] (2003).
[5] Hawking, Stephen W., Hartle, James B., *Wave function of the Universe.* Phys. Rev. D 28, 2960 (1983).
[6] Hawking, Stephen W., *A Brief History of Time.* Bantam (1988).
[7] *The End of Time; A talk with Julian Barbour,* Edge Foundation. [edge.org/documents/archive/edge60.html] (1999).
[8] Greene, Brian, *The Elegant Universe; Hidden Dimensions and the Quest for the Ultimate Theory.* Vintage Books, New York (2000).
[9] Greene, Brian, *The Fabric of the Cosmos; Space, Time, and the Texture of Reality.* Vintage Books, New York (2004).

Chapter Two

[10] Parmenides, *On Nature.* [elea.org/Parmenides/] (1988).

Chapter Three

[11] Davies, Paul, *Space-time Singularities in Cosmology and Black Hole Evaporation.* (1978).
[12] Seife, Charles, *Zero; The Biography of a Dangerous Idea.* Penguin (2000).
[13] Seife, Charles, *Alpha and Omega; The Search for the Beginning and End of the Universe.* Viking (2003).
[14] Caldwell, Robert R., Kamionkowski, Mark., Weinberg, Nevin N., *Phantom Energy and Cosmic Doomsday.* (Big Rip Theory) [arxiv.org/abs/astro-ph/0302506] (2003).
[15] Anderson, Mike H., et al. . *Observation of Bose-Einstein condensation in a dilute atomic vapor.* Science 269, July 14, Pg. 198-201 (1995) .
[16] Hut, Piet., *Structuring Reality; The Role of Limits.* [.ids.ias.edu/~piet/publ/abisko/ab.html] (1996).
[17] Hut, Piet, *As in a Dream.* [ids.ias.edu/~piet/publ/other/dream.html] (2000).

Chapter Four

[18] Guth, Alan, *A Golden Age of Cosmology.* Edge Foundation. [edge.org/documents/day/day_guth.html] (2001).
[19] Misner, Charles W., Thorne, Kip S., Wheeler, John A., *Gravitation.* Freeman (1973).
[20] Boltzmann, Ludwig, *On the relation between the second law of the mechanical theory of heat and the probability calculus with respect to theorems of thermal equilibrium.* Sitzungsber. Kais. Akad. Wiss. Wien, Math. Naturwiss. Classe 76, 373 (1877).
[21] Barbour, Julian, *The End of Time; The Next Revolution in Physics.* Oxford University Press, (1999). [Figure 4.1 adapted from Pg. 312-321].
[22] Barbour, Julian, *The timelessness of quantum gravity. Classical and Quantum Gravity.* 11, 2853 (1994).
[23] Boltzmann, Ludwig, *On Certain Questions of the Theory of Gases.* Nature 51, 413 (1895).

Chapter Five

[24] Beck, Charles, Artist. [crk.umn.edu/campusinfo/tour/BerglandLab/Art/]
[25] Crystalline Structure see: cst-www.nrl.navy.mil/lattice/

Chapter Six

[26] Raymond, George Lansing, *The Genesis of Art Form; An Essay in Comparative Aesthetics.* C.P. Putman's Sons (1893).

Chapter Seven

[27] Bohm, David, *Wholeness and the Implicate Order.* Routledge & Kegan Paul, (1980).
[28] Bohm, David, Peat, David F., *Science, Order & Creativity.* Bantam Books, (1987).
[29] *Interview with David Bohm.* Omni Magazine, Jan. (1987).
See: www.fdavidpeat.com
[30] Talbot, Michael, *The Holographic Universe.* Harper Collins (1991).
[31] John Fudjack, Patricia Dinkelaker, *The Enneagram and the MBTI, About Face; Part three* [www.tap3x.net/EMBTI/]
[32] Bergson, Henri, *Creative Evolution.* Random, [Pg. 233-261] (1944).
[33] Yeats, William, *A Vision.* The MacMillan Company, New York [Pg. 67-68] (1956).

Chapter Eight

[34] Emerson, Ralph Waldo, *Essays and English Traits (Compensation).* P.F Collier & Son, New York (1909).
[35] Lin, Shu-Kun, see www.mdpi.org/lin/
[36] Lin, Shu-Kun *Correlation of Entropy with Similarity and Symmetry.* Journal of Chemical Information and Computer Sciences, 36, 367-376 (1996).
[37] Lin, Shu-Kun *The Nature of the Chemical Process. 1. Symmetry Evolution –Revised Information Theory, Similarity Principle and Ugly Symmetry.* Int. J. Mol. Sci. 10-39 (2001).

Chapter Nine

[38] Darling, David, *On creating something from nothing.* New Scientist, 14 Sept. (1996).
[39] Lucretius, Titus, *On the Nature of Things.* W.W. Norton & Company, New York (1977).
[40] Tegmark, Max, *Does the universe in fact contain almost no information?* Foundations of Physics Letters. 25-42, [arxiv.org/abs/quant-ph/9603008] (1996).

Chapter Ten

[41] de Bernardis, P., et al., Nature 404, 955, (CMB-Boomerang) [arxiv.org/abs/astro-ph/0105296] (2000).
[42] Brougham, Henry, *Review of Thomas Young's article The Bakerian Lecture on the Theory of Light and Colors; published anonymously in the Edinburgh Review* (Jan. 1803).
[43] Herbert, Nick. *Quantum Reality; Beyond the New Physics.* Doubleday (1985).
[44] Everett, Hugh, *On the Foundations of Quantum Mechanics,* thesis submitted to Princeton University. (1957); *Relative State' Formulation of Quantum Mechanics,* Reviews of Modern Physics, (1957).
[45] Gribbin, John, *In Search of Schrödinger's Cat; Quantum Physics and Reality.* Bantam (1984).
[46] Barrow, John D., Tipler, Frank J., *The Anthropic Cosmological Principle.* Oxford University (1988).
See www.anthropic-principle.com
[47] Tegmark, Max, *Parallel Universes,* [arxiv.org/abs/astro-ph/0302131], published in *Science and Ultimate Reality: From Quantum to Cosmos,* J. D. Barrow, P.C.W. Davies, & C.L. Harper eds. Cambridge University Press (2003).
[48] Tegmark, Max, *Is 'the Theory of Everything' merely the Ultimate Ensemble Theory.* [arxiv.org/abs/astro-ph/9704009], published in Annals of Physics 270 1-51 (1998).
[49] Deutsch, David, *The Fabric of Reality,* Penguin (1997).
[50] Davies, Paul, *Other Worlds; Space, Superspace, and the Quantum Universe.* Penguin (1997).
[51] Wolf, Fred Alen, *The Dreaming Universe; A mind expanding journey into the realm where psyche and physics meet.* Simon and Shuster (1994).
[52] Phillips, J. B., Your God is Too Small; A guide for Believers and Skeptics Alike. Touchstone (2004).

Chapter Eleven

[53] Davies, Paul, *The Mysterious Flow of Time.* Scientific American Magazine (Sept. 2002).

Chapter Twelve

[54] Kauffman, Stuart, *Live moderated chat at International Society for Complexity, Information, and Design.* [iscid.org/stuartkauffman-chat.php] (2002).
[55] Kauffman, Stuart, *At Home in the Universe; The Search for the Laws of Self-organization.* Oxford University Press (1996).
[56] Kauffman, Stuart, *Origins of Order; Self Organization and Selection in Evolution.* Oxford University Press (1993).
Term 'adjacent possible' coined by Stuart Kauffman

Chapter Thirteen

[57] Hawking, Stephen W.; Ellis G.F.R., *The large scale structure of space-time.* Cambridge University Press (1973).
[58] King, Chris C., *Dual-Time Supercausality*, Physics Essays 2/2 128-151 (1989).
[59] Pitts, Trevor, *Dark Matter, Antimatter, and Time-Symmetry.* [arxiv.org/abs/physics/9812021] (1998).
[60] Stenger, Victor. J., *Time's Arrows Point Both Ways.* Skeptic, vol. 8, no. 4, 92 (2001).

Chapter Fourteen

[61] Price, Hue, *Time's Arrow and Archimedes' Point: New Directions for the Physics of Time.* Oxford (1997).

Chapter Fifteen

[62] Hoyle, Fred, *The Asymmetry of Time*, Third annual lecture to the research student's association, Camberra (1962).

Chapter Sixteen

[63] Cramer, John, *The Transactional Interpretation of Quantum Mechanics.* Reviews of Modern Physics 58, 647-688, [www.npl.washington.edu/ti/TI_toc.html] (1986).
[64] Cramer, John, *Generalized absorber theory and the Einstein-Podolsky-Rosen paradox.* Physical Review D 22, 362-376 [http://mist.npl.washington.edu/npl/int_rep/gat_80] (1980).
[65] Cramer, John, *An Overview of the Transactional Interpretation of Quantum Mechanics.* International Journal of Theoretical Physics 27, 227
[http://mist.npl.washington.edu/npl/int_rep/ti_over/ti_over.html] (1988).
[66] Cramer, John, *Velocity Reversal and the Arrow of Time.* Foundations of Physics 18, 1205 [http://mist.npl.washington.edu/npl/int_rep/VelRev/VelRev.html] (1988).

Chapter Seventeen

[67] Weinberg, Steven, *The First Three Minutes; A Modern View of the Origin of the Universe.* Basic Books (1993).
[68] Klyce, Brig, *The Second Law of Thermodynamics.* [www.panspermia.org/seconlaw.htm] (1988).
[69] Feynman, Richard P., *The Feynman Lectures on Physics vol.1*, 44-3. Addison Wesley Publishing (1963).
[70] Thaxton, Charles; Bradley, Walter; Olsen, Roger, *The Mystery of Life's Origin: Reassessing current theories.* Philosophical Library (1984).

Chapter Eighteen

[71] Krauss, Lawrence M. *The Physics of Star Trek.* Harper Perennial (1995).
[72] Larson, Dewey, *The Structure of the Physical Universe.* North Pacific Publishing (1979).

Chapter Nineteen

[73] Einstein, Albert, *Relativity; The Special and the General Theory.* (Appendix Five: Relativity and the Problem of Space. Pg. 155-172) Crown Trade Paperbacks (1961).
[74] Dirac, Paul, *The Principles of Quantum Mechanics.* Oxford University Press (1930).
[75] Dirac, Paul, *Lectures on Quantum Mechanics.* Dover (2001).
[76] Czajko, Jakub, *On conjugate complex time I. Chaos, Solitons & Fractals,* Vol. 11 (13) p.1983 (2000); On conjugate complex time II. Chaos, Solitons & Fractals, Vol. 11 p.2001 (2000). Term 'Multispatiality' was coined by Jakub Czajko.
[00] Bryanton, Rob, *Imagining the Tenth Dimension: A New Way of Thinking about Time and Space* Talking Dog Studios (2006).

Chapter Twenty

[77] Landis, Geoffrey, *Impact Parameter and other Quantum Realities.* Golden Gryphen Press (1988).
[78] Tipler, Frank J., *Refereed Journals: Do They Insure Quality or Enforce Orthodoxy?* International Society for Complexity, Information, and Design, Vol. 2.1 [www.iscid.org/pcid/2003/2/1-2/tipler_refereed_journals.php] (2003).
[79] Tipler, Frank J., *The Physics of Immortality: Modern Cosmology, God, and the Resurrection of the Dead.* Doubleday (1994).
[80] Bruno, Giordano, *Cause, Principle and Unity; And Essays on Magic.* Cambridge University Press (1998).
[81] Bruno, Giordano, *The Expulsion of the Triumphant Beast.* Bison Books (1992).
[82] Bruno, Giordano, *The Ash Wednesday Supper.* University of Toronto Press (1995).

Part Six

[83] Moody, Raymond, *Life after Life; The investigation of a phenomenon – Survival of Bodily Death.* Harper (2001).

Chapter Twenty Three

[84] Keirsey, David, *Please Understand Me; Character and Temperament Types.* Prometheus Nemesis Book Company (1984).
[85] Jung, Carl, *The Development of Personality; Collected Works Vol. 17.* Princeton University Press (1981).
[86] Jung, Carl, *Psychological Types; Collected Works Vol. 6.* Princeton University Press (1976).
[87] Jung, Carl, *The Archetypes and the Collective Unconscious; Collected Works Vol. 9 Part 1.* Princeton University Press (1981).
[88] Sperling, John; Helburn, Suzanne; Samuel, George; Morris, John; Hunt, Carl, *The Great Divide, Retro versus Metro in America.* Polipoint Press (2004).

Chapter Twenty Four

[89] Harris, Thomas *I'm Okay, You're Okay.* Galahad (2004).
[90] Fromm, Erich, *On Disobedience and other Essays.* Seabury Press (1981).
[91] Horney, Karen, *Neurosis and Human Growth; The Struggle Toward Self Realization.* W.W. Norton (1991).
[92] Horney, Karen, *Our Inner Conflicts; A Constructive Theory of Neurosis.* W.W. Norton (1992).
[93] Rumi, Jalal al-Din, *The Rumi Collection.* Shambala (2005).

Part Seven

[94] Bois, Samuel, Epistemics; *The Science-Art of Innovating.* International Society for General Semantics (1972).
[95] Tolle, Eckhart, *The Power of Now; A Guide to Spiritual Enlightenment.* New World Library (1999).

Chapter Twenty Seven

[96] *Interview with David Bohm.* Omni Magazine, Jan. (1987).
See: www.fdavidpeat.com
[97] Donald, Matthew J., *Finitary and Infinitary mathematics, the Possibility of Possibilities and the Definition of Probabilities.* [http://philsci-archive.pitt.edu/archive/00001245/] (2003).

General

[98] Misner, Charles W., *Absolute Zero of Time.* 10.1103/PhysRev.186.1328 [www.prola.aps.org/abstract/PR/v186/i5/p1328_1] (1969).
[99] Woodward, J. F., *Killing Time.* Foundations of Physics Letters, Vol. 9, No. 1, [www.chaos.fullerton.edu/~jimw/kill-time/] (1996).
[100] Stenger, Victor, *Timeless Reality; Symmetry, Simplicity, and Multiple Universes.* Prometheus (2000).
[101] Asimov, Isaac, *What is Beyond the Universe?* Science Digest, vol. 69, 69-70 (1974).
[102] Spergel, D. N. et al., (WMAP) [www.arxiv.org/abs/astro-ph/0302209] (2003).
[103] Page, L. et al., (WMAP) [www.arxiv.org/abs/astro-ph/0302220] (2003).
[104] Caldwell, Robert R., *A Phantom Menace?* Phys. Lett. B 545, 23, [www.arxiv.org/abs/astro-ph/9908168] (2002).
[105] Boltzmann, Ludwig, *Lectures on Gas Theory.* University of California Press (1964).
[106]Davies, Paul, *Superforce; The Search for the Grand Unified Theory of Nature.* Touchstone (2002).

Books and Essays by this Author:

[107] Giorbran, Gevin, *The Superstructure of an Infinite Universe.* (1994); *At the Shore of an Infinite Ocean.* (1996); *Exploring a Many Worlds Universe.* (1997). (copyrighted unpublished evolving editions of this book)
[108] Giorbran, Gevin, *Modeling the Macrocosmic Structure of State Space.* (Essay 2001).
[109] Giorbran, Gevin, *Modeling the Aggregate Structure of Configuration Space.* (Essay 2003).
[110] Giorbran, Gevin, *Omega Zero; The Influence of the Future on Cosmic Evolution.* (Essay 2004).

Index

Recommended reading:

Zero; The Biography of a Dangerous Idea - Charles Seife
The Power of Now - Eckhart Tolle
Quantum Reality - Nick Herbert
A Brief History of Time – Stephen Hawking
Please Understand Me – David Keirsey
Neurosis and Human Growth - Karen Horney
The Earthsea Trilogy - Ursula K. LeGuin
Many Worlds in One – Alex Vilenkin

Lightning Source UK Ltd.
Milton Keynes UK

175600UK00002B/34/A

9 780979 186103